AF795872

Molecular Biology Intelligence Unit

p53

Ayeda Ayed, PhD
Ontario Cancer Institute
University of Toronto
Toronto, Ontario, Canada

Theodore Hupp, PhD
Institute of Genetics and Molecular Medicine
CRUK p53 Signal Transduction Laboratories
University of Edinburgh
Edinburgh, Scotland, UK

Landes Bioscience
Austin, Texas
USA

Springer Science+Business Media
New York, New York
USA

p53

Molecular Biology Intelligence Unit

Landes Bioscience
Springer Science+Business Media, LLC

ISBN: 978-1-4419-8230-8 Printed on acid-free paper.

Copyright ©2010 Landes Bioscience and Springer Science+Business Media, LLC

All rights reserved. This work may not be translated or copied in whole or in part without the written permission of the publisher, except for brief excerpts in connection with reviews or scholarly analysis. Use in connection with any form of information storage and retrieval, electronic adaptation, computer software, or by similar or dissimilar methodology now known or hereafter developed is forbidden.
The use in the publication of trade names, trademarks, service marks and similar terms even if they are not identified as such, is not to be taken as an expression of opinion as to whether or not they are subject to proprietary rights.
While the authors, editors and publisher believe that drug selection and dosage and the specifications and usage of equipment and devices, as set forth in this book, are in accord with current recommendations and practice at the time of publication, they make no warranty, expressed or implied, with respect to material described in this book. In view of the ongoing research, equipment development, changes in governmental regulations and the rapid accumulation of information relating to the biomedical sciences, the reader is urged to carefully review and evaluate the information provided herein.

Springer Science+Business Media, LLC, 233 Spring Street, New York, New York 10013, USA.
http://www.springer.com

Please address all inquiries to the Publishers:
Landes Bioscience, 1806 Rio Grande, Austin, Texas 78701, USA.
Phone: 512/ 637 6050; FAX: 512/ 637 6079
http://www.landesbioscience.com

9 8 7 6 5 4 3 2 1

Library of Congress Cataloging-in-Publication Data

p53 / [edited by] Ayeda Ayed, PhD, Ontario Cancer Institute, University of Toronto, Toronto, Ontario, Canada, Theodore Hupp, PhD, Institute of Genetics and Molecular Medicine, CRUK p53 Signal Transduction Laboratories, University of Edinburgh, Edinburgh, Scotland, UK.
p. ; cm. -- (Molecular biology intelligence unit)
Includes bibliographical references and index.
ISBN 978-1-4419-8230-8 (alk. paper)
1. p53 antioncogene. 2. p53 protein. I. Ayed, Ayeda, 1962- editor. II. Hupp, Theodore, 1969- editor. III. Series: Molecular biology intelligence unit (Unnumbered : 2003)
[DNLM: 1. Cell Transformation, Neoplastic--genetics. 2. Genes, p53. QZ 202]
RC268.44.P16P15 2011
572.8'6--dc22

2010046107

About the Editors...

AYEDA AYED is on leave from the Ontario Cancer Institute at the University of Toronto where she worked as a Postdoctoral Fellow and Associate Scientist on structural aspects of p53 using nuclear Magnetic Resonance Spectroscopy. She was the recipient of the Governor General's Award in Leukemia Research and a National Cancer Institute Fellowship. She obtained her PhD in Chemistry at the University of Manitoba in Winnipeg and currently resides in Toronto, Canada.

About the Editors...

THEODORE HUPP was trained in Chemistry as an undergraduate at Bowling Green State University in Ohio working with Bill Scovell and applied developing interests in life sciences towards a PhD degree at Michigan State University under the mentorship of Jon Kaguni. Interests in enzymology was applied to the cancer field working with Sir David Lane during the time when the p53 field was discovering key p53-inducible genes like *p21*, transgenic technologies showed the key role of p53 as a tumor suppressor, and the p53 protein was found to amenable to activation by post-translational modifications such as phosphorylation, ubiquitination, peptide ligands, or small molecules. The Hupp lab is now based at the University of Edinburgh (UK) funded by the Cancer Research UK charity, where enthusiastic students and colleagues continue to study fundamental enzymological aspects of p53 control by ubiquitination, acetylation, and phosphorylation with the hope of developing novel therapeutics for activating the p53 pathway in human cancers.

CONTENTS

EDITORS

Ayeda Ayed
Ontario Cancer Institute
University of Toronto
Toronto, Ontario, Canada

Theodore Hupp
Institute of Genetics and Molecular Medicine
CRUK p53 Signal Transduction Laboratories
University of Edinburgh
Edinburgh, Scotland, UK
Chapter 4

CONTRIBUTORS

Osnat Alsheich-Bartok
Lautenberg Center, IMRIC
The Hebrew University Hadassah Medical School
Jerusalem, Israel
Chapter 5

Laura D. Attardi
Department of Radiation Oncology
Division of Radiation and Cancer Biology
and
Department of Genetics
Stanford University School of Medicine
Stanford, California, USA
Chapter 2

Georgy Bakalkin
Department of Clinical Neuroscience
Karolinska Institute
Stockholm, Sweden
Chapter 9

Kathryn L. Ball
Cell Signalling Unit
Division of Cancer Biology, IGMM
University of Edinburgh
Edinburgh, UK
Chapter 12

Jeremy Blaydes
University of Southampton School of Medicine
Southampton General Hospital
Southampton, UK
Chapter 6

Claude Caron de Fromentel
INSERM U590
Centre Léon Bérard
Lyon, France
Chapters 1, 3

Stéphanie Courtois
Molecular Carcinogenesis Group
International Agency for Research on Cancer
Lyon, France
Chapter 3

Jennifer A. Fraser
Cell Signalling Unit
Edinburgh Cancer Research Centre
University of Edinburgh
Edinburgh, Scotland, UK
Chapter 7

David Givol
Department of Molecular Cell Biology
Weizmann Institute of Science
Rehovot, Israel
Chapter 10

Zehavit Goldberg
Lautenberg Center, IMRIC
The Hebrew University Hadassah Medical School
Jerusalem, Israel
Chapter 5

Pierre Hainaut
Molecular Carcinogenesis Group
International Agency for Research on Cancer
Lyon, France
Chapters 1, 3

Sue Haupt
Lautenberg Center, IMRIC
The Hebrew University Hadassah Medical School
Jerusalem, Israel
Chapter 5

Ygal Haupt
Research Division
The Peter MacCallum Cancer Centre
East Melbourne, Victoria, Australia
Chapter 5

Nobuo Horikoshi
Department of Radiation Oncology
Division of Molecular Radiation Biology
University of Texas Southwestern Medical School
Dallas, Texas, USA
Chapter 11

Dadi Jiang
Department of Radiation Oncology
Division of Radiation and Cancer Biology
Stanford University School of Medicine
Stanford, California, USA
Chapter 2

Karuppiah Kannan
Cancer Pharmacology
Millennium Pharmaceuticals Inc.
Cambridge, Massachusetts, USA
Chapter 10

Yaara Levav-Cohen
Lautenberg Center, IMRIC
The Hebrew University Hadassah Medical School
Jerusalem, Israel
Chapter 5

Nicola J. Maclaine
Institute of Genetics and Molecular Medicine
CRUK p53 Signal Transduction Laboratories
University of Edinburgh
Edinburgh, Scotland, UK
Chapter 4

Carl G. Maki
Department of Radiation and Cellular Oncology
University of Chicago
Chicago, Illinois, USA
Chapter 8

Sarah E. Meek
Cell Signalling Unit
Division of Cancer Biology, IGMM
University of Edinburgh
Edinburgh, UK
Chapter 12

Vikram Narayan
Cell Signalling Unit
Division of Cancer Biology, IGMM
University of Edinburgh
Edinburgh, UK
Chapter 12

Magali Olivier
Molecular Carcinogenesis Group
International Agency for Research on Cancer
Lyon, France
Chapter 1

Audrey Petitjean
Molecular Carcinogenesis Group
International Agency for Research on Cancer
Lyon, France
Chapter 1

Gideon Rechavi
Chaim Sheba Medical Center
Tel Aviv University
Tel Aviv, Israel
Chapter 10

Vladana Vukojevic
Department of Clinical Neuroscience
Karolinska Institute
Stockholm, Sweden
Chapter 9

Tatiana Yakovleva
Department of Clinical Neuroscience
Karolinska Institute
Stockholm, Sweden
Chapter 9

Valentina Zuckerman
Lautenberg Center, IMRIC
The Hebrew University Hadassah Medical School
Jerusalem, Israel
Chapter 5

PREFACE

Our understanding of human cancer in the past 40 years has been driven by linking innovative concepts and cutting edge technologies to key problems identified by clinical research. Some of the successes in cancer genetics identified from clinical work have been the identification of specific gene deletions in human chromosomes, the use of PCR-based cloning methodologies to identify and clone human cancer genes, the validation of the human cancer genes using transgenetic technologies in the mouse, and the ability to sequence whole genomes that has recently allowed a collation of all somatic and germline mutations in a human genome. In the same generation, entirely different disciplines involved in basic life science research have used model organisms like yeast, flies, worms, and cancer causing animal viruses as tools to develop windows to see into the machinery of the cell life cycle. The discoveries of pro-apoptotic genes, oncogenes, and covalent control mechanisms like phosphorylation and ubiquitination using the tools of science and technology have all been awarded Nobel prizes for their contribution to our understanding of how cells work. The discovery of p53 using the tumor causing animal virus SV40 falls into this pioneering period of biological and medical research. Now, at the 30th year anniversary following the discovery of p53, the international community has demonstrated the fundamental role of p53 in cancer suppression, reproduction, ageing, and anti-viral immunity, further cementing the fundamental role of p53 as a key gene maintained by natural selection to contribute to fitness and health. Although knowledge on p53 continues to advance in leaps and bounds, and is all revealed in international journals, it is relatively difficult for students entering the cancer field or p53 field to get a historical or practical grasp on fundamentals of p53.

This volume, *p53*, was developed primarily as a resource for students to have access to key ideas in the field that have developed over the years including how transgenics have been used to study p53, how clinical genetics have identified and studied mutations in p53 found in human cancers, how p53 can be regulated by post-translational modification, and how key drug targets have been defined, namely MDM2, which has provided fundamental approaches for defining how p53 can be activated with potentially therapeutic effect. This book is by no means comprehensive and the large number of reviews published in peer-review journals always provide the cutting edge ideas developing in the field. However, the key concepts in the chapters included provide a perspective on key paradigms in the p53 field.

Theodore Hupp, PhD

CHAPTER 1

TP53 Mutations in Human Cancers:

Selection versus Mutagenesis

Magali Olivier,* Audrey Petitjean, Claude Caron de Fromentel and Pierre Hainaut

Abstract

The tumor suppressor gene *TP53* differs from most other cancer-related genes by the very high prevalence of missense mutations which result in the expression of a mutant protein. Considerable variations are observed between mutation patterns from different types of cancer and from different population groups, reflecting both mutagenesis and selection processes. These mutations are compiled in a database which includes information on tumor histology and patient characteristics, allowing the analysis of *TP53* mutation patterns according to various parameters (http://www-p53.iarc.fr/). *TP53* mutations are also observed in the germline and are associated with a syndrome of early onset cancers, the Li-Fraumeni syndrome. Germline and somatic mutations are very similar and affect codons located in the DNA-binding domain of the protein. Six major hotspot codons account for 30% of all mutations. Most mutations lead to proteins with impaired transactivation activities. However, all mutations are not equivalent. In addition to the loss of wild-type activity, some mutants exert dominant-negative effects and/or acquire new pro-oncogenic activities. Our understanding of the behavior of mutant p53 functions is expanding and holds promises for applications to cancer risk assessment, early diagnosis, prediction of disease outcome, as well as for development of new therapeutic strategies.

Introduction

Cancer growth involves the sequential accumulation of genetic alterations in genes controlling cell proliferation, lifespan, responses to stress, relationships with neighbours and gene homeostasis.[1] Amongst these alterations, the *TP53* tumor suppressor gene (OMIM #191170) represents a focal point, irrespective of the tissue and cellular origin of the tumor.[2] *TP53* encodes the p53 protein, a transcription factor that controls the expression of several proteins involved in cell-cycle control, DNA-repair, apoptosis and differentiation. The p53 protein acts by inhibiting the growth of cells exposed to chemical or physical stress, including cancer cells. Thus, loss of p53 functions promotes cell growth under conditions which suppress the proliferation of normal cells. The special role of *TP53* in cancer protection is also illustrated by the fact that Li-Fraumeni Syndrome (LFS), a familial syndrome of predisposition to multiple cancers, is caused by germline *TP53* mutation.[3]

*Corresponding Author: Magali Olivier—International Agency for Research on Cancer, 150 Cours Albert Thomas, 69372 Lyon, Cedex 8, France. Email: molivier@iarc.fr

p53, edited by Ayeda Ayed and Theodore Hupp. ©2010 Landes Bioscience and Springer Science+Business Media.

TP53 alterations typically include loss of alleles, gene mutations and inactivation of the protein by sequestration by viral or cellular proteins.[4] A database of mutations reported in human cancers is maintained at the International Agency for Research on Cancer (http://www-p53.iarc.fr/).[5] The nature and distribution of mutations vary among cancer types and population groups. Two main factors contribute to the shaping of a tumor-specific "mutation pattern". The first is mutagenesis: the type of damage caused by a mutagen can be specific in its nature and DNA sequence context, and the rate of mutation formation is limited by the cell's capacity to repair DNA lesions. The second is biological selection: only those mutants that have significant changes in their functional properties will induce a proliferative advantage and contribute to cancer. Weighting the contribution of these two factors provides interesting clues on the molecular mechanisms involved in the etiology and pathogenesis of human cancers. Mutations are also useful biomarkers in epidemiological and clinical studies and for patient management.

TP53 Alterations in Human Cancers

TP53 *Mutation Databases*

Soon after the identification of *TP53* as a frequent target gene for mutation in cancer,[6] it became evident that mutation patterns could significantly differ from one cancer to an other.[7,8] This observation led to the compilation of computerized lists of mutations that have now evolved into complex databases. The *TP53* database, maintained and developed at the International Agency for Research on Cancer (IARC *TP53* Database, http://www-p53.iarc.fr/), includes somatic and inherited mutations or variations that have been reported in the literature since 1989. Independent datasets or mirror datasets are maintained by other groups, providing a variety of analysis tools for data mining (see list at http://www-p53.iarc.fr/p53databases.html).

The information compiled in the IARC *TP53* database includes precise identification of the mutation, detailed description of tumor specimen, patient demographics and, when available, individual risk factors, genetic background and clinical parameters. It is thus possible to search for associations between mutation patterns and individual risk factors. Curated data also include information on biological activities and structural properties of p53 mutant proteins and provide a list of mouse models with engineered *TP53* (Fig. 1).

It should be noted that the database is affected by several intrinsic biases.[9] First, only a minority of publications describes molecular epidemiological studies with adequate controls and exposure groups. Second, as the database is exclusively based on peer-reviewed literature, it reflects changing trends in reporting and publishing of mutations. Other biases may result from the use of different methods for mutation detection. Despite these limitations, the database is a powerful tool to retrieve and analyze large sets of mutation data and generate hypotheses about their causes and consequences.

Sequence Variations

Several polymorphisms in the coding and noncoding regions of the *TP53* gene have been identified in human populations (see list at http://www.p53.iarc.fr/PolymorphismView.asp). Most polymorphisms are located in introns, outside consensus splicing sites, and the functional consequences of these variations remain largely unknown. With the recent discovery of p53 isoforms that are generated by use of an alternative promoter or alternative splicing,[10] some of these variations may affect the production or stability of some isoforms. Indeed, an intronic polymorphism which consists of a 16 bp duplication in intron 3, p53PIN3 (rs17878362; A1: nonduplicated; A2: duplicated), has been shown to affect p53 mRNA levels, with the presence of the A2 allele being correlated with lower p53 mRNA levels and lower p53 activity.[11] Adding to the complexity of p53 regulation, a recent study showed that the p53 regulatory protein MDM2, which mainly regulates p53 through protein-protein interactions, is also able to bind p53 mRNA and facilitate its translation, and that silent mutations within the N-terminus of p53 can abrogate this effect.[12] Thus, synonymous polymorphisms may affect p53 function through this new mechanism.

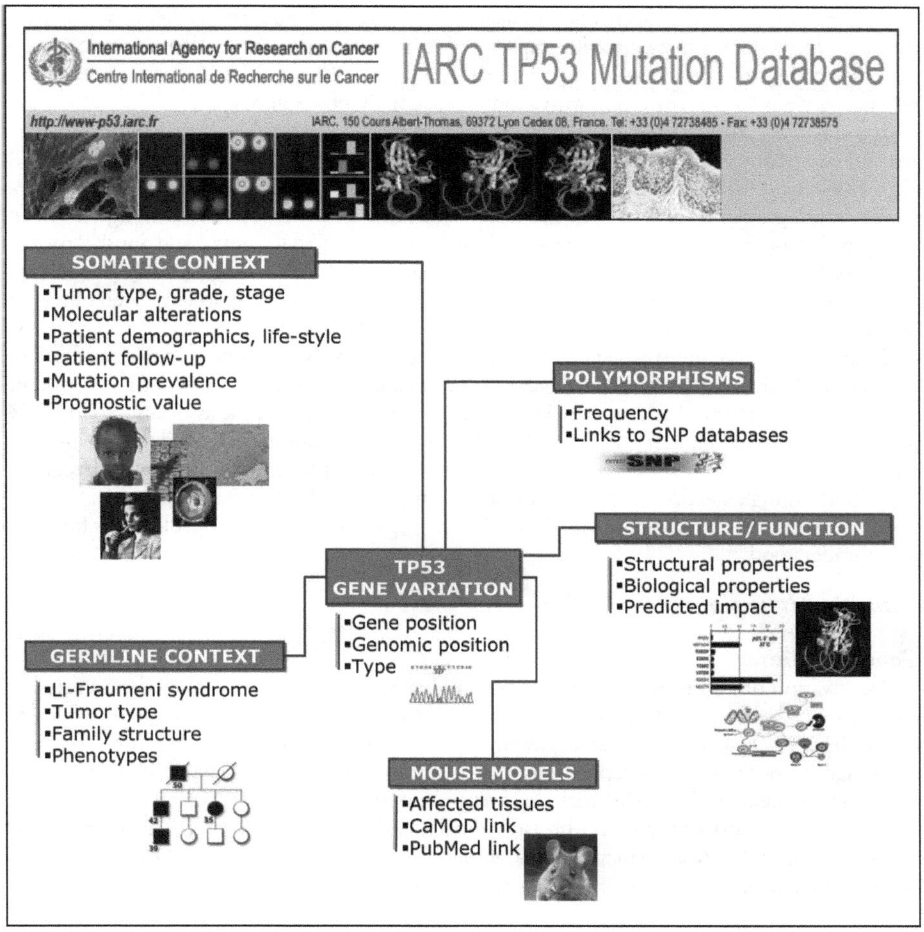

Figure 1. IARC TP53 Database online search system. The IARC *TP53* database can be searched and analyzed through a web interface (http://www-p53.iarc.fr/). Entire datasets, or sets of data selected according to user's queries, can be displayed and downloaded in tabular as well as graphical formats. A user guide is available that describes database and web site contents.

In the coding sequence, four polymorphisms alter the amino acid sequence of p53. There is sufficient molecular evidence that p53 function is affected by these polymorphisms for two of them only. One is a nonsynonymous variation in exon 4 (rs1042522; G/C) that leads to an arginine (R) to proline (P) amino-acid substitution at codon 72 (p53R72P). This residue is located in the proline-rich domain that is thought to be essential for a full p53 apoptotic response. p53 proteins containing a R or P allele display subtle changes in biochemical and functional properties that result in a more potent capacity of the R allele to induce apoptosis, while the P allele is more efficient in inducing cell cycle arrest.[13-15] However, the tissue specificity of these functional differences and their in vivo significance remains to be demonstrated. The other functional polymorphism is a rare C/T variation reported in African populations (rs1800371) that leads to a proline (P) to serine (S) amino-acid substitution at codon 47 (p53P47S). This codon is located in the transactivation domain and is close to a serine residue important for p53-dependent apoptosis induced by DNA damage as well as cellular senescence induced by oncogenic stress.[16] P47S was shown to be a poorer substrate for phosphorylation of serine 46 by p38 MAPK. However, the consequence on protein transactivation capacity has shown conflicting results. While yeast assays showed a more potent transactivation capacity of 47S compared to 47P, other assays showed a decreased ability to transactivate two p53 target-genes, p53AIP1 and PUMA, but not other p53 response genes, which is correlated with a lower capacity to induce apoptosis.[17,18]

Residue 72, although not conserved, is located within the proline-rich region and may affect the structure of the putative SH3-binding domain. Sharp ethnic differences in codon 72 allele frequencies have been observed.[19] In the Northern hemisphere, the Pro72 allele shows a North-South gradient, from 0.17 in Swedish Saamis to 0.63 in African Blacks (Nigerians). In Western Europe (France, Sweden, Norway), North America (USA), Central and South America (Mexico, Costa-Rica, Peru) and Japan, the most common allele is Arg72, with frequencies ranging from 0.60 to 0.83. However, frequencies of Pro72 superior to 0.40 have been observed in African-American and Chinese populations.

Many studies have investigated the association of *TP53* polymorphisms with increased risk of cancer. p53R72P and p53PIN3 have been the most extensively studied, however no consistent results have been found. For example, a meta-analysis of 13 studies on the association of 3 *TP53* polymorphisms, including p53R72P and p53PIN3, and lung cancer risk failed to find any significant association.[20] In breast cancer, a large recent study has found that none of the frequent *TP53* SNPs (Single Nucleotide Polymorphisms) were associated with breast cancer risk.[21]

Overall, the functional significance and clinical impact of *TP53* polymorphisms is far from being understood.[22]

Somatic Alterations

Gene Mutations

Inactivation of p53 tumor suppressor functions by gene mutations is one of the most frequent alterations found in human cancers. Mutations are found in almost every type of cancer by the time the capacity for invasive growth has been acquired. The overall mutation frequencies range from 5% to 50% depending on the tumor type (Fig. 2). Malignancies with high mutation frequencies (40-55%) include ovarian, esophageal, colorectal, head and neck and lung cancers. Tumors of the brain, breast, stomach and liver show an intermediate mutation frequency (20-35%). Malignancies with low mutation frequency include cervical cancer, neuroblastoma, leukemia, sarcoma, testicular cancer and malignant melanoma.

Protein Interactions

In several cancers that do not carry *TP53* mutations, inactivation of p53 occurs by protein-protein interactions that either promote p53 degradation or inhibit its activity. In Human Papilloma Virus (HPV) related cervical cancers, the *TP53* gene is often wild-type, but the protein is inactivated by the HPV protein E6. E6 binds p53 in association with the cellular protein E6AP and targets it for proteasome-mediated degradation.[23,24] In soft tissue sarcomas, the *HDM2* gene is amplified and over expressed without evidence of *TP53* gene mutation in about 30% of the cases, leading to destabilization and inactivation of p53.[25] Amplification of *HDM2* is also observed in other tumors, but not always correlated with the presence of wild-type *TP53*.[26] In retinoblastoma, caused by Rb1 deficiency, a recent study showed that amplification of the *HDMX* gene and increased expression of HDMX protein was responsible for the suppression of the p53 apoptotic response triggered by Rb1 deficiency.[27] In neuroblastoma, Twist1, a protein involved in development, has been found to interact with and inhibit p53 activities.[28-30]

Other Alterations

Other modes of inactivation of structurally normal p53 have been proposed in testicular cancers, but the mechanism is unknown.[31] Loss of function through cytoplasmic retention has been observed in neuroblastoma and in inflammatory breast cancer.[32,33] A similar phenomenon has been proposed in some hepatocellular carcinomas associated with Hepatitis B Virus (HBV) infection, since transgenic expression of the HBx protein in mouse liver blocks p53

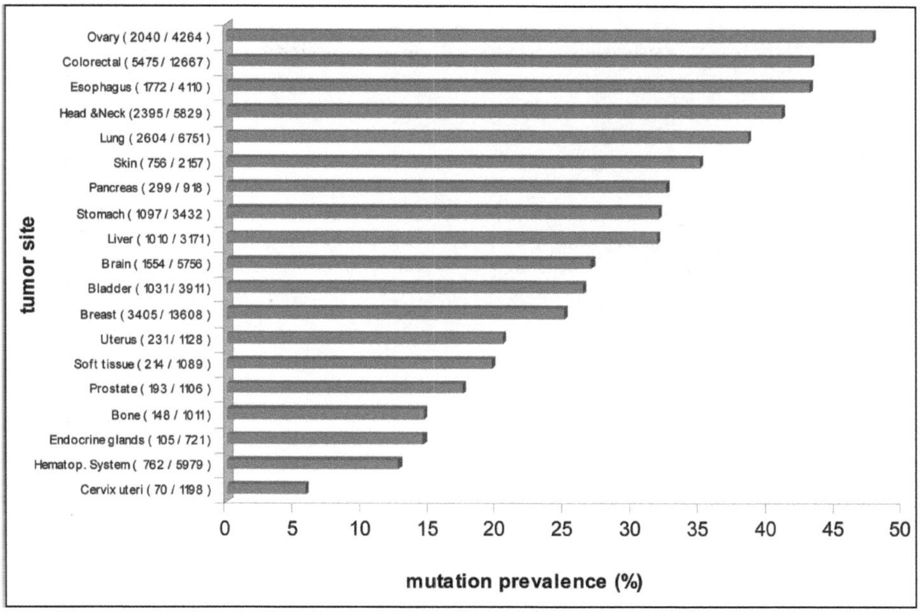

Figure 2. TP53 mutation prevalence in sporadic cancers. The proportion of tumors with somatic *TP53* mutations is indicated. Data from IARC *TP53* Database (R13, November 2008).[9]

entry into the nucleus.[34] Mutations in p53 downstream effectors have been searched for, but have produced largely negative results.

Germline Mutations

Inherited *TP53* mutations are associated with a rare autosomal dominant disorder, the Li-Fraumeni syndrome (LFS). LFS is characterized by familial clustering of tumors diagnosed before 45 years of age, mostly sarcomas, breast, brain and adrenocortical cancers.[35] Families with incomplete features of LFS are referred to as Li-Fraumeni-like Syndrome (LFL), for which several clinical definitions have been proposed.[36] In LFS/LFL patients, normal cells are heterozygous (*TP53* wild-type/mutant), but in cancer cells the wild-type allele is usually lost or inactivated by somatic mutation.

Although breast cancers, sarcomas (soft tissue sarcomas and osteosarcomas), brain tumors and adrenocortical carcinomas account for about 80% or all tumors arising in *TP53* germline mutation carriers, the spectrum of tumors observed in mutation carriers is wide (Fig. 3). This heterogeneous tumor patterns in LFS/LFL families may be explained in part by differences in *TP53* mutation types and their functional consequences.[36] In addition, polymorphisms in the *TP53* pathway have been shown to have modifier effects on *TP53* germline mutations.[37,38]

Types of Mutations

The type and distribution of inherited and somatic *TP53* mutations are very similar (Fig. 4). In contrast to many tumor suppressors such as *RB1*, *APC* or *BRCA1*, which are often inactivated by deletion, frameshift or nonsense mutations, most *TP53* alterations are missense mutations (73%) (Fig. 4A). About 35% of them fall within five "hotspot" codons (Fig. 4B) detectable in almost every types of cancer (codons 175, 245, 248, 273, 282). The corresponding residues are located in the DNA-binding domain of the protein. This domain has a complex structure made of two beta-sheets (forming a sandwich) bridged by flexible loops and helixes.[39] These loops are kept in place by the binding of an atom of zinc. The hotspot residues play important roles either

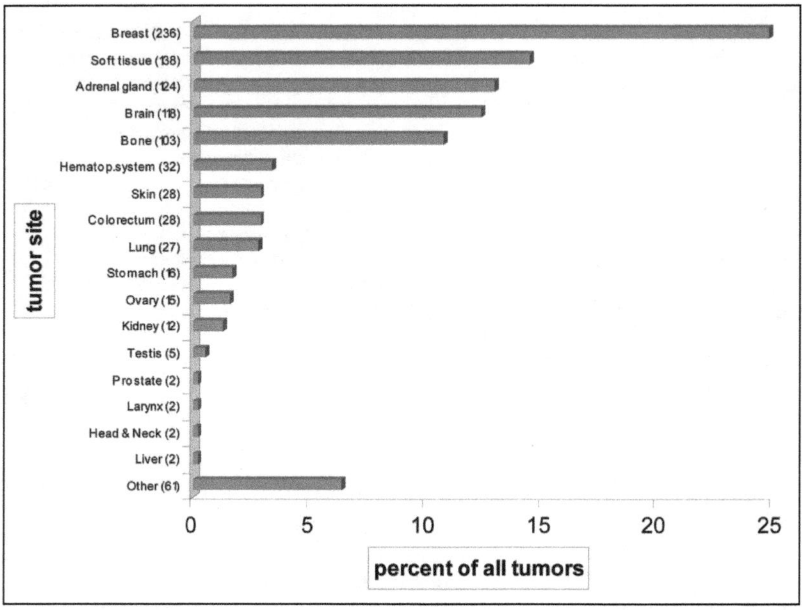

Figure 3. Tumor spectrum in individuals with a germline TP53 mutation. The proportion of specific tumor types among all tumors reported in confirmed *TP53* germline mutation carriers is indicated. Data from IARC *TP53* Germline Database (R13, November 2008, http://www-p53.iarc.fr/Germline.html).[9]

in protein-DNA contacts (codons 248 and 273) or in maintaining the conformation of the protein (175, 245, 282) (Fig. 5), explaining their high mutation frequency in cancer. However, all codons within the DNA-binding domain have been reported to be mutated in cancer and 80% of all mutations fall within this domain, reflecting its importance in the tumor suppressor function of p53. The main function of this domain is to interact with specific DNA sequences that regulate the transcription of p53 target genes. The main consequence of these mutations is thus a loss of p53 capacity to regulate its target genes. However, different types of mutations show different degrees of loss of function. Kato et al[17] have performed, in yeast assays, a systematic analysis of the transactivation capacity of all possible point mutants on several p53 responsive-elements. They showed that mutants that are found in cancer display severe loss of function, while mutants that retain some activity are rarely found in human tumors. These results show the importance of p53 transactivation capacity in its role as a tumor suppressor.

Sequence Variation and Phenotype

Several lines of evidence suggest that germline mutations may illicit tissue specific effects. The most striking example is the R to H mutation at codon 337 (R337H) that has been found in the Brazilian population and shown to predispose preferentially (although not exclusively) to childhood adrenocortical carcinoma.[40,41] Functional analysis revealed that this mutant is pH-sensitive, i.e., inactive (mutant-like) at pH>7.7 and active (wild-type-like) at pH<7.7.[42] The protein may thus adopt a mutant or unfolded conformation only under particular physiological conditions. Although this effect does not explain the tissue-specificity of the R337H mutant, this example illustrates the fact that mutant p53 protein function may depend on the cellular context.

Other evidence come from genotype-phenotype analysis of a large dataset of *TP53* germline carriers.[36] These analyses have shown that missense mutations affecting residues located in the L2 and L3 loops of the p53 structure, that bind to the minor groove of DNA, were preferentially associated with brain tumors, whereas those outside the DNA-binding surface (in the non-DNA-binding loops, beta-sheets and oligomerization domain) were associated with

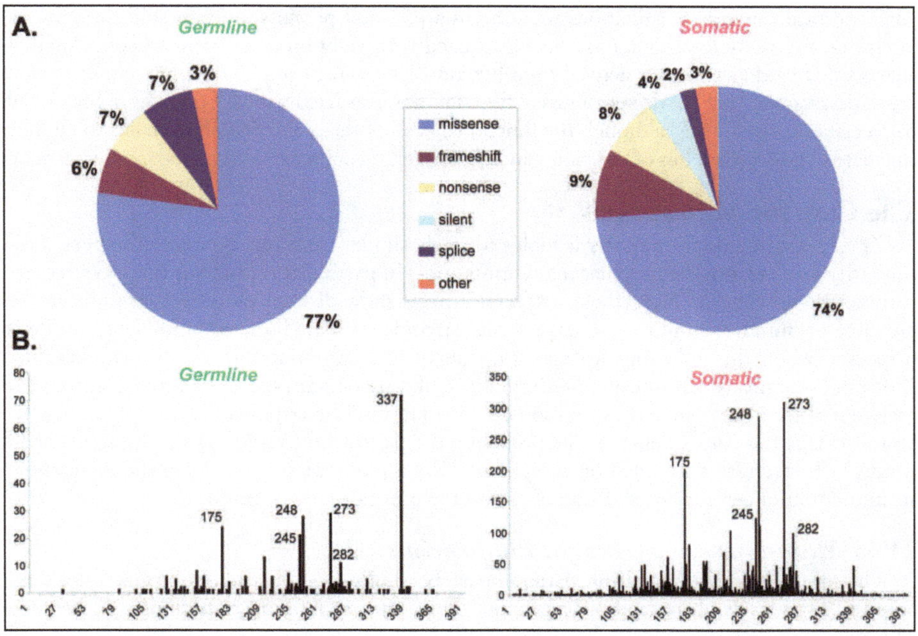

Figure 4. Comparison of germline and somatic TP53 mutations. A) Pie charts showing the proportion of the different types of *TP53* germline and somatic mutations. B) Histograms displaying the position of the germline and somatic point mutations in the coding sequence of the *TP53* gene. Data from the IARC *TP53* Database (R13, November 2008).[9]

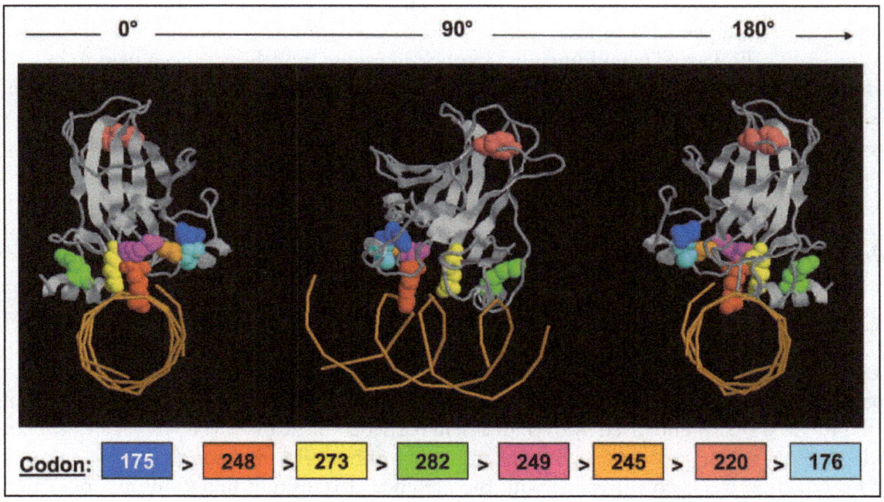

Figure 5. Structural localization of the most frequent TP53 mutations. 3D view of the core domain of the p53 protein in complex with DNA. This domain has a complex structure made of two beta-sheets (forming a sandwich) bridged by sets of loops and helixes. These loops are kept in place by the binding of an atom of zinc (zinc coordination by codons 176, 179, 238 and 242), which is essential for the stability of the whole structure. Codon 248 makes contact with DNA in the minor groove of the helix, whereas codon 273 makes contact in the major groove. Codons 175, 245, 249 and 282 play important roles in the conformation of the protein. Structure from Cho et al.[39]

adrenocortical carcinoma. Mutations resulting in a p53-null phenotype (frameshift deletions or insertions and nonsense mutations) were associated with early onset of brain tumors. Another analysis that used annotations derived from functional assessment of p53 mutants transactivation capacities, showed that the degree of loss of function was associated with age of onset of breast and colorectal cancers.[43] Thus, although the functional basis of these observations remains to be fully elucidated, the degree of loss of function may affect mutation penetrance in a tissue specific manner.

The Case for Mutagenesis

TP53 somatic mutation patterns look extremely similar from one cancer to the next. This similarity results from the fact that many mutations, in particular transitions at CpG sites, are common in all cancers. Nevertheless, several cancers show distinct patterns that indicate the presence of mutations induced by exogenous carcinogens. An "induced" mutation profile is suspected when the following features are present: (1) tumor-specific or exposure-specific "hotspot" mutations; (2) unusual predominance of a particular type of base substitution; (3) preferential accumulation of the mutation on the nontranscribed strand of DNA (strand bias).[44] Strand bias is the consequence of the preferential repair of DNA adducts on the transcribed strand by transcription-coupled repair systems.[45] This phenomenon results in the preferential accumulation of certain types of mutations on the nontranscribed strand.

TP53 *Mutations as Carcinogen Fingerprints*

The most distinctive mutation patterns have been observed in studies on populations exposed to high levels of mutagens. In general, the spectrum of *TP53* mutations is in keeping with mutation patterns generated experimentally by the suspected agents. A well-documented example is that of lung cancer, where a high prevalence of G>T mutations at specific residues have been correlated with exposure to tobacco and with a site of DNA adduct by benzo(a)pyrene, a major carcinogen contained in tobacco smoke.[46] Other examples include hepatocellular carcinoma (dietary aflatoxins) and nonmelanoma skin cancers (solar UVs). In several other cancers, such as bladder and esophageal carcinomas, specific mutation profiles have been observed, but the mutagens have not been clearly identified. These questions have been extensively addressed in recent reviews.[46-48]

Recently, cells derived from human p53 knock-in mouse models have been used to examine induced human p53 gene mutations in cell cultures exposed to mutagenic factors.[49] Mutations observed in these models were very similar to the ones observed in human tumors. Thus, these models provide a basis for generating experimental mutation patterns in human p53 and, together with the analysis of mutation patterns in human tumors, may help to identify carcinogens and mutagenic processes involved in the development of cancer with specific mutation patterns.

Mechanisms of Mutation

Table 1 presents a simple key for the the interpretation of different types of mutations found in sporadic tumors. The most frequent mutations (25%) are transitions (purine to purine or pyrimidine to pyrimidine) at cytosines within CpG sites. These transitions can, in the first instance, be considered as resulting from an endogenous mutagenic process. Spontaneous deamination of methylated cytosine occurs frequently at CpG sites, leading to a substitution to thymine. This process is greatly enhanced by oxyradicals, in particular nitric oxide (NO), that are generated endogenously during processes such as inflammation or bacterial infection.[50,51] In colon cancer, NO production has been correlated with the presence of transition mutations at CpG sites in *TP53*.[52] In contrast, transversions (purine to pyrimidine or vice-versa) at G bases (G:C to T:A or G:C to C:G) are often caused by bulky carcinogens in various experimental systems. G:C to A:T transitions at non-CpG sites can be induced by many different agents, in particular N-nitroso compounds, oxidizing agents and alkylating agents (for review see ref. 53). Altogether, non-CpG transitions and transversions at G bases represent around 40% of all mutations. About 10% of mutations are deletions with the majority being small deletions. Micro deletions, in particular in CG base repeats, are thought to primarily result from polymerase slippage during replication.[54]

Table 1.* TP53 *mutations in sporadic cancers and suspected mechanisms of mutagenesis

Mutation Type	Cancer with High Prevalence	Suspected Agents or Mechanisms
Insertions	Head and neck, Esophagus	
Deletions	Head and neck	Polymerase slippage; Irradiation?
CC tandem	Skin (other than melanoma)	UV
A:T bases	Esophagus (SCC), Head and neck	Acetaldehyde?
A:T>T:A	Liver (Hemangiosarcoma)	Vinyl chloride
G:C>A:T	Bladder, Many other cancers	Alkylating agents? Aromatic amines? Radiations?
G:C>A:T at CpG	Colon, Brain, Stomach, other cancers	Spontaneous deamination of methylated cytosines
G:C>C:G		
G:C>T:A	Lung	PAH (Benzo(a)pyrene)
	Hepatocellular carcinoma (hotspot at codon 249)	Aflatoxin B1

Data from IARC *TP53* database and references 44,53,91. SCC= squamous cell carcinoma.

Nucleotide substitution rates[55] derived from human-mouse aligned sequence of chromosomes 21 and 10 have been applied to *TP53* wild-type and mutated sequences to estimate the propensity of each mutation to occur as a neutral process from replication error or endogenous mutagenesis.[5] The comparison of these mutation rates with frequency of occurrence in cancers shows that rare mutants have the lowest median nucleotide substitution rates while frequent mutants have the highest rates (Fig. 6). Thus, underlying substitution rates are highly associated with mutation frequency, showing that mutagenesis (spontaneous or carcinogen-induced) plays a major role in shaping mutation patterns.

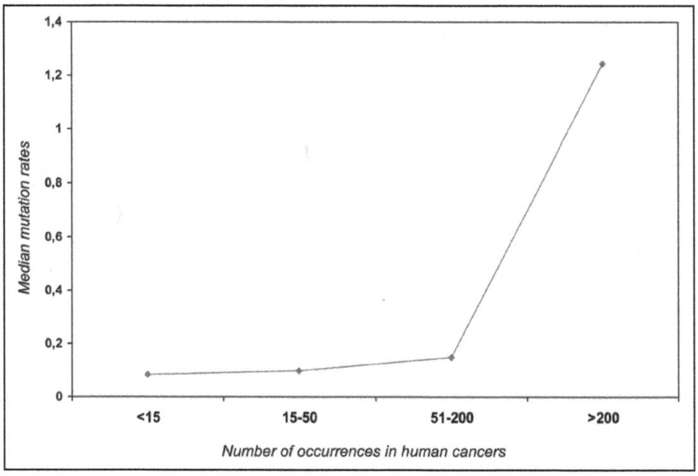

Figure 6. *TP53* mutation rates and frequency in cancer. Single amino-acid substitutions were grouped into four categories according to their frequency in the somatic dataset of the IARC TP53 database (R13, November 2008). The median mutation rates were calculated for each group of mutants. These rates were derived from dinucleotides substitution rates calculated for all point mutations according to Lunter et al.[55] Only mutations detected by DNA sequencing and located within the DNA-binding domain were included. Data from the IARC *TP53* Database (R13, November 2008).[9]

The Case for Selection

Structural and Functional Properties of Mutations

TP53 mutations cluster within exons 5 to 8, encoding the DNA-binding domain of the protein. After initial reports that mutations cluster in this central portion of the coding sequence, many studies were limited to exons 5 to 8, resulting in an over-representation of these mutations in databases. Nonetheless, when taking into account only studies that have screened the entire coding sequence, 70% (3326/4768) of the mutations are located within exons 5-8 (Fig. 7). Other major functional domains such as the oligomerization domain, and regulatory sites located in the N-terminus and C-terminus of the protein are rarely mutated in cancer (Fig. 7). For example, only 1% of reported mutations fall within the transactivation domain (TA) which also contains the binding site for Mdm2, the main regulator of p53 stability. The C-terminus, which participates in the regulation of DNA-binding activity, contains less than 5% of all mutations. Moreover, mutations at regulatory sites, such as Ser 15 and 37 (phosphorylation by ATM and/or DNA-PK), Phe 19, Leu 22, Trp 23 and Leu 26 (interactions with Mdm-2 and TBP), Ser 315 (phosphorylation by cyclin-dependent kinases), Ser 376 and 378 (interactions with 14-3-3 sigma), Lys 370, 372, 373, 381 and 382 (sites of acetylation) and Ser 392 (phosphorylation by CKII), are extremely rare. These observations suggest that mutation of any of these residues is not sufficient in itself to fully inactivate p53. In the case of residues of the TA domain, studies have shown that mutation of at least two of the residues is required to inactivate p53 transcriptional activity.[56] In fact, post-translational modifications of these residues have only modulatory roles and blocking single regulatory modifications of p53 often lead to subtle effects on protein activities in in vitro assays and have even less phenotypic impact in mouse-models.[16,57]

In the DNA-binding domain, missense mutations have been reported at all residues, but with striking variations in prevalence. About 35% of missense mutations in this domain fall at 6 hotspot codons (175, 245, 248, 220, 273, 282). Four of these codons correspond to Arg residues (175, 248, 273 282) involved in protein-DNA interactions, either by direct contact with DNA (residue 248 and 273) or by stabilization of the DNA-binding surface (residues 175, 282).[39]

Recent structural analysis of p53 wild-type and mutants proteins have shown that the stability of full-length p53 is highly dependent on its core domain, which is highly structured but poorly stable and melts at slightly above body temperature.[58] The N- and C-terminal domains are largely unstructured apart from small regions that exhibit helical structures. Common cancer mutants exhibit a variety of distinct local structural changes and can be grouped in distinct classes, which often coincide with the location of the mutations in the structure. DNA-contact mutants, such as R273H, have only a small effect on the thermodynamic stability of the protein, and the overall structural scaffold is largely preserved. Thus they merely remove the DNA-contact residue without inducing structural perturbations in neighbouring residues. In contrast, mutations in the DNA-binding surface show different degrees of destabilization (G245S<R249S<R282W). Another class correspond to mutations that affect residues involved in zinc binding (C242S) or in structuring the zinc-binding site (R175H). These mutations are highly destabilizing, even at low temperatures. All tested beta-sandwich mutants (e.g., V143A, F270C, Y220C) are also highly destabilized. Overall, these structural analyses show the diversity of the structural impacts of the mutants observed in cancer (see Joerger et al[59] for review).

Biological Properties of p53 Mutants

Loss of Function

The main activity of the p53 protein is to regulate gene expression (activation or repression) through specific DNA-binding to responsive-elements located in p53 target genes.[60] A systematic assessment of the transactivation capacity of all possible mutants on eight different p53 responsive-elements has been performed in yeast assays.[17] A large number of mutants have also

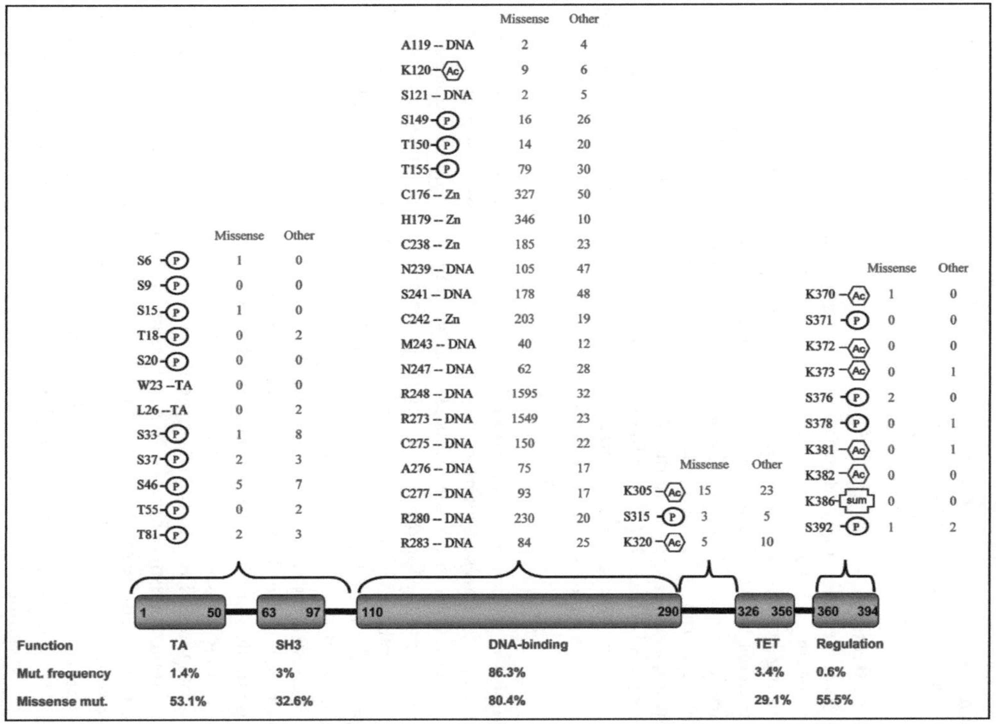

Figure 7. Mutations at known structural and regulatory sites in p53 protein. A linear representation of the p53 protein is shown with its main regulatory domains (TA, transactivation domain; SH3, SH3-like motif; DNA-binding, core domain involved in sequence specific DNA binding; TET, tetramerization domain; Regulation, regulatory domain). Bottom: distribution of the mutations in the different functional domains of the protein (Mut. Frequency), and proportion of missense mutations in each domain (missense mut.). Percent is calculated from the total number of mutations reported in the IARC *TP53* database, taking into account only studies that have screened the entire coding region of the *TP53* gene (exon 2 to 11). Top: number of missense and other mutations reported at functionally and/or structurally important residues, as recorded in the somatic dataset of the IARC *TP53* Database (R13, November 2008).[9] Residues are identified with the wild-type amino-acid (one letter code), codon number and attributed function (P, phosphorylation site; TA, transactivation; DNA, DNA binding; Ac, acetylation; Sum, sumoylation; Zn, zinc binding).

been tested in human cell assays for their capacity to regulate target genes or to induce apoptosis or growth arrest. These analyses showed that, while all hotspot mutations exhibit complete loss of function, other mutations exhibit a wide diversity of loss of function. Some mutants retain an activity on some p53 responsive-elements and completely lose their activity on other elements, while some mutants retain a partial activity on most response-elements. Different amino-acid substitutions at the same residues may have a very different impact. For example, the hotspot mutant R175H is a complete loss of function mutant, while the R175P mutant retains the capacity to transactivate *WAF1* but is defective for *BAX* activation and fails to induce apoptosis.[61] Missense mutants outside the DNA-binding domain are more likely to retain some activity compared to missense mutants within the DNA-binding domain, which certainly explains the fact that missense mutations cluster in this domain, while truncating mutations are more frequent in the N- and C-termini.

Dominant-Negative Effects

p53 transcriptional activity relies on the formation of tetramers (dimers of dimers). Since most mutant proteins retain an intact oligomerization domain, mutants may interfere with wild-type p53 by forming hetero-oligomers less competent for specific DNA binding. The capacity of mutant proteins to interfere with the wild-type form (dominant-negative effect, DNE) has been studied in yeast and human cell assays in various settings.[62-64] Data have been produced on more than 200 mutants. These data show that DNE may be promoter- and cell type dependent. However, a recent analysis of the IARC *TP53* database confirmed initial studies that showed a correlation between DNE and frequency of occurrence in cancer.[5] DNE may thus play a significant role in the selection of mutation in addition to loss of function and may explain why missense mutations are more frequent than truncating mutations.

Gain of Function

Over 20 years ago, it was proposed that mutant p53 may exert pro-oncogenic effects, and that mutation was turning p53 into some kind of oncogene.[65] Indeed, most cancer cells accumulate mutant p53 protein, even in distant metastases, suggesting that mutation does more than inactivate protein function. Moreover, several mutant proteins have been shown to display molecular and biological activities that are independent of wt p53 and that promote tumor growth. These properties are refereed to as gain of function (GOF). They include activation of genes normally unaffected or repressed by wild-type p53, interference with other transcription factors and resistance to specific drugs. It is only recently that in vivo evidence of GOF have been demonstrated and that some clues about the mechanisms involved started to emerge. Indeed, mouse models of LFS using knock-in of R175H and R273H mutants clearly showed that tumors in these knock-in mice were more metastatic and occurred in different tissues than the ones arising in p53 KO mice.[66,67] It is only recently that progress in the understanding of the molecular mechanism behind GOF has been made. It appears that p53 mutant proteins have a global impact on the transcriptome by coordinating oncogenic transcriptional responses. While mutant proteins have lost sequence-specific DNA binding activity, they are still able to associate with specific loci on DNA by utilizing different mechanisms. Thus, alterations of target DNA selectivity may be the driving force of mutant p53 specific transcription underlying the growth-promoting effects of mutant p53.[68]

TP53 Mutations as Biomarkers

TP53 *Mutations as Markers of Exposure*

As discussed before, specific *TP53* mutations are observed in some types of cancer that are caused by known carcinogen exposure. An example for such a "mutagen fingerprint" is *TP53* mutation at codon 249 in hepatocellular carcinoma in regions of the world characterized by high levels of the mutagen aflatoxin B1 and endemic HBV infection. Recently, *TP53* mutations have been detected in surrogate sources of genetic material such as free circulating DNA isolated from plasma.[69] Plasma *TP53* mutations can be detected in the blood of precancer and cancer patients, with potential application for early cancer detection.[70] Thus, *TP53* mutations have multiple applications as markers of mutagenic exposure, or as intermediate end-points in assessment of cancer occurrence and progression.[71]

TP53 *Mutations as Markers of Clonality*

Mutations in *TP53* are useful markers of tumor clonality. They have been used as such to compare, in individual patients, separate clusters of tumor cells from the same lesion, or multiple lesions arising in the same tissue.[72-74] Mutations are also useful for the follow-up of patients for detection of minimal residual disease, in comparison between primary and recurrent tumors and in tracing the origin of distant metastases.[72] However, it should be kept in mind that in many cancers, *TP53* mutations appear to be a relatively late event in the sequence of genetic alterations that lead to tumor progression. Thus, the finding of different *TP53* mutations in separate clusters of a tumor does not exclude the possibility of a clonal origin and merely indicates that the two clusters might have diverged at some point in tumor progression.

TP53 *Mutations for the Detection and Follow-Up of Cancer Lesions*

Mutations in Precancerous Lesions

Identification of *TP53* mutations may be of interest to identify early lesions that are at a high risk of malignant evolution, particularly in tumors where *TP53* mutation is considered as an early event. For example, in esophageal and endometrial cancers, detection of a mutation in a dysplastic lesion may be considered as an indicator of high risk of malignant transformation.[75,76]

Circulating Antibodies

Circulating p53 auto-antibodies have been found in a subset of cancer patients.[77-79] In rare instances, these antibodies have been found in blood samples collected months to years before cancer diagnosis. The proportion of cancer patients with circulating p53 auto-antibodies is usually low (3-15%) but for some types of cancer this proportion is clearly higher. While mutations affect codons that are mainly located in the DNA-binding domain, p53 auto-antibodies are directed against the N- and C-terminal domains of the protein. The biological mechanisms leading to antibody production, the basis of inter-individual differences and the prognostic significance, if any, are far from being understood.

TP53 Mutations as Prognostic and Predictive Factors

Several studies have investigated the possible value of *TP53* gene mutation as a prognostic or predictive factor in cancer. It is of note that a large number of studies that have used immuno-histochemistry (IHC) as a surrogate marker for *TP53* status have failed to provide consistent results. This is due to the high rate of false positive (over expression of p53 wild-type protein is frequent) and false negative cases (a significant number of inactivating mutation are nonsense, frameshift or splice site mutations that stain negative) obtained with IHC. Hence, the use of IHC leads to an unacceptable number of misclassified cases and to a greater inter-study variability. When only studies that have used gene sequencing to assess *TP53* mutation status are taken into account, the presence of *TP53* mutations has been repeatedly associated with a poor prognosis in several types of cancers such as breast, colorectal, head and neck and leukemia (Table 2).[43] In breast cancer, *TP53* mutation was found to be an independent prognostic factor,[80,81] and has been associated with poor response to different treatment regimens.[81-83]

However, in a recent study, the presence of a *TP53* mutation was associated with good response to treatment.[84] In fact, in this study, patients were treated with a dose-dense epirubicin-cyclophosphamide regimen that targets highly proliferating tumors, a hallmark of *TP53* mutated tumors. In other types of cancer such as brain and pancreas tumors, mutations were found to be associated with both poor and good prognosis depending on the study (Table 2). These results show that the tissue type and the type of treatment may be an

Table 2. **TP53** *mutation status as prognostic factor in human cancer*

Tumor Site	Studies Reporting an Association of Mutation with Poor Prognosis	Studies Reporting an Association of Mutation with Good Prognosis	Studies Reporting No Association
Bladder	3	-	3
Bones	1	-	1
Brain	3	2	4
Breast	21	-	5
Colorectum	15	-	8
Esophagus	2	-	2
Head & Neck	6	-	2
Hematol.	11	-	-
Larynx	-	-	1
Liver	2	-	-
Lung	8	-	6
Ovary	7	1	2
Pancreas	1	1	-
Prostate	1	-	1
Soft Tissues	2	-	-
Stomach	1	-	2
Uterus	2	-	-

The number of studies is indicated. Data from the IARC *TP53* Database (R13, November 2008).[5]

important determinant of the prognostic and predictive value of *TP53* mutations. This can be explained by experimental data that show that different types of treatment activate p53 through different pathways and cause different types of post-translational modifications of p53.[85] These modifications, together with the cellular context, will influence the type of response that can range from senescence, cell-cycle arrest to apoptosis. The tissue specific effects of wild-type p53 are thus a crucial determinant of the ultimate outcome of any anti-cancer treatment that activates p53. Current knowledge of p53 biology and function is still not sufficient to accurately predict the response of a specific tumor type to a specific type of treatment according to *TP53* status. Thus, large clinical studies in the context of controlled treatment regimens are needed to define in which context *TP53* mutations may be useful to predict outcome.

Therapies Targeting TP53 *Mutations*

Several strategies that target mutant p53 have been developed to restore normal p53 function. They include molecules that reactivate mutant p53 suppressive functions in tumor cells (PRIMA, RITA, scFv)[86-88] or that target specific point mutations to restore wild-type-like structure (Phikan059 targeting R220C).[89] Peptides have also been generated that interact specifically and strongly with several p53 mutants and block their non specific transactivation capacities, inhibiting their GOF properties. These approaches lead to cell death exclusively in cells expressing mutant p53.

These therapies complement other approaches such as gene therapy that target tumors that do not express p53 mutants.[90] Indeed, reintroducing p53 by gene therapy would work best in cells that express wild-type p53 or do not express p53 due to the presence of a truncating mutation.

Conclusion and Perspectives

After 30 years of p53 investigations and the recent development of global approaches to analyze the complexity of the interactions between *TP53* mutations and other cellular factors, our knowledge of the clinical impact of *TP53* mutations is increasing and *TP53* translational research efforts are finally advancing to clinical applications.

In the future, it is predicted that good quality mutation knowledge bases will provide the first line of information on the clinical significance of mutations, and, as such, will be frequently accessed by physicians and patients alike in the process of medical decision. This trend is already very significant in the current mutation databases, which are more and more frequently consulted by physicians in search of clues for developing individualized treatments. Meeting these expectations requires a large, concerted effort to develop the power, impact and flexibility of mutation knowledge bases. The first step in this process is to develop awareness towards these issues, and lead all actors to embrace common resolutions to avoid the loss of extremely valuable, and potentially life-saving, information. In the long-term, meeting this challenge will fulfill the essential ethical requirement of returning to the largest community the benefits of years and years of basic and clinical research.

References

1. Hanahan D, Weinberg RA. The hallmarks of cancer. Cell 2000; 100:57-70.
2. Vogelstein B, Lane D, Levine AJ. Surfing the p53 network. Nature 2000; 408:307-310.
3. Malkin D, Li FP, Strong LC et al. Germ line p53 mutations in a familial syndrome of breast cancer, sarcomas, and other neoplasms. Science 1990; 250:1233-1238.
4. Harris CC. Structure and function of the p53 tumor suppressor gene: clues for rational cancer therapeutic strategies. J Natl Cancer Inst 1996; 88:1442-1455.
5. Petitjean A, Mathe E, Kato S et al. Impact of mutant p53 functional properties on TP53 mutation patterns and tumor phenotype: lessons from recent developments in the IARC TP53 database. Hum Mutat 2007; 28:622-629.
6. Baker SJ, Fearon ER, Nigro JM et al. Chromosome 17 deletions and p53 gene mutations in colorectal carcinomas. Science 1989; 244:217-221.
7. Caron de Fromentel C, Soussi T. TP53 tumor suppressor gene: a model for investigating human mutagenesis. Genes Chromosomes Cancer 1992; 4:1-15.
8. Hollstein M, Sidransky D, Vogelstein B, Harris CC. p53 mutations in human cancers. Science 1991; 253:49-53.
9. Olivier M, Eeles R, Hollstein M et al. The IARC TP53 database: new online mutation analysis and recommendations to users. Hum Mutat 2002; 19:607-614.
10. Bourdon JC. p53 and its isoforms in cancer. Br J Cancer 2007; 97:277-282.
11. Gemignani F, Moreno V, Landi S et al. A TP53 polymorphism is associated with increased risk of colorectal cancer and with reduced levels of TP53 mRNA. Oncogene 2004; 23:1954-1956.
12. Candeias,M.M. et al. P53 mRNA controls p53 activity by managing Mdm2 functions. Nat Cell Biol 2008; 10:1098-1105.
13. Dumont P, Leu JI, Della PA et al. The codon 72 polymorphic variants of p53 have markedly different apoptotic potential. Nat Genet 2003; 33:357-365.
14. Matlashewski GJ, Tuck S, Pim D et al. Primary structure polymorphism at amino acid residue 72 of human p53. Mol Cell Biol 1987; 7:961-963.
15. Thomas M, Kalita A, Labrecque S et al. Two polymorphic variants of wild-type p53 differ biochemically and biologically. Mol Cell Biol 1999; 19:1092-1100.
16. Feng L, Hollstein M, Xu Y. Ser46 phosphorylation regulates p53-dependent apoptosis and replicative senescence. Cell Cycle 2006; 5:2812-2819.
17. Kato S, Han SY, Liu W et al. Understanding the function-structure and function-mutation relationships of p53 tumor suppressor protein by high-resolution missense mutation analysis. Proc Natl Acad Sci USA 2003; 100:8424-8429.
18. Li X, Dumont P, Della PA et al. The codon 47 polymorphism in p53 is functionally significant. J Biol Chem 2005; 280:24245-24251.
19. Beckman G, Birgander R, Själander A et al. Is p53 polymorphism maintained by natural selection? Hum Hered 1994; 44:266-270.
20. Matakidou A, Eisen T, Houlston RS. TP53 polymorphisms and lung cancer risk: a systematic review and meta-analysis. Mutagenesis 2003; 18:377-385.

21. Baynes C, Healey CS, Pooley KA et al. Common variants in the ATM, BRCA1, BRCA2, CHEK2 and TP53 cancer susceptibility genes are unlikely to increase breast cancer risk. Breast Cancer Res 2007; 9:R27.
22. Whibley C, Pharoah PD, Hollstein M. p53 polymorphisms: cancer implications. Nat Rev Cancer 2009; 9:95-107.
23. Scheffner M, Werness BA, Huibregtse JM et al. The E6 oncoprotein encoded by human papillomavirus types 16 and 18 promotes the degradation of p53. Cell 1990; 63:1129-1136.
24. Scheffner M, Huibregtse JM, Howley PM. Identification of a human ubiquitin-conjugating enzyme that mediates the E6-AP-dependent ubiquitination of p53. Proc Natl Acad Sci USA 1994; 91:8797-8801.
25. Oliner JD, Kinzler KW, Meltzer PS et al. Amplification of a gene encoding a p53-associated protein in human sarcomas. Nature 1992; 358:80-83.
26. Momand J, Jung D, Wilczynski S et al. The MDM2 gene amplification database. Nucleic Acids Res 1998; 26:3453-3459.
27. Laurie NA, Donovan SL, Shih CS et al. Inactivation of the p53 pathway in retinoblastoma. Nature 2006; 444:61-66.
28. Ansieau S, Bastid J, Doreau A et al. Induction of EMT by twist proteins as a collateral effect of tumor-promoting inactivation of premature senescence. Cancer Cell 2008; 14:79-89.
29. Shiota M, Izumi H, Onitsuka T et al. Twist and p53 reciprocally regulate target genes via direct interaction. Oncogene 2008; Epub ahead of print.
30. Valsesia-Wittmann S, Magdeleine M, Dupasquier S et al. Oncogenic cooperation between H-Twist and N-Myc overrides failsafe programs in cancer cells. Cancer Cell 2004; 6:625-630.
31. Lutzker SG, Levine AJ. A functionally inactive p53 protein in teratocarcinoma cells is activated by either DNA damage or cellular differentiation. Nat Med 1996; 2:804-810.
32. Moll UM, Riou G, Levine AJ. Two distinct mechanisms alter p53 in breast cancer: mutation and nuclear exclusion. Proc Natl Acad Sci USA 1992; 89:7262-7266.
33. Moll UM, LaQuaglia M, Benard J, Riou G. Wild-type p53 protein undergoes cytoplasmic sequestration in undifferentiated neuroblastomas but not in differentiated tumors. Proc Natl Acad Sci USA 1995; 92:4407-4411.
34. Ueda H, Ullrich SJ, Gangemi JD et al. Functional inactivation but not structural mutation of p53 causes liver cancer. Nat Genet 1995; 9:41-47.
35. Li FP, Fraumeni JF Jr, Mulvihill JJ et al. A cancer family syndrome in twenty-four kindreds. Cancer Res 1988; 48:5358-5362.
36. Olivier M, Goldgar DE, Sodha N et al. Li-Fraumeni and related syndromes: correlation between tumor type, family structure, and TP53 genotype. Cancer Res 2003; 63:6643-6650.
37. Bond GL, Hu W, Bond EE et al. A single nucleotide polymorphism in the MDM2 promoter attenuates the p53 tumor suppressor pathway and accelerates tumor formation in humans. Cell 2004; 119:591-602.
38. Bougeard G, Baert-Desurmont S, Tournier I et al. Impact of the MDM2 SNP309 and p53 Arg72Pro polymorphism on age of tumor onset in Li-Fraumeni syndrome. J Med Genet 2006; 43:531-533.
39. Cho Y, Gorina S, Jeffrey PD, Pavletich NP. Crystal structure of a p53 tumor suppressor-DNA complex: understanding tumorigenic mutations. Science 1994; 265:346-355.
40. Achatz MI, Olivier M, Le Calvez F et al. The TP53 mutation, R337H, is associated with Li-Fraumeni and Li-Fraumeni-like syndromes in Brazilian families. Cancer Lett 2007; 245:96-102.
41. Ribeiro RC, Sandrini F, Figueiredo B et al. An inherited p53 mutation that contributes in a tissue-specific manner to pediatric adrenal cortical carcinoma. Proc Natl Acad Sci USA 2001; 98:9330-9335.
42. DiGiammarino EL, Lee AS, Cadwell C et al. A novel mechanism of tumorigenesis involving pH-dependent destabilization of a mutant p53 tetramer. Nat Struct Biol 2002; 9:12-16.
43. Petitjean A, Achatz MI, Borresen-Dale AL et al. TP53 mutations in human cancers: functional selection and impact on cancer prognosis and outcomes. Oncogene 2007; 26:2157-2165.
44. Hainaut P, Hollstein M. p53 and human cancer: the first ten thousand mutations. Adv Cancer Res 2000; 77:81-137.
45. Scicchitano DA, Hanawalt PC. Intragenomic repair heterogeneity of DNA damage. Environ Health Perspect 1992; 98:45-51.
46. Pfeifer GP, Denissenko MF, Olivier M et al. Tobacco smoke carcinogens, DNA damage and p53 mutations in smoking- associated cancers. Oncogene 2002; 21:7435-7451.
47. Olivier M, Hussain SP, Caron de Fromentel C et al. TP53 mutation spectra and load: a tool for generating hypotheses on the etiology of cancer. IARC Sci Publ 2004; 157:247-270.
48. Hussain SP, Schwank J, Staib F et al. TP53 mutations and hepatocellular carcinoma: insights into the etiology and pathogenesis of liver cancer. Oncogene 2007; 26:2166-2176.

49. Liu Z, Hergenhahn M, Schmeiser HH et al. Human tumor p53 mutations are selected for in mouse embryonic fibroblasts harboring a humanized p53 gene. Proc Natl Acad Sci USA 2004; 101:2963-2968.
50. Ambs S, Hussain SP, Marrogi AJ, Harris CC. Cancer-prone oxyradical overload disease. IARC Sci Publ 1999; 150:295-302.
51. Ohshima H, Bartsch H. Chronic infections and inflammatory processes as cancer risk factors: possible role of nitric oxide in carcinogenesis. Mutat. Res 1994; 305:253-264.
52. Ambs S, Bennett WP, Merriam WG et al. Relationship between p53 mutations and inducible nitric oxide synthase expression in human colorectal cancer. J Natl Cancer Inst 1999; 91:86-88.
53. Dogliotti E, Hainaut P, Hernandez T et al. Mutation spectra resulting from carcinogenic exposure: from model systems to cancer-related genes. Recent Results Cancer Res 1998; 154:97-124.
54. Greenblatt MS, Grollman AP, Harris CC. Deletions and insertions in the p53 tumor suppressor gene in human cancers: confirmation of the DNA polymerase slippage/misalignment model. Cancer Res 1996; 56:2130-2136.
55. Lunter G, Hein J. A nucleotide substitution model with nearest-neighbour interactions. Bioinformatics 2004; 20(Suppl 1):I216-I223.
56. Lin J, Chen J, Elenbaas B, Levine AJ. Several hydrophobic amino acids in the p53 amino-terminal domain are required for transcriptional activation, binding to mdm-2 and the adenovirus 5 E1B 55-kD protein. Genes Dev 1994; 8:1235-1246.
57. Toledo F, Wahl GM. Regulating the p53 pathway: in vitro hypotheses, in vivo veritas. Nat. Rev. Cancer 2006; 6:909-923.
58. Bullock AN, Henckel J, Fersht AR. Quantitative analysis of residual folding and DNA binding in mutant p53 core domain: definition of mutant states for rescue in cancer therapy. Oncogene 2000; 19:1245-1256.
59. Joerger AC, Fersht AR. Structure-function-rescue: the diverse nature of common p53 cancer mutants. Oncogene 2007; 26:2226-2242.
60. Pietenpol JA, Tokino T, Thiagalingam S et al. Sequence-specific transcriptional activation is essential for growth suppression by p53. Proc Natl Acad Sci USA 1994; 91:1998-2002.
61. Ryan KM, Vousden KH. Characterization of structural p53 mutants which show selective defects in apoptosis but not cell cycle arrest. Mol Cell Biol 1998; 18:3692-3698.
62. Brachmann RK, Vidal M, Boeke JD. Dominant-negative p53 mutations selected in yeast hit cancer hot spots. Proc Natl Acad Sci USA 1996; 93:4091-4095.
63. Dearth LR, Qian H, Wang T et al. Inactive full-length p53 mutants lacking dominant wild-type p53 inhibition highlight loss-of-heterozygosity as an important aspect of p53 status in human cancers. Carcinogenesis 2007; 28(2):289-298.
64. Marutani M, Tonoki H, Tada M et al. Dominant-negative mutations of the tumor suppressor p53 relating to early onset of glioblastoma multiforme. Cancer Res 1999; 59:4765-4769.
65. Halevy O, Michalovitz D, Oren M. Different tumor-derived p53 mutants exhibit distinct biological activities. Science 1990; 250:113-116.
66. Lang GA, Iwakuma T, Suh YA et al. Gain of function of a p53 hot spot mutation in a mouse model of Li-Fraumeni syndrome. Cell 2004; 119:861-872.
67. Olive KP, Tuveson DA, Ruhe ZC et al. Mutant p53 gain of function in two mouse models of Li-Fraumeni syndrome. Cell 2004; 119:847-860.
68. Kim E, Deppert W. Interactions of mutant p53 with DNA: guilt by association. Oncogene 2007; 26:2185-2190.
69. Kirk GD, Camus-Randon AM, Mendy M et al. Ser-249 p53 mutations in plasma DNA of patients with hepatocellular carcinoma from The Gambia. J Natl Cancer Inst 2000; 92:148-153.
70. Gormally E, Vineis P, Matullo G et al. TP53 and KRAS2 mutations in plasma DNA of healthy subjects and subsequent cancer occurrence: a prospective study. Cancer Res 2006; 66:6871-6876.
71. Le Roux E, Gormally E, Hainaut P. Somatic mutations in human cancer: applications in molecular epidemiology. Rev Epidemiol Sante Publique 2005; 53:257-266.
72. Franklin WA, Gazdar AF, Haney J et al. Widely dispersed p53 mutation in respiratory epithelium. A novel mechanism for field carcinogenesis. J Clin Invest 1997; 100:2133-2137.
73. Pontén F, Berg C, Ahmadian A et al. Molecular pathology in basal cell cancer with p53 as a genetic marker. Oncogene 1997; 15:1059-1067.
74. Dix BR, Robbins PD, Spagnolo DV et al. Clonal analysis of colorectal tumors using K-ras and p53 gene mutations as markers. Diagn Mol Pathol 1995; 4:261-265.
75. Montesano R, Hainaut P. Molecular precursor lesions in oesophageal cancer. Cancer Surv 1998; 32:53-68).
76. Jia L, Liu Y, Yi X et al. Endometrial glandular dysplasia with frequent p53 gene mutation: a genetic evidence supporting its precancer nature for endometrial serous carcinoma. Clin Cancer Res 2008; 14:2263-2269.

77. Soussi T. p53 Antibodies in the sera of patients with various types of cancer: a review. Cancer Res 2000; 60:1777-1788.
78. Caron de Fromentel C, May-Levin F, Mouriesse H et al. Presence of circulating antibodies against cellular protein p53 in a notable proportion of children with B-cell lymphoma. Int J Cancer 1987; 39:185-189.
79. Crawford LV, Pim DC, Bulbrook RD. Detection of antibodies against the cellular protein p53 in sera from patients with breast cancer. Int J Cancer 1982; 30:403-408.
80. Langerød A, Zhao H, Borgan Ø et al. TP53 mutation status and gene expression profiles are powerful prognostic markers of breast cancer. Breast Cancer Res 2007; 9:R30.
81. Olivier M, Langerød A, Carrieri P et al. The clinical value of somatic TP53 gene mutations in 1,794 patients with breast cancer. Clin Cancer Res 2006; 12:1157-1167.
82. Aas T, Børresen AL, Geisler S et al. Specific P53 mutations are associated with de novo resistance to doxorubicin in breast cancer patients. Nat Med 1996; 2:811-814.
83. Bergh J, Norberg T, Sjogren S et al. Complete sequencing of the p53 gene provides prognostic information in breast cancer patients, particularly in relation to adjuvant systemic therapy and radiotherapy. Nat Med 1995; 1:1029-1034.
84. Bertheau P, Turpin E, Rickman DS et al. Exquisite sensitivity of TP53 mutant and basal breast cancers to a dose-dense epirubicin-cyclophosphamide regimen. PLoS Med 2007; 4:e90.
85. Bode AM, Dong Z. Post-translational modification of p53 in tumorigenesis. Nat Rev Cancer 2004; 4:793-805.
86. Issaeva N, Bozko P, Enge M et al. Small molecule RITA binds to p53, blocks p53-HDM-2 interaction and activates p53 function in tumors. Nat Med 2004; 10:1321-1328.
87. Bykov VJ, Issaeva N, Zache N et al. Reactivation of mutant p53 and induction of apoptosis in human tumor cells by maleimide analogs. J Biol Chem 2005; 280:30384-30391.
88. Caron de Fromentel C, Gruel N, Venot C et al. Restoration of transcriptional activity of p53 mutants in human tumor cells by intracellular expression of anti-p53 single chain Fv fragments. Oncogene 1999; 18:551-557.
89. Boeckler FM, Joerger AC, Jaggi G et al. Targeted rescue of a destabilized mutant of p53 by an in silico screened drug. Proc Natl Acad Sci USA 2008; 105:10360-10365.
90. Senzer N, Nemunaitis J. A review of contusugene ladenovec (Advexin) p53 therapy. Curr Opin Mol Ther 2009; 11:54-61.
91. Pfeifer GP. p53 mutational spectra and the role of methylated CpG sequences. Mutat Res 2000; 450:155-166.

Chapter 2

Lessons on *p53* from Mouse Models

Dadi Jiang and Laura D. Attardi*

Introduction

Mutations in the *p53* gene are implicated in the development of at least half of all human cancers, of a wide variety of types.[1,2] This high incidence of mutations suggests that there exists a strong selection pressure for *p53* inactivation during tumorigenesis. The idea that *p53* mutations are important for tumor development in humans has also been supported by the finding that in the Li-Fraumeni syndrome, individuals inherit a mutant *p53* allele and show a predisposition to developing a wide variety of cancers.[3] An unambiguous cause and effect relationship, however, between *p53* mutation and tumorigenesis has been clearly provided through the generation and analysis of *p53* knockout mice.[4-6] *p53* null mice are subject to tumorigenesis at 100% frequency, indicating that the presence of *p53* is crucial for preventing cancer development. In addition to this initial observation, significant understanding of the role of *p53* as a tumor suppressor has come through further analysis of the *p53* knockout mouse as well as other versions of mice with altered *p53* genes. In this chapter, we will summarize various *p53* knockout and knock-in models and how these models have helped us to understand *p53* function in vivo.

Early Studies of *p53* in the Mouse

Initial evidence from mouse models suggesting a role for *p53* in tumor suppression came from analysis of mice infected with Friend leukemia virus. These mice developed erythroleukemia, and cell lines derived from their spleens expressed either a truncated version of p53 or no p53 protein.[7] Moreover, the *p53* genomic locus was shown to be rearranged in these tumors. In cases in which there was no obvious rearrangement of the locus, the *p53* gene was shown to have sustained a point mutation.[8] These studies suggested that *p53* inactivation might play a role in leukemia development induced by Friend virus.

In addition, transgenic mice expressing *p53* point mutants, either the $p53^{Pro193}$ or the $p53^{Val135}$ allele, displayed a propensity to developing tumors.[9,10] In these transgenic mice, the *p53* alleles were expressed in the lung, spleen, lymph nodes, thymus, ovary, and testes. These mice developed tumors, particularly lung adenocarcinomas, osteosarcomas, or lymphomas at an incidence of 20-30% by 18 months of age.[9,11] These findings provided the first direct evidence that *p53* mutations cause tumor development in vivo. This oncogenic activity was attributed to a dominant negative effect of the p53 mutant proteins on the native *p53* gene product, supporting the role of wild-type p53 as a tumor suppressor.

*Corresponding Author: Laura D. Attardi—CCSR-South, Room 1255 , 269 Campus Drive , Stanford, California 94305-5152, USA. Email: attardi@stanford.edu

p53, edited by Ayeda Ayed and Theodore Hupp. ©2010 Landes Bioscience and Springer Science+Business Media.

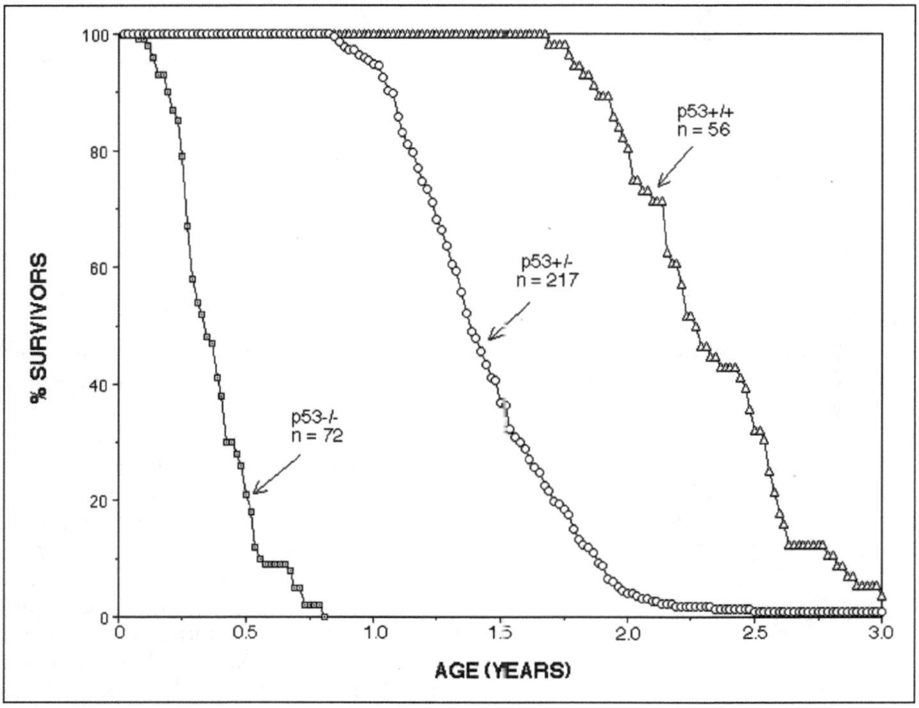

Figure 1. Kaplan Meyer curve illustrating the survival of $p53^{-/-}$, $p53^{+/-}$ and wild-type mice on a mixed genetic background (C57BL/6 X 129/Sv) with increasing age. Reprinted with permission from: Venkatachalam S et al. Toxicol Pathol 2001; 29:147-54;[15] ©2001 SAGE Publications.

Generation of *p53* Knockout Mice

p53$^{-/-}$ *Mice*

An unequivocal demonstration of the role of *p53* in tumor suppression came through the use of gene targeting to inactivate the *p53* gene in the mouse. In 1992, the first report describing the creation and analysis of *p53* knockout mice was published.[4] This was followed by reports from other laboratories, which made similar observations.[5-6] In all these cases, homologous recombination was used to disrupt the *p53* gene by introducing a selectable marker into the *p53* locus. Importantly, slightly different strategies were used. In the case of the Donehower group, parts of intron 4 and exon 5 were replaced,[4] whereas the Jacks and Purdie groups deleted exons 2-6.[5,6] Both of these approaches interrupted the DNA binding domain of p53, rendering the protein inactive. Surprisingly, upon intercrossing heterozygous mutant mice, it was found that *p53* null mice are viable, indicating that p53 is not crucial for normal cell cycles during development, as was previously believed. However, 100% of these mice went on to rapidly succumb to tumors, within several months after birth, indicating a crucial role for p53 in preventing cancer (Fig. 1). The majority of these mice (>70%), which were on a mixed genetic background (75% C57BL/6 and 25% 129/Sv), developed lymphoma, either CD4, CD8 double positive T-cell or generalized B-cell lymphomas.[4,5,12] In addition, these *p53* null mice developed various sarcomas (osteosarcomas, hemangiosarcomas, fibrosarcomas, rhabdomyosarcomas, anaplastic sarcomas), testicular tumors, and occasional carcinomas or nervous system tumors (Fig. 2). These experiments demonstrated a clear cause and effect relationship between *p53* inactivation and tumorigenesis. Thus, p53 is definitively a tumor suppressor, and this is true in many different tissues.

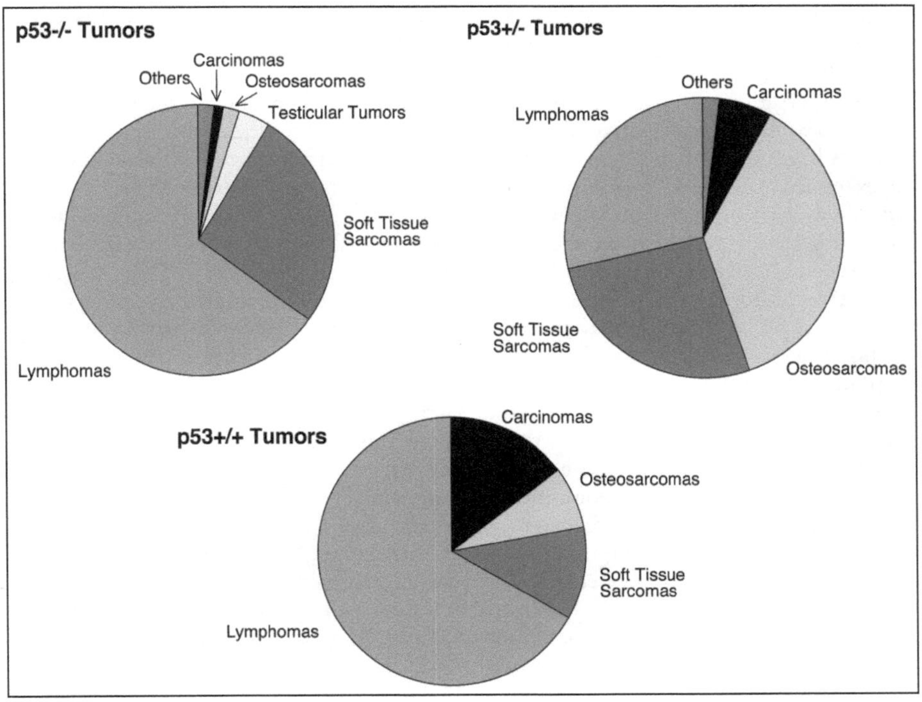

Figure 2. Tumor spectra of $p53^{-/-}$, $p53^{+/-}$ and wild-type mice. Reprinted with permission from: Venkatachalam S et al. Toxicol Pathol 2001; 29:147-54;[15] ©2001 SAGE Publications.

There were several additional lessons derived from these studies. First, p53 is one of the few dedicated "tumor suppressors" known in the sense that it is dispensable for normal development but plays a fundamental role in protecting an organism from tumor development. In addition, deletion of *p53* is sufficient to allow tumorigenesis, indicating that the point mutant versions so commonly observed in human cancers are not necessary to promote tumorigenesis. Finally, the fact that there is some lag time before tumor development in the *p53* null animals indicates that other mutations are required for tumorigenesis.

$p53^{+/-}$ *Mice*

p53$^{+/-}$ mice were also tumor prone, but they showed a longer latency to tumor development and developed a different tumor spectrum than did *p53*$^{-/-}$ mice (Fig. 1). While no obvious phenotype was observed before 9 months of age, half of the mice developed tumors by approximately 18 months of age.[4,13] These mice succumbed primarily to sarcomas of a variety of types—including hemangiosarcomas, anaplastic sarcomas, osteosarcomas, fibrosarcomas, and rhabdomyosarcomas—but also developed lymphomas and some carcinomas (Fig. 2). The other studies of *p53* heterozygotes showed similar tumor spectra and latencies.[5-6] In all of these studies, at least half of these heterozygous mice displayed loss of heterozygosity (LOH) at the *p53* locus, suggesting that complete loss of *p53* is important for tumor development in many, but not all cases (see below).

Although lymphomas dominated the tumor spectrum of *p53*$^{-/-}$ mice, in *p53* heterozygotes sarcomas were the predominant tumors; the incidence of sarcomas of some type was ~58% and the incidence of lymphomas was 25-32% in *p53*$^{+/-}$ mice.[4,5,13] An explanation for this difference in tumor spectrum in the heterozygotes compared to the null animals has been proposed.[5] In the heterozygous mice, there is limited time to lose the wild-type *p53* allele in the thymus because of thymic involution several weeks after birth, when the number of thymocytes decreases substantially. Thus there is a restricted developmental window for LOH before the target cell size decreases dramatically, lessening the chance for LOH.

The observation that tumors from *p53* heterozygous mice did not always show LOH of the *p53* gene was further pursued, to test the idea that *p53* is haploinsufficient for tumor suppression.[14] A large colony of $p53^{+/-}$ mice was generated, and the frequency of LOH in these mice was examined in detail. LOH was observed in 50% of tumors in mice under 18 months of age and in only 15% of mice that developed tumors after 18 months of age. That the wild-type allele left in these tumors in older mice was functional was demonstrated by showing that p53 DNA binding, transcriptional activation and apoptotic activities were intact. The authors concluded that in addition to complete loss of *p53* playing a role in tumorigenesis, a reduction in the dosage of *p53* can contribute to tumorigenesis as well. This role for *p53* haploinsufficiency in tumorigenesis is consistent with cell culture studies, in which *p53* heterozygous cells have apoptotic and arrest responses intermediate between wild-type and *p53* null cells.[15]

The *p53* heterozygote provides an example of a tumor suppressor gene knockout mouse that is a reasonably accurate model for the analogous human disease, in this case Li-Fraumeni syndrome. In Li-Fraumeni syndrome, *p53* heterozygous individuals develop a host of tumors, that commonly include bone and soft tissue sarcomas, breast cancers, brain tumors, and leukemias.[3] Similarly, the *p53* heterozygous mice develop osteosarcomas and soft tissue sarcomas. In addition, depending on the genetic background, the *p53* heterozygous mice can also develop breast cancer (see below). Unlike Li-Fraumeni kindred, the $p53^{+/-}$ mice do not typically develop brain cancers or leukemia, which could be accounted for by such factors as species differences, genetic background differences, or differences in the types of *p53* mutations in the two cases.

Effects of Genetic Background on p53 Knockout Phenotypes

The genetic background has been shown to influence the tumor spectra and latencies of *p53*-deficient mice. The original *p53* knockout mice analyzed for tumor phenotypes were of a mixed C57BL/6 and 129/Sv background (approximately 75% and 25%, respectively;[4,5]). C57BL/6 wild-type mice have a propensity to developing lymphoma, so it was possible that *p53* loss was simply accentuating the natural predisposition of that strain. To test this idea, the propensity of *p53*-deficient mice to tumorigenesis on a pure 129/Sv genetic background was also examined.[16] This strain is not prone to developing lymphoma, but rather testicular tumors. On a pure 129/Sv background, lymphomagenesis still occurred, with similar latency and with similar characteristics to those on the mixed background, indicating that *p53*-deficiency itself leads to lymphoma development. In addition, the natural predisposition of 129/Sv mice to developing testicular tumors was enhanced, with incidence going from 1-2% in 129/Sv wild-type males to 50% in 129/Sv *p53* null males. Thus, while lymphoma development is due to *p53* loss, testicular tumor development is due to *p53*-deficiency accentuating an inherent predisposition of the background strain. In addition, tumors generally developed faster on a 129/Sv background than on a mixed background, suggesting that there are modifiers in 129/Sv mice that cooperate with *p53*-deficiency.

Another example of *p53* loss exacerbating an inherent predisposition of a strain comes from analysis of BALB/c mice. Untreated wild-type BALB/c mice are not inherently more susceptible than wild-type C57BL/6 mice to breast cancer, but radiation treatment exposes their susceptibility.[17] *p53*-deficient mice were bred onto the BALB/c background to determine if, on this sensitized background, they would now develop mammary cancer.[18] $p53^{-/-}$ mice on the BALB/c background developed a similar spectrum of lymphomas and sarcomas, with similar latencies, to $p53^{-/-}$ mice of other backgrounds, but almost no mammary cancer. That mammary cancer did not develop was thought to be due to the early mortality from other tumors. Indeed, $p53^{-/-}$ BALB/c breast epithelial tissue was competent to develop into breast cancer, as shown by transplanting this tissue into the mammary gland of a wild-type host, where it frequently developed into cancer. In contrast, BALB/c *p53* heterozygotes, while developing the sarcomas and lymphomas typical of *p53* heterozygotes, were also highly predisposed to mammary tumors, with 55% of females developing this cancer. In fact, mammary carcinomas were the most common cancer, occurring in 42% of the mice, and all females showed either mammary hyperplasia or overt cancer. LOH was frequently observed. Thus, the specific modifiers in the BALB/c strain revealed the potential of *p53* loss to promote mammary tumorigenesis, as predicted from the role for *p53* mutation in both sporadic and hereditary human breast cancers. As mentioned above, this BALB/c *p53* heterozygous mouse is a good model for Li-Fraumeni syndrome, as it recapitulates the predisposition of female Li-Fraumeni patients to breast cancer—the most common cancer in these women.

Response of p53 Mutant Mice to DNA Damaging Agent Treatment

p53 plays a crucial role in responses to DNA damage. Specifically, DNA damage stimuli cause cells to either undergo cell cycle arrest or apoptosis, in a p53-dependent manner.[19] In the absence of p53, cells that have sustained DNA damage proliferate or survive inappropriately, allowing the propagation of cells with potentially oncogenic mutations. Mouse models have been used to demonstrate that loss of this p53 DNA damage response can contribute to tumorigenesis. Irradiation of either transgenic mice expressing p53 dominant negative proteins ($p53^{Pro193}$ or $p53^{Val135}$) or *p53* heterozygotes showed that these mice have a shorter lifespan due to tumorigenesis than both untreated counterparts or irradiated wild-type mice.[10,20] In fact, treated wild-type mice failed to develop tumors at all in these two studies. The rates of LOH were near 100% in the *p53* heterozygous mice, suggesting that *p53* is a target for inactivation by irradiation. These studies support a role for p53 in limiting tumorigenesis through its response to DNA damage.

Carcinogenesis Studies

Many studies have examined the role of *p53* mutation in tumors induced by carcinogens. One type of study involves treating wild-type mice with carcinogens and examining the fate of *p53* alleles during cancer development. If *p53* is mutated, it suggests that *p53* loss plays a role in the development of those particular types of cancer. The *p53* mutations observed may reveal hallmarks of particular carcinogens, suggesting a clear molecular basis for tumorigenesis. Another type of study has entailed utilizing the *p53* heterozygous or null mice to determine the effects of the absence of *p53* on carcinogenesis.

For example, one set of studies aimed to define the role of *p53* in multistep carcinogenesis using a well-established skin cancer progression model. This model involves initiation of tumorigenesis through dimethylbenzanthracene (DMBA) treatment and promotion of tumorigenesis through repeated 12-O-tetradecanoyl-phorbol-13-acetate (TPA) treatments. Through this treatment, papillomas develop and after some time, a small percentage of these papillomas progress to malignant carcinomas. Analysis of the stages at which *p53* was mutated indicated that *p53* mutations occurred in carcinomas but not papillomas, suggesting a role for *p53* loss in tumor progression rather than initiation.[21,22] Through DMBA/TPA treatment of wild-type *p53* heterozygous and *p53* null mice, the step at which p53 acts in this multistep cancer model was further defined.[23] *p53*-deficiency did not enhance tumor initiation, as assessed by numbers and latencies of papillomas developing. However, the frequency and rate of progression to carcinomas were dramatically increased in the absence of *p53*, further supporting a role for *p53* loss in progression.

p53 heterozygotes are in fact considered a good model for carcinogenicity testing by the United States Food and Drug Administration.[15] Candidate pharmaceuticals need to be tested for carcinogenicity on two rodent models. The FDA allowed replacement of one of the two required standard two-year rodent bioassays on wild-type mice and rats with a 6-month assay on $p53^{+/-}$ mice. Besides the fact that they are compromised for their p53 function, making them sensitive to tumor development, they do not develop spontaneous cancers before 9 months of age, providing a good time frame to treat and assay the effects of carcinogens on the mice. These mice are also susceptible to a broad array of tumor types, making them useful for testing many types of carcinogens.

Crosses to p53-Deficient Mice

A vast number of tumor-prone strains have been crossed to the *p53*-deficient mice to determine if *p53* loss can alter either the latency of tumorigenesis or the tumor spectrum in such compound mutant mice. By revealing genetic interactions between p53 and other proteins involved in tumorigenesis, such crosses will help to define the pathways governing growth control.

Why Are p53-Deficient Mice Lymphoma Prone?

One class of crosses utilizing the *p53* knockout mice has aimed to understand why these mice develop thymic lymphoma so readily. p53 is known to play a role in inducing apoptosis of thymocytes in response to DNA damaging agent treatment, and *p53*$^{-/-}$ cells are resistant to this cell death.[24,25] It was hypothesized that this p53-dependent apoptotic response activated by double strand breaks may also keep tumor growth in check in wild-type mice. It was speculated that the double strand breaks arising during V(D)J recombination could activate p53, leading it to eliminate, through apoptosis, those cells that had sustained aberrant, potentially oncogenic recombination events. In the absence of p53, there would be a failure of apoptosis, leading to the inappropriate survival of potentially defective cells that were lymphomagenic. To test this idea, *p53* null mice were crossed with mouse strains defective in different steps of V(D)J recombination.

Scid (Severe combined immunodeficient) mice are deficient in the DNA-dependent protein kinase (DNA-PK) catalytic subunit, which is involved in joining the DNA ends generated during V(D)J recombination.[26] This recombination defect leads to blocks at early stages of T and B cell development in *Scid* mice. In a *p53* null background, T cell, but not B cell, development in *Scid* mice was rescued.[27-29] Moreover, tumor latency was reduced in *Scid/Scid;p53*$^{-/-}$ mice relative to either single mutant, indicating that these lesions cooperated to induce cancer.[28-30] However, because most of the observed tumors in the double mutants were B-cell lymphomas, it appeared that *Scid* and *p53* loss cooperated in B-cell but not T-cell lymphomagenesis. Indeed, similar results were obtained by crossing *p53* null mice with mice lacking other non-homologous end-joining (NHEJ) components involved in V(D)J recombination: XRCC4, DNA ligase IV and Ku80.[31-34] Rapid, widely disseminated pro-B-cell lymphomas developed in XRCC4-, Ligase IV- or Ku80-deficient mice also null for *p53*. In the case of Ku80, it was hypothesized that these lymphomas developed because the high levels of pro-B cell apoptosis seen in *Ku80*$^{-/-}$ mice were diminished in the absence of p53, perhaps facilitating tumorigenesis.[31]

Rag-1 and Rag-2 are involved in generating the double strand breaks critical for V(D)J recombination.[35] It was predicted that mice deficient for either one of these proteins might suppress lymphoma development by blocking the initiation of recombination at antigen receptor loci. In contrast, *p53*$^{-/-}$;*Rag-2*$^{-/-}$ mice developed thymic lymphoma with similar latencies to *p53*$^{-/-}$ mice, and *p53*$^{-/-}$;*Rag-1*$^{-/-}$ mice developed thymic lymphoma with shorter latencies.[36] Moreover, there was no evidence of V(D)J recombination, suggesting further that it is not crucial for T-cell lymphoma development in *p53*-deficient mice. Together, these studies suggest that aberrant recombination events are not the cause of thymic lymphomagenesis in *p53*$^{-/-}$ mice, although in mice with impaired NHEJ and loss of p53, impaired resolution of RAG-initiated breaks can lead to oncogene amplification and B-cell lymphomagenesis.[37]

Crosses with Oncogene-Expressing or Tumor Suppressor Knockout Strains

Transgenic strains expressing any of a variety of oncogenes, including *c-Myc*, *Wnt*, *Neu*, *Ras* and *Scl*, among others, have been bred to *p53*-deficient mice to assess potential cooperativity in tumor development.[38-41] Most often, tumors in these models form more rapidly in the absence of p53 than in its presence, suggesting that the lesions act in different pathways that cooperate in tumorigenesis. One notable exception was the cross between Ig-*Bcl-2* transgenic mice and *p53* knockout mice.[42] These compound mice showed no enhancement of tumorigenesis, suggesting that these two genetic lesions act in the same pathway.

Tumor suppressor knockout strains have also been bred to *p53*-deficient mice to assess cooperativity. These include mice with mutations in *Retinoblastoma* (*Rb*), *Adenomatous polyposis coli* (*APC*), *Neurofibromatosis Type I* (*Nf1*), *Neurofibromatosis Type II* (*Nf2*), *Patched* (*Ptc*) and the Breast Cancer Susceptibility Genes *Brca1* and *Brca2*.[43-51] Cooperativity between these lesions and *p53* deficiency was often manifested as the appearance of novel tumor types in compound mutants. For example, *Rb* heterozygous mice typically developed pituitary and thyroid tumors, but on a $p53^{-/-}$ background, they now became susceptible to novel tumor types, including pancreatic islet cell carcinomas and pinealoblastomas.[43,44] Similarly, $Nf1^{+/-}$ mice also heterozygous for *p53* developed malignant peripheral nerve sheath tumors and astrocytomas, neither of which was observed in single heterozygous parents.[46,47] In addition, $Apc^{+/-};p53^{-/-}$ mice developed novel tumors not seen in the individual mutants—desmoid tumors and pancreatic acinar tumors—as well as displaying a role for *p53* loss in progression of the characteristic adenomas seen in $Apc^{+/-}$ mice.[45,52]

In fact, breeding some of these tumor suppressor knockout models to *p53*-deficient mice has been shown to be critical for recapitulating particular types of cancer seen in humans. For example, both the Brca1 and Brca2 tumor suppressors play a role in human hereditary breast cancer and *p53* loss is implicated in these cases. However, when either *Brca1* or *Brca2* was conditionally deleted in the mouse mammary gland, their loss was not sufficient to cause frequent mammary tumorigenesis.[50,51] Upon breeding to either a constitutive or a conditional *p53* null allele, however, mammary tumors developed at a high frequency, resulting in an accurate model for inherited human breast cancer.

Crosses with Telomerase-Deficient Mice

The *p53* knockout mouse has also been instrumental in demonstrating a role for telomere dysfunction in the genesis of epithelial cancers. Telomerase is an RNA-protein complex involved in maintaining the ends of eukaryotic chromosomes, the telomeres.[53] Mice have very long telomeres, and hence mice deficient in the RNA component of telomerase, *mTERC*, do not show obvious phenotypes until they have been bred for several generations, at which time their telomeres have shortened significantly. Critical telomere shortening leads to deprotection or uncapping of the chromosome ends, activating p53-dependent responses.[54] For example, $mTERC^{-/-}$ mice at generation 5-6 showed increased apoptosis in their germ cells, and by breeding these mice with $p53^{-/-}$ mice, this cell death was rescued. The consequence of rescuing cell death was seen in late generation (G6-8) $mTERC^{-/-};p53^{+/-}$ mice, which were highly prone to developing cancers.[55] They developed cancer with a shorter latency than $mTERC^{-/-};p53^{+/+}$ mice, and also developed cancers of various epithelia, including the colon, breast and skin. Thus, telomere attrition in the absence of p53 led to survival of cells with dysfunctional telomeres, providing a permissive environment for chromosomal fusion-bridge-breakage cycles, and ultimately resulting in non-reciprocal translocations. In contrast, G1 or G2 $mTERC^{-/-};p53^{+/-}$ mice did not show an increased susceptibility to tumorigenesis relative to *p53*-deficient mice. This finding suggested that it is telomere dysfunction, not telomerase loss per se, that leads to tumorigenesis. Moreover, epithelial cancers were only seen in these late generation mice. These events are thought to model a progression occurring in human epithelial cancers, in which epithelial cell renewal during a lifetime leads to telomere dysfunction similar to that seen in the *mTERC*-deficient mouse. If p53 is inactivated, cells with such telomere dysfunction can survive, and non-reciprocal translocations can fuel the tumorigenic process.[53] In support of this hypothesis, chromosomal instability and non-reciprocal translocations in this $mTERC^{-/-};p53^{+/-}$ model result in gene copy number aberrations that closely resemble those seen in human carcinomas.[56]

p53 Conditional Knockout Mice

One of the major limitations in studying *p53* knockout mice is the fact that they succumb to lymphomas at an early age, thus making it difficult to examine tumorigenesis occurring with longer latency in other tissues. To study the role of *p53* loss in other tissues, a *p53* conditional knockout mouse was generated.[57] This model was used originally to show a role for combined *Rb*- and *p53*-deficiency in promoting medulloblastoma. The utility of *p53* conditional knockout mice was also demonstrated in the generation of the *Brca2*-associated breast cancer model mentioned above.[51] Conditional inactivation of *Brca2* in the mammary gland through expression of Cre under a K14 promoter failed to result in mammary tumors. However, when both *p53* and *Brca2* were conditionally inactivated in the mammary gland, tumors developed. Squamous cell carcinomas (SCCs) also formed in the skin, another site where K14-Cre is active. Breast and skin cancer development occurred with *p53* either in the homozygous or heterozygous state, with the latter case showing LOH. Thus, inactivation of the p53 pathway was crucial to the development of these cancers and was instrumental for creating an accurate model for the human disease. This approach has since been used to elucidate the role of p53 in tumor development in multiple tissues. For example, homozygous deletion of both *Pten* and *p53* in the prostate epithelium using the probasin-Cre strain resulted in lethal invasive prostate cancer which was absent in mice lacking either gene alone.[58] In another example, conditional homozygous inactivation of both *Rb* and *p53* in lung epithelial tissue through intratracheal administration of Adenoviral-Cre promoted the development of aggressive lung tumors with features very similar to small cell lung cancer (SCLC).[59] Similar strategies have also provided evidence for p53 loss in causing ovarian cancer and pancreatic cancer.[60-62] A demonstration of the role of p53 in these cancer models would have been impossible without the conditional knockout because of the early mortality from lymphomas in the constitutive knockout mice.

p53 Knock-In Mutant Mice

In recent years, knock-in technology has been utilized to probe p53 function in vivo. The advantage of this approach, in which the endogenous wild-type *p53* gene is replaced with a subtle variant, is that it allows the in vivo function of a particular allele to be probed in a physiological way (Fig. 3). Such an allele is expressed at normal levels from the native *p53* promoter, with the correct spatial and temporal regulation. This strategy has been utilized both to mimic *p53* mutations found in human cancers as well as to test the role of specific post-translational modifications or activities of p53 for its in vivo function.

p53 *Tumor-Derived Mutants*

During tumorigenesis *p53* is typically altered by missense mutations within the DNA binding domain rather than by nonsense mutations, suggesting that there may be some selective advantage conferred by missense mutations.[63] Some of these mutations have been identified in human cancers at much higher frequencies than others and are therefore termed "hotspot" mutations. To understand the tumor-promoting effects of such *p53* missense mutations, and to develop accurate models of human cancer, knock-in mice replacing the endogenous *p53* allele with tumor-derived *p53* mutants have been generated. For example, $p53^{R172H/+}$ mice expressing $p53^{R172H}$, the analogue of the human $p53^{R175H}$ hotspot mutant, were generated.[64] Although there was no significant difference in the tumor spectrum between $p53^{R172H/+}$ and $p53^{+/-}$ mice, osteosarcomas and carcinomas in $p53^{R172H/+}$ mice had the propensity to metastasize, whereas none of the $p53^{+/-}$ mice exhibited metastasis. In addition, both $p53^{R172H/R172H}$ and $p53^{R172H/+}$ mouse embryo fibroblasts (MEFs) grew faster and reached a higher saturation density than $p53^{-/-}$ MEFs, and $p53^{R172H/R172H}$ MEFs were transformed by oncogenic Ras more efficiently than $p53^{-/-}$ MEFs. Concomitantly, another group generated knock-in mouse lines expressing two different *p53* hotspot mutations.[65] In this study, the authors found that $p53^{R172H/+}$, $p53^{R270H/+}$ (encoding the equivalent of the human $p53^{R273H}$ hotspot mutant), and $p53^{+/-}$ mice differed significantly in their tumor spectra. Compared to

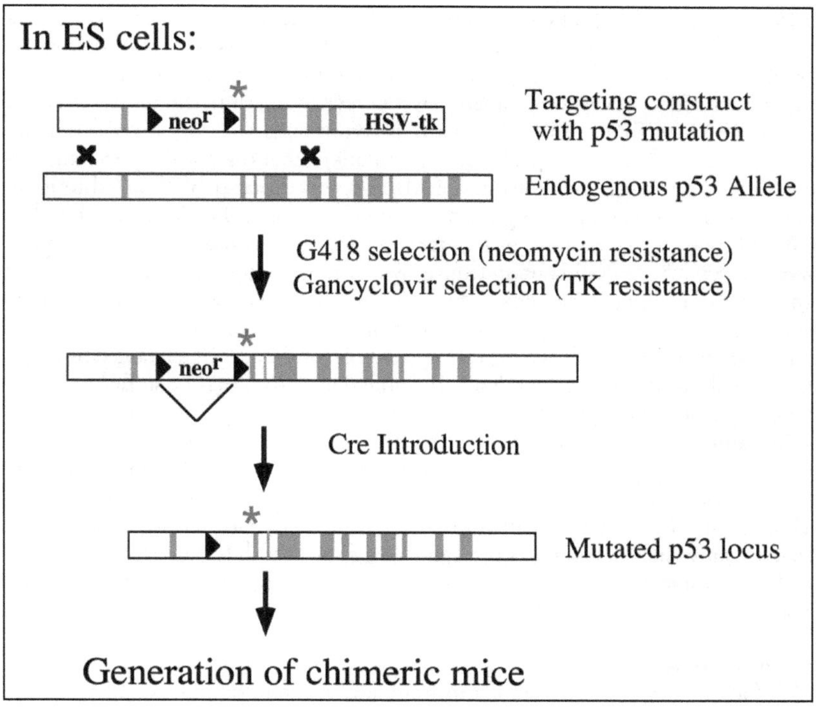

Figure 3. Schematic of a hypothetical strategy utilized to make a *p53* knock-in mouse strain. A targeting vector consisting of the *p53* genomic region with the introduced mutation (indicated by an asterisk) and bearing both a positive selection cassette (e.g., neomycin resistance) and a negative selection cassette (e.g., thymidine kinase) is generated. The 11 exons of the *p53* locus are indicated by gray (green online) boxes. The targeting vector is introduced into ES cells, cells are subjected to selection, and clones surviving selection are screened for homologous recombination. Cre is transiently introduced to remove the selection cassette. ES cells are then used to make mice stably propagating the allele through the germline. A color version of this figure is available online at www.landesbioscience.com/curie.

$p53^{+/-}$ mice, $p53^{R270H/+}$ mice developed more carcinomas, especially lung carcinomas, with a higher overall tumor burden. On the other hand, $p53^{R172H/+}$ mice developed osteosarcomas with the highest incidence, and they were more metastatic than osteosarcomas arising in either $p53^{R270H/+}$ or $p53^{+/-}$ mice. Furthermore, in contrast to $p53^{-/-}$ mice, which predominantly developed lymphomas and sarcomas, $p53^{R270H/-}$ and $p53^{R172H/-}$ mice developed significantly more carcinomas of various tissue origins. The discrepancies between the two studies, with a difference in metastatic potential observed in one study versus a difference in tumor spectra in the other, may be caused by differences in genetic background. The underlying mechanism for the observed gain-of-function phenotypes with the hotspot mutations still remains elusive, although it was suggested that it may be due to interference with the p53 family members, p63 and p73, by the mutant p53 protein.[64,65]

Mutants to Elucidate Mechanisms of p53 Action

Knock-in mice also provide a powerful tool to study the mechanisms of p53 regulation and the activities of p53 important for its function. Knock-in mouse strains have been generated for studies of the functional significance of a variety of post-translational modifications clustered at both the N- and C-termini of the p53 protein,[66-76] the importance of the proline-rich domain of p53 for its function,[77,78] and the contribution of the N-terminal transactivation domains to p53 function.[79-82] In the latter case, the transactivation mutant mice were made because despite being well-recognized as a transcriptional activator of a large number of target genes, it still was unclear whether p53 transactivation function was necessary and sufficient for its tumor suppression function in vivo. To determine whether transactivation is necessary for p53 function, our laboratory generated knock-in mice harboring mutations that disrupt the p53 transactivation domain (L25Q and W26S).[79,80] Analysis of MEFs revealed that $p53^{25,26}$ failed to induce the expression of most canonical p53 target genes in response to acute DNA damage, but did retain the ability to transactivate a subset of p53 target genes, such as *Bax*. At the phenotypic level, homozygous $p53^{25,26}$ MEFs did not undergo growth arrest or apoptosis in response to treatment with the DNA damaging agent doxorubicin but did undergo apoptosis upon exposure to non-genotoxic stresses such as hypoxia and serum starvation. These findings suggest either that mechanism(s) other than transactivation are responsible for the activities retained by $p53^{25,26}$, or that the residual transactivation function (e.g., induction of *Bax*) may be sufficient for its activity. As a complementary approach, we also generated a chimeric p53 protein in which the 80 N-terminal amino acids were replaced by transactivation sequences from the Herpes Simplex Virus VP16 protein to generate a p53 mutant capable of transcriptional activation but lacking sequences involved in other p53 activities.[81] Although $p53^{VP16}$ was capable of inducing a strong cell cycle arrest in MEFs associated with features of cellular senescence, it failed to induce apoptosis in oncogene-expressing MEFs. This observation suggests that while transactivation function of p53 is sufficient for regulating cell cycle arrest and senescence, transactivation-independent functions may contribute to the p53-dependent apoptotic response.

The Hupki Mouse

One knock-in strain was developed to humanize the mouse p53 to make a more exact model for determining the etiology of human *p53* mutations in vivo and for developing therapeutics for human cancer. In this model, known as the hupki model (human *p53* knock-in), exons 4-9 of mouse *p53* were replaced with the corresponding exons from human *p53*.[83] $p53^{KI/KI}$ cells derived from these mice were essentially indistinguishable from wild-type cells in their expression and induction of p53 upon DNA damage and in their p53 activities, including p53 target gene activation and induction of apoptosis. These mice have indeed provided a useful experimental model system for mutagenesis studies of human *p53* in vivo. Because the mouse *p53* does not have the same nucleotide sequence as human *p53*, the wild-type mouse cannot be used as a model to show the connection between a carcinogen and specific *p53* mutations in human cancer. Using the hupki mouse, spontaneous and carcinogen-induced mutational spectra in mouse tumors can be examined and compared with human *p53* tumor mutation databases to identify potential causes of human cancers. In fact, chronic UVB-treated hupki epidermis displayed specific p53 DNA binding-domain mutations typical of sun-exposed human skin and sunlight-associated human tumors.[84] The hupki mouse will additionally be a useful system to test cancer therapeutics.

A Role for p53 in Aging?

Another model aimed to recapitulate a human tumor allele of *p53*, at residue 248, but inadvertently revealed a role for p53 in organismal aging.[85] Specifically, upon targeting, an aberrant integration event occurred, resulting in the deletion of the first 6 exons of *p53* and placement of the last 5 exons in proximity to an unknown upstream promoter. This mutant allele, termed the "m" allele, encoded a truncated C-terminal version of p53 that is capable of activating the endogenous wild-type allele. Thus, $p53^{+/m}$ mice had a greater p53 response to DNA damage than $p53^{+/-}$ mice, or even, in some cases, than wild-type mice. In cell culture assays, the m allele stimulated p53 function in transactivation and growth suppression. This activation of the wild-type *p53* allele by the m allele was consistent with previous results showing that C-terminal fragments of p53 could activate wild-type p53 DNA binding and transactivation. The idea that p53 is hyperactive in these mice was supported by the finding that as these mice aged, they were highly resistant to developing cancer, with a mere 6% of these mice developing cancer during their lifespan. This is in contrast with almost all *p53* heterozygotes and half of aging wild-type mice developing cancer during their lifetimes. Nonetheless, these $p53^{+/m}$ mice had a shorter lifespan than wild-type mice, and this was accompanied by a set of dramatic phenotypes. These mice were prone to premature aging, manifesting a host of symptoms including osteoporosis, lordokyphosis (hunchbacked spine), decreased body weight, atrophy of various tissues, reduced hair growth, decreased cellularity of organs, and poor wound healing. These observations implicated hyperactive p53 in organismal aging. The caveat to these experiments is the unknown nature of the deletion in the upstream part of the *p53* gene, which could have resulted in an alteration of another locus with the consequence of causing this aging phenotype. To implicate *p53* more specifically, these results were supported through analysis of the *pL53* transgenic mouse strain that overexpresses a temperature sensitive *p53* allele.[85] In these mice, some of the same features of aging were observed, including lordokyphosis, weight loss, reduced hair growth, osteoporosis, and slow wound healing. These data supported the idea that hyperactive p53, while preventing cancer, is detrimental because of its ability to promote premature aging. In further support of this notion, a more recent study showed that transgenic mice expressing p44, a naturally occurring amino-terminal p53 deletion mutant,[86] displayed similar increased protection against cancer and accelerated aging.[87] Finally, telomere shortening in late generation $mTERC^{-/-}$ mice led to a similar constellation of premature aging phenotypes.[88] Activation of p53 by telomere shortening may also underlie the organismal aging phenotypes seen in this model.

As suggested from the studies just described, activated p53 can promote cancer resistance and induce aging. To determine if high levels of wild-type p53 can also confer tumor-resistance and promote aging, mice bearing supernumerary copies of *p53* were generated.[89] These mice were created with BACs bearing the *p53* genomic region to allow normal *p53* expression and regulation. Two transgenics were generated, *p53*-tg and *p53*-tg^b, with one or two BACs inserted, respectively. As proof that this allele encoded normal p53 activity, it was examined first in a *p53* null background. In these $p53^{-/-}$;tg mice, the *p53* BAC transgene rescued all p53 activities tested, including apoptosis, target gene activation, G_1 checkpoint induction, and suppression of Ras-induced transformation. In terms of tumors, $p53^{-/-}$;tg mice lived longer than $p53^{-/-}$ mice, but not as long as $p53^{+/-}$ mice. The finding that they did not live as long as *p53* heterozygotes is attributed to the fact that these transgenes are telomeric and may more easily undergo allele loss than the endogenous gene. Thus except for the difference in tumor latency, the $p53^{-/-}$;tg cells behaved identically to *p53* heterozygous mice, indicating that the transgene behaves like wild-type *p53*. $p53^{+/+}$;tg mice, known as "super p53" mice, were then examined for their resistance to tumorigenesis. These mice clearly had an enhanced p53 response to DNA damage compared to wild-type mice, based on both p53 target gene expression and thymocyte apoptosis. This apoptotic response to gamma rays was even more pronounced when the *p53*-tg^b line with two copies of the transgene was examined. When the $p53^{+/+}$;tg mice were subjected to carcinogen treatment, either 3MC to induce fibrosarcomas or BBN to induce bladder carcinomas, they showed increased resistance to tumor development relative to treated wild-type mice. As for the mice that died of natural causes, they had fewer tumors than wild-type mice as well, although the sample size was limited. In addition, there were neither signs of a shorter lifespan nor overt aging phenotypes. It is thought that because normally-regulated p53 is upregulated in this model, these mice show an increased resistance to tumor development without accelerated aging. The enhanced tumor suppression observed is likely due to both augmented p53 responses keeping tumors in check as well as the increasing difficulty of inactivating *p53* when more copies are present. Modulation of other components of the p53 pathway, through insertion of an additional copy of the $p19^{ARF}$ locus or decreasing *Mdm2* dosage, both of which efficiently increased p53 levels, elicited similar anti-tumor effects without affecting the normal aging process.[90,91] These studies indicate that properly regulated p53 activation does not have the deleterious effects of premature aging.

How p53 Acts as a Tumor Suppressor

From cell culture model systems, several mechanisms have been proposed to explain how p53 restricts tumorigenesis.[19] In some cases, p53 activation by DNA damage causes G_1 arrest, thus preventing progression of cells with potentially oncogenic lesions through the cell cycle. Failure of this checkpoint is proposed to lead to genomic instability, a characteristic feature of many tumors lacking p53. In addition, the induction of a permanent cell cycle arrest response, known as cellular senescence, by p53 may also be important for tumor suppression. In other cases, activation of p53 by any of a number of stresses, including DNA damage, hypoxia and hyperproliferative signals, leads to apoptosis. From analysis of mouse models, it seems that any of these mechanisms can be operational in tumor suppression in vivo.[92]

In some tumor models, p53 appears to play a role in a cell cycle checkpoint function. For example, in an MMTV *Wnt-1* transgenic mammary tumor model, tumors formed more quickly on a *p53*-deficient background than on a wild-type background.[39] However, apoptotic levels were low in these tumors and they did not clearly correlate with *p53* status.[93] Instead, there was accelerated cell proliferation in the absence of p53, with accompanying genomic instability. A similar scenario was observed in the salivary gland tumors that developed in MMTV-*H-Ras* transgenic mice.[40] Furthermore, knock-in mice expressing $p53^{R172P}$ (equivalent to human $p53^{R175P}$), a rare mutant form of p53 which still induces cell cycle arrest but is completely defective in inducing apoptosis, escaped the early-onset of lymphomagenesis seen with *p53* null mice, indicating that p53 cell cycle arrest activity is important for suppressing T-cell lymphoma development.[94] These cases provide examples of how p53's function as a cell cycle regulator may play a role in tumor suppression.

Oncogene-induced cellular senescence as a tumor suppression mechanism in vivo has been established by studies utilizing a variety of mouse tumor models and analyzing human tumor samples.[58,95-97] The dependence on p53 in this tumor barrier mechanism was clearly demonstrated in one study, where acute deletion of *Pten* in the prostate stabilized p53 and induced senescence, while loss of p53 rescued this phenotype.[58] The importance of senescence as a tumor suppressor mechanism was also demonstrated by studies using the $p53^{R172P/R172P}$ knock-in mice mentioned above, which are defective in p53-dependent apoptosis, but retain intact cell-cycle arrest and cellular senescence pathways.[94,98] When crossed to mice deficient in the RNA subunit of telomerase, spontaneous tumorigenesis was prevented through induction of senescence by $p53^{R172P}$ in response to telomere dysfunction.[99] Thus, senescence is another important means through which p53 limits tumorigenesis.

That the induction of apoptosis by p53 is a relevant mechanism for tumor suppression was found by studying other oncogene-expressing transgenic mice. For example, Eμ-*myc* mice developed B-cell lymphoma that was dramatically accelerated upon *p53* loss, concomitant with a drastic reduction in apoptotic levels in these tumors.[100] This finding suggested that apoptosis normally limits tumor development. Furthermore, upon loss of p53, these tumors became refractory to chemotherapies, which act in this mouse model by stimulating the DNA damage-activated p53 apoptotic response. These results have been bolstered through analysis of the Eμ-*myc* mice on a background deficient for the caspase 9 apoptotic pathway component, which has further demonstrated that defects in apoptosis promote tumorigenesis.[101] Much of the understanding of the specific molecular function of p53 involved in tumor suppression has also come from studies of a transgenic brain cancer model. In this model, a fragment of Large T-Antigen (TgT121), which sequesters the Rb family of proteins, was expressed in the choroid plexus epithelium, leading to inappropriate cell proliferation.[102] This hyperproliferation activated p53 to induce apoptosis, resulting in slow-growing tumors. When the TgT121 transgenics were crossed to *p53* null mice, however, the brain tumors were not held in check by apoptosis, and they grew very rapidly. These studies provided strong support for the idea that activation of apoptosis plays a major role in p53-dependent tumor suppression.

The Therapeutic Value of p53 Restoration in Cancer Therapy

It has long been wondered whether, during multi-stage tumor development, p53 loss is only relevant in the early stages to facilitate aberrant proliferation and permit the accumulation of mutations, or whether its absence is actually required throughout the tumor lifespan. If p53 loss is indispensable for maintaining tumor cell growth and survival, reactivating wild-type p53 function in tumor cells holds promise as a potent therapeutic approach against malignancies. A few groups have addressed this question independently utilizing different mouse tumor models. One study demonstrated that restoration of wild-type p53 function in lymphomas from Eμ-*myc* mice—by addition of tamoxifen to activate a p53-ER fusion—triggered massive tumor cell death and tumor regression.[103] In another model system of *H-ras*-driven invasive hepatocellular carcinoma, activation of p53 through modulation of a tetracycline-responsive shRNA directed against p53 led to complete tumor regression mediated by senescence and clearance of the senescent cells by the innate immune system.[104] In a third study, p53 expression was silenced by virtue of incorporating a floxed (flanked by loxP recombination sites) transcriptional stop element into the *p53* locus, leading mice to develop lymphomas and sarcomas as *p53* null mice do.[105] When widespread Cre recombinase activity was induced in these mice to excise the transcriptional stop cassette, activation of p53 expression resulted in apoptosis in the lymphomas and senescence in the sarcomas. Together with the absence of deleterious effects in normal tissues with p53 reactivation, these findings not only revealed the reliance of established tumors on the continuous absence of p53, but demonstrated the value of p53 reactivation as a promising strategy for human cancer therapy.

Conclusion

In the nearly two decades since the *p53* knockout mouse was first published, many advances have been made using mouse models to understand p53 function. We have had a clear demonstration of the central role *p53* inactivation plays in spontaneous and induced cancers of a variety of types. Through crosses with other mouse strains, we have seen how universally *p53* loss either accelerates tumorigenesis or reveals novel tumor types in combination with other mutations. We have developed sophisticated ways to study *p53*, through conditional knockouts and knock-in models that have helped us to develop more accurate models for human cancer and to identify new in vivo functions of p53. We have started to explore the detailed molecular mechanism of p53 action in in vivo tumor suppression, as well as to appreciate the therapeutic values of p53 restoration as an anti-cancer strategy. However, in some senses, we have just scratched the surface of what we can learn about *p53*. There remain many unanswered questions: How exactly does *p53* loss facilitate tumorigenesis? Why are *p53* null mice lymphoma-prone? What activities of p53 are most crucial for tumor suppression? What is the basis of the tissue-specificity of certain p53 functions? These and other questions will keep researchers busy for many years to come.

Acknowledgments

We would like to thank Larry Donehower, Steven Artandi, Rebecca Ihrie, and Colleen Brady for helpful discussions and critical reading of the manuscript. We are also grateful to Larry Donehower for providing his data to present in Figures 1 and 2.

References

1. Hollstein M, Sidransky D, Vogelstein B, Harris CC. p53 mutations in human cancers. Science 1991; 253(5015):49-53.
2. Greenblatt MS, Bennett WP, Hollstein M, Harris CC. Mutations in the p53 tumor suppressor gene: clues to cancer etiology and molecular pathogenesis. Cancer Res 1994; 54(18):4855-78.
3. Malkin D, Li FP, Strong LC et al. Germ line p53 mutations in a familial syndrome of breast cancer, sarcomas, and other neoplasms. Science 1990; 250(4985):1233-8.
4. Donehower LA, Harvey M, Slagle BL et al. Mice deficient for p53 are developmentally normal but susceptible to spontaneous tumours. Nature 1992; 356(6366):215-21.

5. Jacks T, Remington L, Williams BO et al. Tumor spectrum analysis in p53-mutant mice. Curr Biol 1994; 4(1):1-7.
6. Purdie CA, Harrison DJ, Peter A et al. Tumour incidence, spectrum and ploidy in mice with a large deletion in the p53 gene. Oncogene 1994; 9(2):603-9.
7. Mowat M, Cheng A, Kimura N et al. Rearrangements of the cellular p53 gene in erythroleukaemic cells transformed by Friend virus. Nature 1985; 314(6012):633-6.
8. Ben-David Y, Bernstein A. Friend virus-induced erythroleukemia and the multistage nature of cancer. Cell 1991; 66(5):831-4.
9. Lavigueur A, Maltby V, Mock D et al. High incidence of lung, bone, and lymphoid tumors in transgenic mice overexpressing mutant alleles of the p53 oncogene. Mol Cell Biol 1989; 9(9):3982-91.
10. Lee JM, Abrahamson JL, Kandel R et al. Susceptibility to radiation-carcinogenesis and accumulation of chromosomal breakage in p53 deficient mice. Oncogene 1994; 9(12):3731-6.
11. Harvey M, Vogel H, Morris D et al. A mutant p53 transgene accelerates tumour development in heterozygous but not nullizygous p53-deficient mice. Nat Genet 1995; 9(3):305-11.
12. Donehower L.A, Harvey M, Vogel H et al. Effects of genetic background on tumorigenesis in p53-deficient mice. Mol Carcinog 1995; 14(1):16-22.
13. Harvey M, McArthur MJ, Montgomery CA Jr et al. Spontaneous and carcinogen-induced tumorigenesis in p53-deficient mice. Nat Genet 1993; 5(3):225-9.
14. Venkatachalam S, Shi YP, Jones SN et al. Retention of wild-type p53 in tumors from p53 heterozygous mice: reduction of p53 dosage can promote cancer formation. EMBO J 1998; 17(16):4657-67.
15. Venkatachalam S, Tyner SD, Pickering CR et al. Is p53 haploinsufficient for tumor suppression? Implications for the p53+/- mouse model in carcinogenicity testing. Toxicol Pathol 2001; 29:147-54.
16. Harvey M, McArthur MJ, Montgomery CA Jr et al. Genetic background alters the spectrum of tumors that develop in p53-deficient mice. FASEB J 1993; 7(10):938-43.
17. Ullrich RL, Bowles ND, Satterfield LC, Davis CM. Strain-dependent susceptibility to radiation-induced mammary cancer is a result of differences in epithelial cell sensitivity to transformation. Radiat Res 1996; 146(3):353-5.
18. Kuperwasser C, Hurlbut GD, Kittrell FS et al. Development of spontaneous mammary tumors in BALB/c p53 heterozygous mice. A model for Li-Fraumeni syndrome. Am J Pathol 2000; 157(6):2151-9.
19. Vousden KH, Lu X. Live or let die: the cell's response to p53. Nat Rev Cancer 2002; 2(8):594-604.
20. Kemp CJ, Wheldon T, Balmain A. p53-deficient mice are extremely susceptible to radiation-induced tumorigenesis. Nat Genet 1994; 8(1):66-9.
21. Ruggeri B, Caamano J, Goodrow T et al. Alterations of the p53 tumor suppressor gene during mouse skin tumor progression. Cancer Res 1991; 51(24):6615-21.
22. Ruggeri B, DiRado M, Zhang SY et al. Benzo[a]pyrene-induced murine skin tumors exhibit frequent and characteristic G to T mutations in the p53 gene. Proc Natl Acad Sci USA 1993; 90(3):1013-7.
23. Kemp CJ, Donehower LA, Bradley A, Balmain A. Reduction of p53 gene dosage does not increase initiation or promotion but enhances malignant progression of chemically induced skin tumors. Cell 1993; 74(5):813-22.
24. Clarke AR, Purdie CA, Harrison DJ et al. Thymocyte apoptosis induced by p53-dependent and independent pathways. Nature 1993; 362(6423):849-52.
25. Lowe SW, Schmitt EM, Smith SW et al. p53 is required for radiation-induced apoptosis in mouse thymocytes. Nature 1993; 362(6423):847-9.
26. Bosma MJ, Carroll AM. The SCID mouse mutant: definition, characterization, and potential uses. Annu Rev Immunol 1991; 9:323-50.
27. Bogue MA, Zhu C, Aguilar-Cordova E et al. p53 is required for both radiation-induced differentiation and rescue of V(D)J rearrangement in scid mouse thymocytes. Genes Dev 1996; 10(5):553-65.
28. Guidos CJ, Williams CJ, Grandal I et al. V(D)J recombination activates a p53-dependent DNA damage checkpoint in scid lymphocyte precursors. Genes Dev 1996; 10(16):2038-54.
29. Nacht M, Strasser A, Chan YR et al. Mutations in the p53 and SCID genes cooperate in tumorigenesis. Genes Dev 1996; 10(16):2055-66.
30. Gurley KE, Vo K, Kemp CJ. DNA double-strand breaks, p53, and apoptosis during lymphomagenesis in scid/scid mice. Cancer Res 1998; 58(14):3111-5.
31. Difilippantonio MJ, Zhu J, Chen HT et al. DNA repair protein Ku80 suppresses chromosomal aberrations and malignant transformation. Nature 2000; 404(6777):510-4.
32. Gao Y, Ferguson DO, Xie W et al. Interplay of p53 and DNA-repair protein XRCC4 in tumorigenesis, genomic stability and development. Nature 2000; 404(6780):897-900.

33. Frank KM, Sharpless NE, Gao Y et al. DNA ligase IV deficiency in mice leads to defective neurogenesis and embryonic lethality via the p53 pathway. Mol Cell 2000; 5(6):993-1002.
34. Lim DS, Vogel H, Willerford DM et al. Analysis of ku80-mutant mice and cells with deficient levels of p53. Mol Cell Biol 2000; 20(11):3772-80.
35. Gellert M. V(D)J recombination: RAG proteins, repair factors, and regulation. Annu Rev Biochem 2002; 71:101-32.
36. Nacht M, Jacks T. V(D)J recombination is not required for the development of lymphoma in p53-deficient mice. Cell Growth Differ 1998; 9(2):131-8.
37. Zhu C, Mills KD, Ferguson DO et al. Unrepaired DNA breaks in p53-deficient cells lead to oncogenic gene amplification subsequent to translocations. Cell 2002; 109(7):811-21.
38. Li B, Rosen JM, McMenamin-Balano J et al. neu/ERBB2 cooperates with p53-172H during mammary tumorigenesis in transgenic mice. Mol Cell Biol 1997; 17(6):3155-63.
39. Donehower LA, Godley LA, Aldaz CM et al. Deficiency of p53 accelerates mammary tumorigenesis in Wnt-1 transgenic mice and promotes chromosomal instability. Genes Dev 1995; 9(7):882-95.
40. Hundley JE, Koester SK, Troyer DA et al. Increased tumor proliferation and genomic instability without decreased apoptosis in MMTV-ras mice deficient in p53. Mol Cell Biol 1997; 17(2):723-31.
41. Curtis DJ, Robb L, Strasser A, Begley CG. The CD2-scl transgene alters the phenotype and frequency of T-lymphomas in N-ras transgenic or p53 deficient mice. Oncogene 1997; 15(24):2975-83.
42. Marin MC, Hsu B, Meyn RE et al. Evidence that p53 and bcl-2 are regulators of a common cell death pathway important for in vivo lymphomagenesis. Oncogene 1994; 9(11):3107-12.
43. Harvey M, Vogel H, Lee EY et al. Mice deficient in both p53 and Rb develop tumors primarily of endocrine origin. Cancer Res 1995; 55(5):1146-51.
44. Williams BO, Remington L, Albert DM et al. Cooperative tumorigenic effects of germline mutations in Rb and p53. Nat Genet 1994; 7(4):480-4.
45. Halberg RB, Katzung DS, Hoff PD et al. Tumorigenesis in the multiple intestinal neoplasia mouse: redundancy of negative regulators and specificity of modifiers. Proc Natl Acad Sci USA 2000; 97(7):3461-6.
46. Cichowski K, Shih TS, Schmitt E et al. Mouse models of tumor development in neurofibromatosis type 1. Science 1999; 286(5447):2172-6.
47. Reilly KM, Loisel DA, Bronson RT et al. Nf1;Trp53 mutant mice develop glioblastoma with evidence of strain-specific effects. Nat Genet 2000; 26(1):109-13.
48. McClatchey AI, Saotome I, Mercer K et al. Mice heterozygous for a mutation at the Nf2 tumor suppressor locus develop a range of highly metastatic tumors. Genes Dev 1998; 12(8):1121-33.
49. Wetmore C, Eberhart DE, Curran T. Loss of p53 but not ARF accelerates medulloblastoma in mice heterozygous for patched. Cancer Res 2001; 61(2):513-6.
50. Xu X, Wagner KU, Larson D et al. Conditional mutation of Brca1 in mammary epithelial cells results in blunted ductal morphogenesis and tumour formation. Nat Genet 1999; 22(1):37-43.
51. Jonkers J, Meuwissen R, van der Gulden H et al. Synergistic tumor suppressor activity of BRCA2 and p53 in a conditional mouse model for breast cancer. Nat Genet 2001; 29(4):418-25.
52. Clarke AR, Cummings MC, Harrison DJ. Interaction between murine germline mutations in p53 and APC predisposes to pancreatic neoplasia but not to increased intestinal malignancy. Oncogene 1995; 11(9):1913-20.
53. Artandi SE. Telomere shortening and cell fates in mouse models of neoplasia. Trends Mol Med 2002; 8(1):44-7.
54. Chin L, Artandi SE, Shen Q et al. p53 deficiency rescues the adverse effects of telomere loss and cooperates with telomere dysfunction to accelerate carcinogenesis. Cell 1999; 97(4):527-38.
55. Artandi SE, Chang S, Lee SL et al. Telomere dysfunction promotes non-reciprocal translocations and epithelial cancers in mice. Nature 2000; 406(6796):641-5.
56. O'Hagan RC, Chang S, Maser RS et al. Telomere dysfunction provokes regional amplification and deletion in cancer genomes. Cancer Cell 2002; 2(2):149-55.
57. Marino S, Vooijs M, van Der Gulden H et al. Induction of medulloblastomas in p53-null mutant mice by somatic inactivation of Rb in the external granular layer cells of the cerebellum. Genes Dev 2000; 14(8):994-1004.
58. Chen Z, Trotman LC, Shaffer D et al. Crucial role of p53-dependent cellular senescence in suppression of Pten-deficient tumorigenesis. Nature 2005; 436(7051):725-30.
59. Meuwissen R, Linn SC, Linnoila RI et al. Induction of small cell lung cancer by somatic inactivation of both Trp53 and Rb1 in a conditional mouse model. Cancer Cell 2003; 4(3):181-9.
60. Morton JP, Klimstra DS, Mongeau ME, Lewis BC. Trp53 deletion stimulates the formation of metastatic pancreatic tumors. Am J Pathol 2008; 172(4):1081-7.
61. Quinn BA, Brake T, Hua X et al. Induction of ovarian leiomyosarcomas in mice by conditional inactivation of Brca1 and p53. PLoS One 2009; 4(12):e8404.

62. Clark-Knowles KV, Senterman MK, Collins O, Vanderhyden BC. Conditional inactivation of Brca1, p53 and Rb in mouse ovaries results in the development of leiomyosarcomas. PLoS One 2009; 4(12):e8534.
63. Soussi T. p53 alterations in human cancer: more questions than answers. Oncogene 2007; 26:2145-56.
64. Lang GA, Iwakuma T, Suh YA et al. Gain of function of a p53 hot spot mutation in a mouse model of Li-Fraumeni syndrome. Cell 2004; 119(5):861-72.
65. Olive KP, Tuveson DA, Ruhe ZC et al. Mutant p53 gain of function in two mouse models of Li-Fraumeni syndrome. Cell 2004; 119(6):847-60.
66. Armata HL, Garlick DS, Sluss HK. The ataxia telangiectasia-mutated target site Ser18 is required for p53-mediated tumor suppression. Cancer Res 2007; 67(24):11696-703.
67. Chao C, Hergenhahn M, Kaeser MD et al. Cell type- and promoter-specific roles of Ser18 phosphorylation in regulating p53 responses. J Biol Chem 2003; 278(42):41028-33.
68. Sluss HK, Armata H, Gallant J, Jones SN et al. Phosphorylation of serine 18 regulates distinct p53 functions in mice. Mol Cell Biol 2004; 24(3):976-84.
69. MacPherson D, Kim J, Kim T et al. Defective apoptosis and B-cell lymphomas in mice with p53 point mutation at Ser 23. EMBO J 2004; 23(18):3689-99.
70. Chao C, Herr D, Chun J, Xu Y. Ser18 and 23 phosphorylation is required for p53-dependent apoptosis and tumor suppression. EMBO J 2006; 25(11):2615-22.
71. Feng L, Hollstein M, Xu Y. Ser46 phosphorylation regulates p53-dependent apoptosis and replicative senescence. Cell Cycle 2006; 5(23):2812-9.
72. Bruins W, Zwart E, Attardi LD et al. Increased sensitivity to UV radiation in mice with a p53 point mutation at Ser389. Mol Cell Biol 2004; 24(20):8884-94.
73. Feng L, Lin T, Uranishi H et al. Functional analysis of the roles of posttranslational modifications at the p53 C terminus in regulating p53 stability and activity. Mol Cell Biol 2005; 25(13):5389-95.
74. Krummel KA, Lee CJ, Toledo F, Wahl GM. The C-terminal lysines fine-tune P53 stress responses in a mouse model but are not required for stability control or transactivation. Proc Natl Acad Sci USA 2005; 102(29):10188-93.
75. Chao C, Wu Z, Mazur SJ et al. Acetylation of mouse p53 at lysine 317 negatively regulates p53 apoptotic activities after DNA damage. Mol Cell Biol 2006; 26(18):6859-69.
76. Tang Y, Zhao W, Chen Y et al. Acetylation is indispensable for p53 activation. Cell 2008; 133(4):612-26.
77. Toledo F, Krummel KA, Lee CJ et al. A mouse p53 mutant lacking the proline-rich domain rescues Mdm4 deficiency and provides insight into the Mdm2-Mdm4-p53 regulatory network. Cancer Cell 2006; 9(4):273-85.
78. Toledo F, Lee CJ, Krummel KA et al. Mouse mutants reveal that putative protein interaction sites in the p53 proline-rich domain are dispensable for tumor suppression. Mol Cell Biol 2007; 27(4):1425-32.
79. Lin J, Chen J, Elenbaas B, Levine AJ. Several hydrophobic amino acids in the p53 amino-terminal domain are required for transcriptional activation, binding to mdm-2 and the adenovirus 5 E1B 55-kD protein. Genes Dev 1994; 8(10):1235-46.
80. Johnson TM, Hammond EM, Giaccia A, Attardi LD. The p53QS transactivation-deficient mutant shows stress-specific apoptotic activity and induces embryonic lethality. Nat Genet 2005; 37(2):145-52.
81. Johnson TM, Meade K, Pathak N et al. Knockin mice expressing a chimeric p53 protein reveal mechanistic differences in how p53 triggers apoptosis and senescence. Proc Natl Acad Sci USA 2008; 105(4):1215-20.
82. Broz DK, Attardi LD. In vivo analysis of p53 tumor suppressor function using genetically engineered mouse models. Carcinogenesis 2010; 31(8):1311-8.
83. Luo JL, Yang Q, Tong WM et al. Knock-in mice with a chimeric human/murine p53 gene develop normally and show wild-type p53 responses to DNA damaging agents: a new biomedical research tool. Oncogene 2001; 20(3):320-8.
84. Luo JL, Tong WM, Yoon JH et al. UV-induced DNA damage and mutations in Hupki (human p53 knock-in) mice recapitulate p53 hotspot alterations in sun-exposed human skin. Cancer Res 2001; 61(22):8158-63.
85. Tyner SD, Venkatachalam S, Choi J et al. p53 mutant mice that display early ageing-associated phenotypes. Nature 2002; 415(6867):45-53.
86. Rovinski B, Munroe D, Peacock J et al. Deletion of 5'-coding sequences of the cellular p53 gene in mouse erythroleukemia: a novel mechanism of oncogene regulation. Mol Cell Biol 1987; 7(2):847-53.

87. Maier B, Gluba W, Bernier B et al. Modulation of mammalian life span by the short isoform of p53. Genes Dev 2004; 18(3):306-19.
88. Rudolph KL, Chang S, Lee HW et al. Longevity, stress response, and cancer in aging telomerase-deficient mice. Cell 1999; 96(5):701-12.
89. Garcia-Cao I, García-Cao M, Martín-Caballero J et al. "Super p53" mice exhibit enhanced DNA damage response, are tumor resistant and age normally. EMBO J 2002; 21(22):6225-35.
90. Matheu A, Maraver A, Klatt P et al. Delayed ageing through damage protection by the Arf/p53 pathway. Nature 2007; 448(7151):375-9.
91. Mendrysa SM, O'Leary KA, McElwee MK et al. Tumor suppression and normal aging in mice with constitutively high p53 activity. Genes Dev 2006; 20(1):16-21.
92. Attardi LD, Jacks T. The role of p53 in tumour suppression: lessons from mouse models. Cell Mol Life Sci 1999; 55(1):48-63.
93. Jones JM, Attardi L, Godley LA et al. Absence of p53 in a mouse mammary tumor model promotes tumor cell proliferation without affecting apoptosis. Cell Growth Differ 1997; 8(8):829-38.
94. Liu G, Parant JM, Lang G et al. Chromosome stability, in the absence of apoptosis, is critical for suppression of tumorigenesis in Trp53 mutant mice. Nat Genet 2004; 36(1):63-8.
95. Braig M, Lee S, Loddenkemper C et al. Oncogene-induced senescence as an initial barrier in lymphoma development. Nature 2005; 436(7051):660-5.
96. Michaloglou C, Vredeveld LC, Soengas MS et al. BRAFE600-associated senescence-like cell cycle arrest of human naevi. Nature 2005; 436(7051):720-4.
97. Collado M, Gil J, Efeyan A et al. Tumour biology: senescence in premalignant tumours. Nature 2005; 436(7051):642.
98. Barboza JA, Liu G, Ju Z et al. p21 delays tumor onset by preservation of chromosomal stability. Proc Natl Acad Sci USA 2006; 103(52):19842-7.
99. Cosme-Blanco W, Shen MF, Lazar AJ et al. Telomere dysfunction suppresses spontaneous tumorigenesis in vivo by initiating p53-dependent cellular senescence. EMBO Rep 2007; 8(5):497-503.
100. Schmitt CA, McCurrach ME, de Stanchina E et al. INK4a/ARF mutations accelerate lymphomagenesis and promote chemoresistance by disabling p53. Genes Dev 1999; 13(20):2670-7.
101. Schmitt CA, Fridman JS, Yang M et al. Dissecting p53 tumor suppressor functions in vivo. Cancer Cell 2002; 1(3):289-98.
102. Symonds H, Krall L, Remington L et al. p53-dependent apoptosis suppresses tumor growth and progression in vivo. Cell 1994; 78(4):703-11.
103. Martins CP, Brown-Swigart L, Evan GI. Modeling the therapeutic efficacy of p53 restoration in tumors. Cell 2006; 127:1323-34.
104. Xue W, Zender L, Miething C et al. Senescence and tumour clearance is triggered by p53 restoration in murine liver carcinomas. Nature 2007; 445:656-60.
105. Ventura A, Kirsch DG, McLaughlin ME et al. Restoration of p53 function leads to tumour regression in vivo. Nature 2007; 445:661-5.

Chapter 3

TP63, *TP73*: The Guardian's Elder Brothers

Stéphanie Courtois, Pierre Hainaut and Claude Caron de Fromentel*

Abstract

TP73 and *TP63* recently emerged as sharing overall architectural similarities with *TP53*. Phylogeny indicates that these three genes derive from a common ancestor, thus defining a new gene family. All three genes bind similar DNA consensus sequences in the promoters of many genes and regulate common generic aspects of growth control, survival, DNA repair or differentiation. However, their regulation patterns are distinct. While p53 is an ubiquitous, stress-response protein regulated at the post-translational level, p63 and p73 are expressed in a tissue and differentiation-specific manner and are also regulated at the transcriptional level. This regulation results in isoforms generated by alternative splicing or by the use of different promoters. They differ from each other in their C-terminus (which contains important regulatory domains) and, most strikingly, in their N-terminus. Thus, the major forms of p63 and p73 in many normal tissues are ΔN isoforms, which lack the transactivation domain, and can behave as repressors of the genes normally regulated by transactivation-competent (TA) forms of the protein. Control of the balance between levels of TA and ΔN forms of p63 and p73 is important in differentiation. Mice lacking *TP63* or *TP73* are not predisposed to cancer, but show developmental defects. In particular, *TP63* knock-out mice have major defects in cranial and limb morphogenesis, and in the formation of squamous epithelia. These defects are partially recapitulated in human subjects with germline mutation in *TP63*. There is growing evidence that p63 and p73 are involved in carcinogenesis through several mechanisms. For instance, amplification of *TP63* in squamous cancers results in overexpression of a ΔN protein that may counteract suppression by p63 as well as other family members. On the other hand, some mutant p53 can bind and inactivate p63 or p73, providing a mechanism for mutant p53 "gain-of-function" effect.

Introduction

In 1993, the p53 protein was awarded the title of "molecule of the year", acknowledging its rise to fame as the "guardian of the genome".[1,2] Four years later, it emerged that the guardian, who was considered as an orphan until now, had two big brothers who were living much less glamorous and dangerous lives, *TP63* and *TP73*.[3-7] Since then, we have become much more familiar with the complex structure, expression patterns and biology of *TP63* (3q27.29) and *TP73* (1p36). As a result, their lives now appear much more exciting, in particular since they are expressed as multiple isoforms that play important roles in differentiation and morphogenesis, and since their interactions with p53 are emerging as potentially important events in tumorigenesis.

Note that all amino acid positions cited in the text for p53, p63 or p73, refer to human sequences, except when the species is mentioned.

*Corresponding Author: Claude Caron de Fromentel—INSERM U590, Centre Léon Bérard, 28 rue Laennec, 69008 Lyon, France. Email: carondef@lyon.fnclcc.fr

p53, edited by Ayeda Ayed and Theodore Hupp. ©2010 Landes Bioscience and Springer Science+Business Media.

The p53 family members have a typical structure of transcription factors, with a N-terminal transactivation domain, a central domain carrying the DNA-binding properties and a C-terminal domain needed for the oligomerization. The *TP53* gene is the most frequently mutated gene in a wide variety of human tumors.[8] The database of published mutations now contains 17,689 somatic mutations and 225 germline mutations (http://www.iarc.fr/p53). Lack of *TP53* function predisposes mice to early and multiple tumors, demonstrating its function as a tumor suppressor.[9,10] By analogy, it was thought that *TP63* and *TP73* would also be frequently altered in human cancers. However, the search for mutations in tumors gave very poor results, and knock out mouse models did not reveal any increase in spontaneous tumorigenesis. In contrast, these mice exhibited a wide spectrum of specific defects in the differentiation and morphogenesis of several tissues, indicating that *TP63* and *TP73* are acting as physiological regulators of developmental processes rather than as tumor suppressors.[3,4,7] This is an unusual situation among gene families, since the members of the same family often have similar, if not overlapping functions. In the p53 family group, the odd one out is clearly p53, which differs from its two brothers by its mode of regulation, its acute inducibility, its ubiquity and its functional restriction to specific aspects of stress response.

In this chapter, we will present and discuss the structural and functional characteristics of the p63 and p73 proteins. In addition, we will review recent evidence on their implication in cancer, focusing on the potential role of particular isoforms, the ΔN isoforms, which lack the transactivation domain, and therefore may have growth-promoting effects.

Structure of the p63 and p73 Isoforms

Structural Homology between the Family Members

TP63 and *TP73* are expressed as multiple isoforms, which all conserve the DNA-binding domain but differ by their N- or C-terminal regions (Fig. 1). The highest percentage of identity

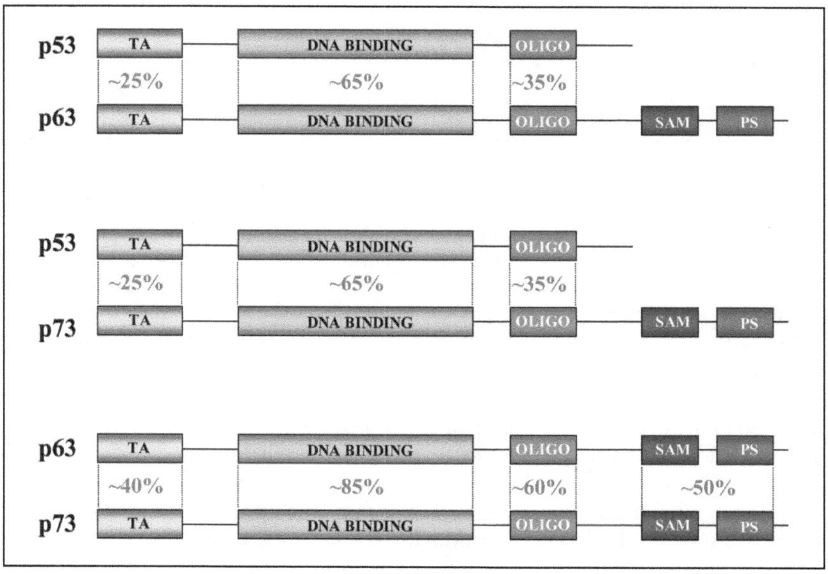

Figure 1. Structural comparison and homology between p53, p63 and p73. The schematic structure of TAα isoforms of p63 and p73 is presented, including the transactivation domain (TA), the DNA-binding domain and the oligomerization domain (oligo). SAM: sterile alpha motif; PS: post-SAM domain. This PS domain has been reported to exert a negative effect on the TA domain. The percentage of identity between p53/p63, p53/p73 and p63/p73 is indicated for each domain.

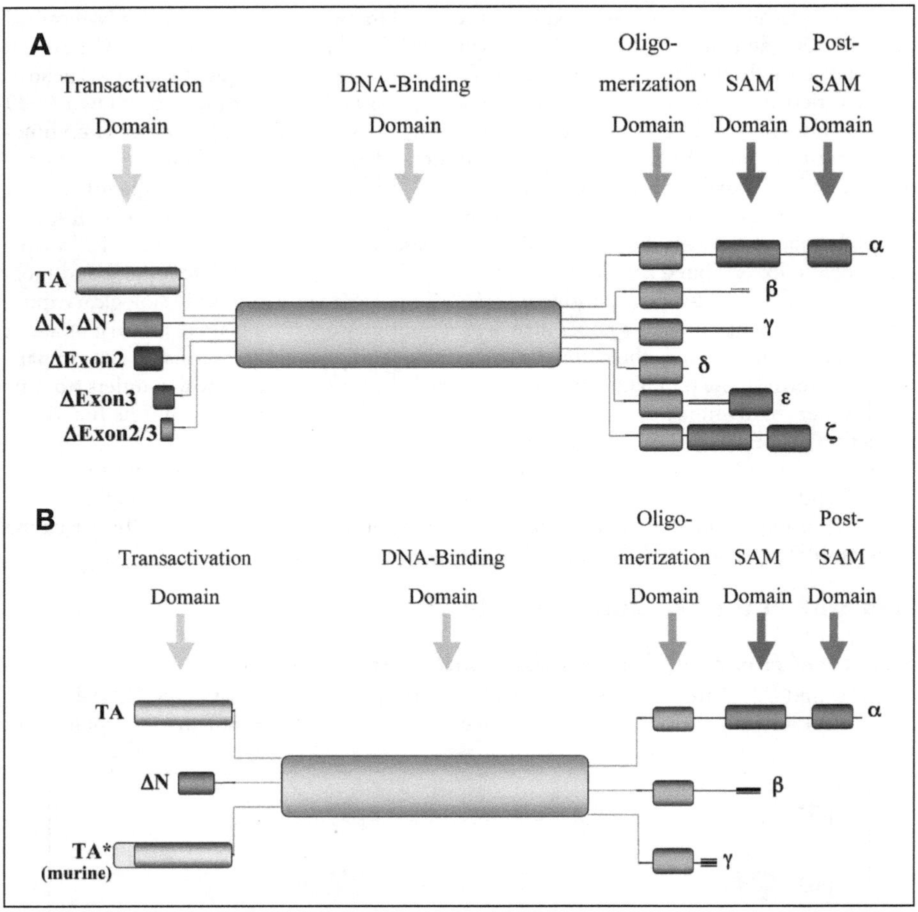

Figure 2. Structure of mammalian p63 and p73 isoforms. The proteins are represented with an invariant, DNA-binding domain, that combines with variable N- (TA or Δ) and C-terminal (greek letters) segments generated by alternative splicing and/or by the use of an alternative promoter. The nomenclature is generated by combining the exact name of each segment actually present in any particular isoform (for example ΔNp73β). A) p73; B) p63. The different styles of black lines indicate different sequences, due to alternative splicing in the C-terminal region. TA*: mouse-specific N-terminal variant with 39 additional amino acids, generated from an AUG located 117 base pairs upstream the main initiation codon.

between these proteins is observed in the central domain (65% between p53 and p63 or p73, 85% between p63 and p73).[3,7] All the residues that play an important role in the folding of the DNA-binding domain and in the contact with DNA are well conserved, suggesting that the overall shape of the domain is similar in the three family members. This similarity accounts for their capacity to bind and to transactivate many of the same promoters as those activated by p53. Furthermore, the DNA-binding domain alone is responsible for the specificity of transactivation, the other domains providing either generic (transactivation, oligomerization) or regulatory functions.[11]

The C-terminal isoforms are generated by an alternative splicing between exons 10 and 15. Six different forms have been identified for p73 (α, β, γ, δ, ε and ζ)[12-14] and three for p63 (α, β and γ)[7] (Fig. 2A,B). The p63α and p73α isoforms contain a sterile alpha motif (SAM) domain, usually implicated in development.[15] This structural feature has no equivalent in p53.

The N-terminal isoforms show differences in the transactivation domain. They are generated either by alternative splicing or by the use of an internal promoter in intron 3 (P2). These N-truncated forms lack the main transactivation domain and are termed "Δ" forms, as opposed to "TA", transactivation-competent forms. However, Δ forms may retain some transcriptional capacity through secondary domains located in the C-terminus or residual N-terminus.[16,17] Several types of Δ forms are distinguished according to their exact divergence with TA counterparts. For p73, Δ2, Δ3, Δ2/3 arise by alternative splicing of, respectively, exons 2, 3, or both. ΔNp73 lacks exons 1, 2 and 3 and is generated from the internal promoter in intron 3 (P2), resulting in the use of an initiation codon located in an alternative exon 3'.[18] ΔN'p73 corresponds to a protein identical to ΔNp73, but translated from a mRNA initiated at the first promoter and incorporating the 3' portion of the alternative exon 3'.[19] For p63, only ΔN isoforms, initiated at the P2 promoter, have been identified so far.[7] However, a different, transcription-competent N-terminal variant, TA*p63, has been characterized in the mouse but not (yet) in other species.[7] This variant contains 39 additional amino acids resulting from the use of an in-frame AUG located upstream of the major initiation site. This particular AUG is not conserved in human *TP63*.

The Proline-Rich Domain in the p53 Family

The N-terminal part of the p53 protein contains a proline-rich region suspected to bind a number of cellular proteins (Fig. 3). Human p53 contains five repeats of PXXP motif (P = proline, X = any amino acid) between amino acids 61 and 94, whereas human TAp63 and TAp73 contain only two repeats (between amino acids 60 and 130 and 80 and 120,

```
Human p53      ---------MEEPQSDPSVEPPLSQETFSDLWKLLPENNVLSPLPSQAMDDLMLSPDDIEQWFTED------------
Squirrel p53   -----------------------------DLWNLLPENNVLSPVLSPPMDDLLLSSEDVENWFDK-------------
Rabbit p53     ---------MEESQSDLSLEPPLSQETFSDLWKLLPENNLLTTSLNPPVDDLLS-AEDVANWLNE-------------
Mouse p53      ------MTAMEESQSDISLELPLSQETFSGLWKLLPPEDILP---SPHCMDDLLLPQDVEEFFEG-------------
Clam p53       MSHEALHKMSQVAIHGTLPNQPMSQETFEYLWHTLEEVTDNVDYTHINTRELDYSYDDSEDGTSLQVEKFRINQHHTD
Clam p73       MSHEALHKMSQVAIHGTPPNQPMSQETFEYLWHTLEEVTDNVDYTHINTRELDYSYDDSEDGTSLQVEKFRINQHHTD
Human TAp63    ------------MSQSTQTNEFLSPEVFQHIWDFLEQPICSVQPIDLNFVDEPSEDGATNKIEISMDCIRMQDSDLSD
Xenopus p63    -------------------------------------------------------------------MLYLENNAQAQ
Human TAp73    -------------MAQSTATSPDGGTTFEHLWSSLEPDSTYFDLPQSSRGNNEVVGGTDSS------MDVFHLEGMTT
Barbel p73     ---------------MSQSSTADEGTTFEHLWSTLEPDSTYFELPQAGHSGDRVASSSLPSNRAEVCMDVYHMRDMRD
Squid p53      ---------------MSQGTSPNSQETFNLLWDSLEQVTANEYTQIHERGVGYEYHEAEPDQTSLEISAYRIAQPDPY
Xenopus p53    ----------MEPSSETGMDPPLSQETFEDLWSLLPDPLQTVTCRLDNLSEFPDYPLAADMTVLQEG-----------
Barbel p53     -----------MADLAENVSLPLSQESFEDLWKMNLNLVAVQPPETESWVGYDNFMMEAPLQVEFDP-----------
Chicken p53    -------------MAEEMEPLLEPTEVFMDLWSMLPYSMQQLPLPEDHSNWQELSPLEPSDPPPPPPPP---------
Fly p53        ------------MYISQPMSWHKESTDSEDDSTEVDIKEDIPKTVEVSGSELTTEPMAFLQGLNSG------------

Human p53      ----------------PGPDEAPRMPEAAPP----VAPAPAAPTPAAPAPAPSWPLSS-SVPSQKTYQG-SYGFRLGFLHS-
Squirrel p53   -----------------GPDEALQMSAAPAP----KAPTPAASTLAAPTPAISWPLSS-SVPSQNTYPG-VYGFRLGFIHS-
Rabbit p53     -----------------DPEEGLRVPAAPAP----EAPAPAAPALAAPAPATSWPLSS-SVPSQKTYHG-NYGFRLGFLHS-
Mouse p53      ------------------PSEALRVSGAPAA----QDPVTETPGPVAPAPATPWPLSS-FVPSQKTYQG-NYGFHLGFLQS-
Clam p53       VSDLLNPIIGTTSSSSMSPDSQTNISGSTASSPY-QEMALTSPPPYSPHTNLTSPIP--TVPSNTNYPG-DYGFEISFATPS
Clam p73       VSDPLNPIIGTTSSSSMSPDSQTNISGSTASSPY-QEMALTSPPPYSPHTNLTSPIP--TVPSNTNYPG-DYGFEISFATPS
Human TAp63    PMWPQYTNLGLLNSMDQQIQNGSSSTSPYNTDHA-QNSVTAPSPYAQPSSTFDALSPSPAIPSNTDYPG-PHSFDVSFQQS-
XenoPus p63    YSEPQYTNLGLLNSMEQQIQNGSSSTSPYANDHA-QNSVTAPSPYAQPSSTFDALSPSPAIPSNTDYPG-PHSFDVSFQQS-
Human TAp73    S---VMAQFNLLSSTMDQ-MSSRAASASPYTPEH-AASVPTHSPYAQPSSTFDTMSPAPVIPSNTDYPG-PHHFEVTFQQS-
Barbel p73     MNDNVMSQYSLLSSSMEQGLGNRAASTSPYSSET-TSNVPTPSPYSQPNSTFEAMSPAPAIPSNTDYPG-PHNFEVTFQQS-
Squid p53      GRSESYDLLNPIINQIPAPMPIADTQNNPLVNHCPYEDMPVSSTPYSPHDHVQSPQPS--VPSNIKYPG-EYVFEMSFAQPS
Xenopus p53    -----------------------------------------------LMGNAVPTVTSCAVPSTDDYAG-KYGLQLDFQQN-
Barbel p53     --------------------------------SLFEVSATEPAPQPSISTLDTGSPPTSTVPTTSDYPG-ALGFQLRFLQS-
Chicken p53    -----------------------------------PLPLAAAAPPPLNPPTPPRAAPSPVVPSTEDYGG-DFDFRVGFVEA-
Fly p53        -----------------------------------NLMQFSQQSVLREMMLQDIQIQANTLPKLENHNIGGYCFSMVL----
```

Figure 3. Comparison between the proline-rich domains of p53, p63 and p73 from various species. The amino acid sequence of the N-terminal domain of p53, p63 and p73 from selected species were aligned using Clustalw program. Proline residues are colored in blue; in the PXXP motifs, nonproline X residues are in purple. Accession numbers: Human p53, Swiss-Prot P04637; Squirrel p53, Swiss-Prot Q64662; Rabbit p53, Swiss-Prot Q95330; Mouse p53, Swiss-Prot P02340; Clam p53, Genbank AAF67733; Clam p73, Genbank AAF67734; Human TAp63, Genbank AAC62635; Xenopus p63, Genbank AAK15622; Human TAp73, Swiss-Prot O15350; Barbel p73, Genbank AAD27752; Squid p53, Genbank AAA98563; Xenopus p53, Swiss-Prot P07193; Barbel p53, Genbank AAD34212; Chicken p53, Swiss-Prot P10360; Fly p53, Genbank AAF75270. A color version of this figure is available online at www.landesbioscience.com/curie.

respectively). The PXXP motifs are known to form a left-handed polyproline Type II helix, which creates a binding site for the SH3 domains present in many signal transduction proteins. Such motifs play a role in p53-mediated signal transduction.[20] In particular, the murine p53 polyproline domain, which contains two PXXP motifs, has been shown to be a docking site in the transmission of Gas1-dependent anti-proliferative signals[21] and to be required to activate apoptosis, but not growth arrest.[22] Although the biological role of the proline-rich domains of p63 and p73 has not been demonstrated experimentally, it is tempting to speculate that they serve as protein-binding sites in p63 or p73-mediated signal transduction.

Phylogeny of the TP53 Family

During the late eighties, the *TP53* gene was identified in all vertebrate species tested, but not in invertebrates or lower organisms, such as yeast or Drosophila. In the early nineties, genes showing homologies with mammalian *TP53* were isolated from mollusks (*Loligo forbesi*, *Mya arenaria*).[23] The later discovery of *TP63* and *TP73* led to the realization that the *TP53*-like genes of mollusks were, in fact, very similar to *TP63/TP73*, in particular with the presence of a SAM domain. The current view on the phylogeny of the family is that all three members derive from a common ancestor, which may correspond to the gene identified in mollusks (Fig. 4). This unique ancestor probably underwent duplication in chordates. The earliest duplication resulted in the specialization into two related genes that have evolved nearly simultaneously, *TP63* and *TP73*. *TP53*, in contrast, may have appeared later through a distinct duplication event that implied the loss of exons coding for the SAM domain. As a result of this evolution, the three family members are present only in vertebrates. It seems counter-intuitive that *TP53*, which is the most recently evolved gene, is much less complex than *TP63* and *TP73*. This paradox is also observed at the level of protein functions. The p53 protein exerts multiple functions in stress response and DNA repair, that are more similar to the ones of the unique, prototypic gene product present in invertebrates, than to the complex, development-related functions of p63 and p73.

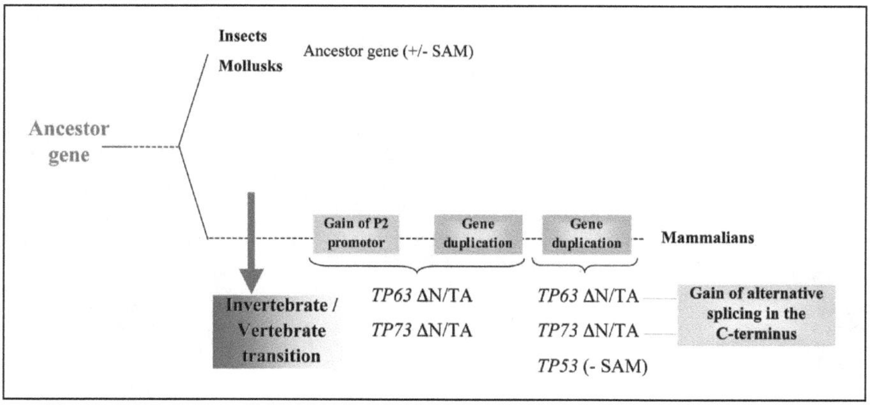

Figure 4. Phylogeny of *TP53*, *TP63* and *TP73*. A simplified view of the family's phylogeny is proposed, based on Clustalw alignment. Although sequence alignment suggests that *TP63* might be the most ancestral gene, there is evidence that the functions of the gene present in invertebrates are more similar to those of *TP53* and of *TP63/TP73*. These considerations led us to suggest the evidence of a common ancestor, from which the present day's genes may have derived. After invertebrate/vertebrate transition, the gain of a second promoter and gene duplication lead to *TP63* and *TP73* genes, which show similar characteristics with their human counterparts. *TP53* appears later, probably resulting from further gene duplication and loss of domains in the C-terminus.

Functions of p63 and p73 Isoforms

Lessons from Knock-Out Mouse Models

Inactivation of *TP53* by homologous recombination in the mouse is developmentally viable, although 25% of female embryos develop exencephaly and die in utero. After birth, *TP53*$^{-/-}$ mice show normal growth, but exhibit a dramatic susceptibility to early tumors and die before one year of age.[9,24]

Mice lacking all p63 isoforms reach term, but die within one day, owing to desiccation and maternal neglect (Fig. 5).[25,26] They show cranofacial abnormalities, absence or truncation of the limbs and absence of an epidermis. Products of epidermal-mesenchymal interactions, such as hair follicles, teeth and mammary glands, are absent. All squamous epithelia show a lack of stratification and mature into a monolayer of epithelial cells that lack the typical squamous differentiation markers. Impaired squamous development results in abnormalities in many organs including skin, tongue, esophagus, cervix and bladder. This defect is interpreted as the consequence of a fundamental role of p63 in regulating the asymmetric division of precursor cells in squamous epithelia (see below). Heterozygous, *TP63*$^{+/-}$ mice are viable and do not show any major developmental defects, or increased susceptibility to tumorigenesis.

The *TP73*$^{-/-}$ mice have a complex phenotype characterized by hippocampal dysgenesis, hydrocephalus, chronic infections and inflammation, gastro-intestinal hemorrhages, as well as abnormalities in the pheromone sensory pathway (Fig. 5).[27] *TP73* deficient males lack both a sexual interest in females and aggressiveness in response to other males, implying an altered hormonal or sensory pathway. *TP73* deficient females do not get pregnant when mated with wild-type males, indicating a defect in conceiving or maintaining embryos. All this data clearly indicates that p73 plays an important role in the development of the brain and many epithelial

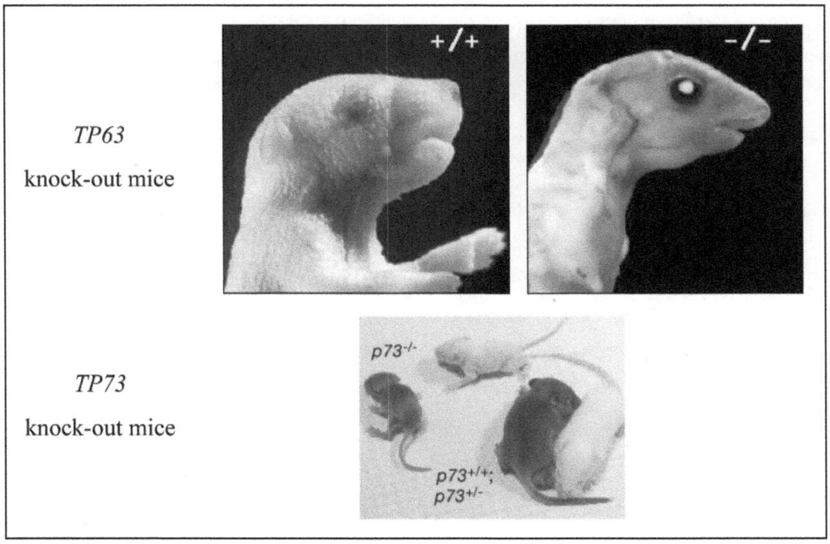

Figure 5. Phenotype of *TP63* and *TP73* knock-out mice. *TP63* and *TP73* knock-out mice are shown, next to their wild-type counterpart. *TP63* knock-out mice (upper panel) exhibit dramatic defects in craniofacial and epidermal development and die within the first day of life. *TP73* knock-out mice (lower panel) have a reduced size and they show numerous abnormalities in brain development or immune response, as well as a high youth mortality. From references 26 and 27 with permission, ©1999 and 2000, Nature.

tissues. However, p73 does not act as a tumor suppressor in mouse, since after 15 months of follow-up, these mice do not develop spontaneous tumors.

Inherited TP63 Mutations in Humans

In humans, *TP63* maps to chromosome 3q27, a region involved in several malformation syndromes, such as EEC (Ectrodactily Ectodermal dysplasia and facial Clefting), SHFM (Split-Hand/split-Foot Malformation), Hay-Wells syndrome, LMS (Limb Mammary Syndrome), ADULT syndrome (Acro-Dermato-Ungual-Lacrimal-Tooth), CLEPD1 (cleft lip/palate ectodermal dysplasia).[28] Genetic analysis reveals that these syndromes, all characterized by limb development and/or ectodermal dysplasia, actually result from heterozygous missense or frameshift mutations in *TP63*.

The clinical features of these syndromes clearly correlate with the phenotype of the *TP63* knock-out mice, therefore implicating p63 in ectodermal development. Interestingly, the mutations associated with the different syndromes cluster in different domains of the p63 protein, resulting in a genotype-phenotype correlation as illustrated in Figure 6.[29]

There is no known human syndrome associated with loss of *TP73* or inheritance of a *TP73* mutation. The chromosomal region that contains *TP73* on 1p36 is the centre of somatic allelic loss in many cancers,[30] but there is no evidence that the *TP73* locus is the specific target of these losses.[31-33] In the case of *TP53*, inherited missense mutations have been linked to a rare autosomal dominant syndrome, Li-Fraumeni syndrome (LFS), characterized by high susceptibility to multiple tumors at an early age, including tumors of the adrenal gland, brain, breast, as well as sarcomas.[34] Although the molecular basis of this tissue-specificity is not well understood, this phenotype is consistent with the impairment of the tumor suppressive properties of p53.

A comparison between the location of "hot spot" *TP53* mutations in human cancer and of *TP63* mutations in malformation syndromes reveals that these mutations often target the same positions (Fig. 7). This striking concordance has a structural basis, as illustrated in Figure 8, which shows the position of these residues in the crystal structure of the DNA-binding domain of the p53 protein. This domain consists of a sandwich of two

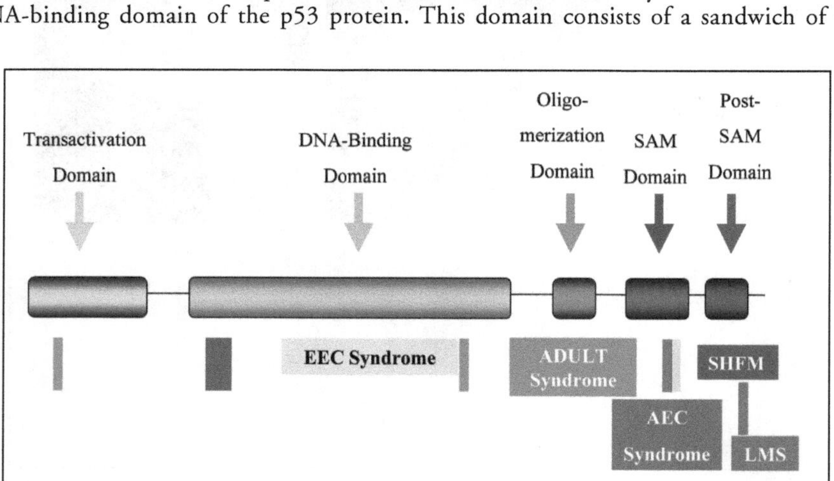

Figure 6. Clustering of *TP63* mutations found in human developmental disorders. The domain structure of p63 is shown and the regions where germline mutations cluster are indicated by boxes, corresponding to different syndromes. EEC, Ectrodactily Ectodermal dysplasia and facial Clefting; AEC, Ankyloblepharon Ectodermal dysplasia Clefting; SHFM, Split-Hand/split-Foot Malformations; LMS, Limb Mammary Syndrome; ADULT, Acro-Dermato-Ungual-Lacrimal-Tooth. For p63 domains, see Figure 2B.

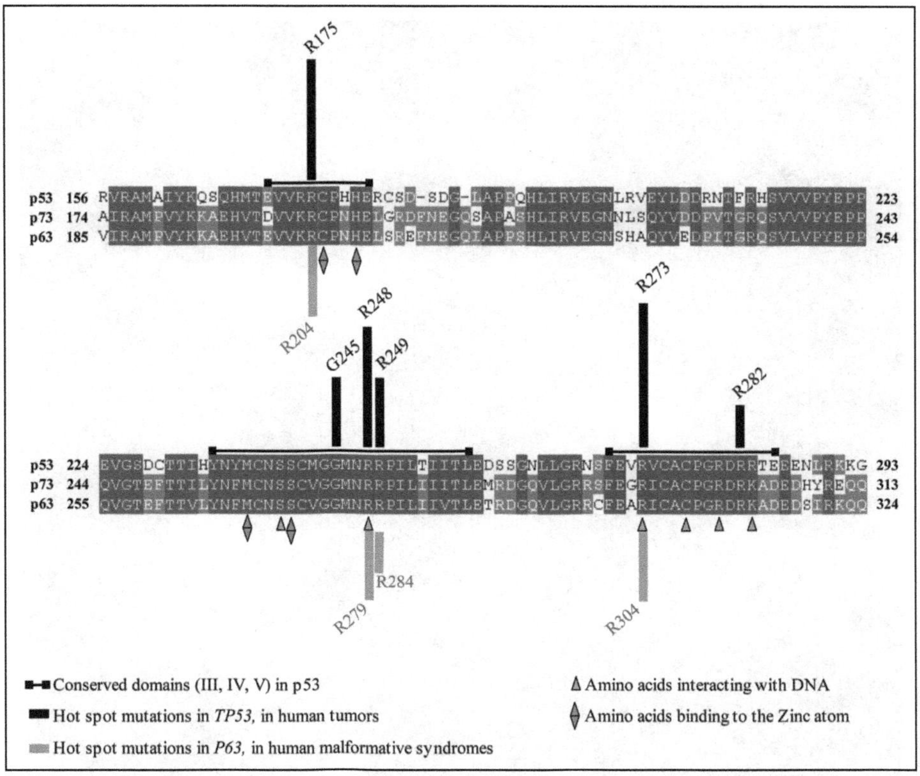

Figure 7. Comparison of "hot spot" Mutation Codons in p53 (from tumors) and in p63 (from Developmental Disorders). A portion of the p53 DNA-binding domain (156-293), containing most of the common mutation hot spots, is aligned with the corresponding portions of p63 and p73 (Edtaln program). The most frequently substituted amino acids are represented by vertical lines, proportional to the frequency of mutation. Black lines, "hot spot" amino acids found in p53 from human tumors; orange lines, "hot spot" amino acids found in p63 from human syndromes. Amino acids conserved in all the three human proteins are colored in navy blue, amino acids only conserved in two of them in gray. The highly conserved domains III, IV and V (Soussi et al, 1987) are represented by ■—■. Note that the frequent targeted amino acids are conserved both in human p53, p63 and p73 proteins and in p53 from vertebrate species. Green triangle, p53 amino acids interacting with DNA; purple lozenge, p53 amino acids interacting with zinc. A color version of this figure is available at www.landesbiosicence.com/curie.

beta-sheets that support several flexible loops and alpha helices. Common *TP53* mutations in cancer and *TP63* mutations in malformation syndromes affect residues in direct contact with DNA (amino acids 248, 273 in p53), as well as residues involved in the folding of the molecule (amino acids 175, 249 in p53). Thus, the folding of the two proteins may depend on the same, conserved amino acids, providing a rationale for the existence of concordant "hot spot" mutations that abrogate sequence-specific DNA-binding. It is quite surprising, therefore, to observe that these mutations have opposite biological effects. While *TP53* mutation results in tumorigenesis due to loss of suppressor function (and perhaps gain of promoting function, see chapter "*TP53* mutations in human cancers: selection versus mutagenesis"), mutation of *TP63* impairs morphogenesis and results in the under-development of specific organs and tissues.

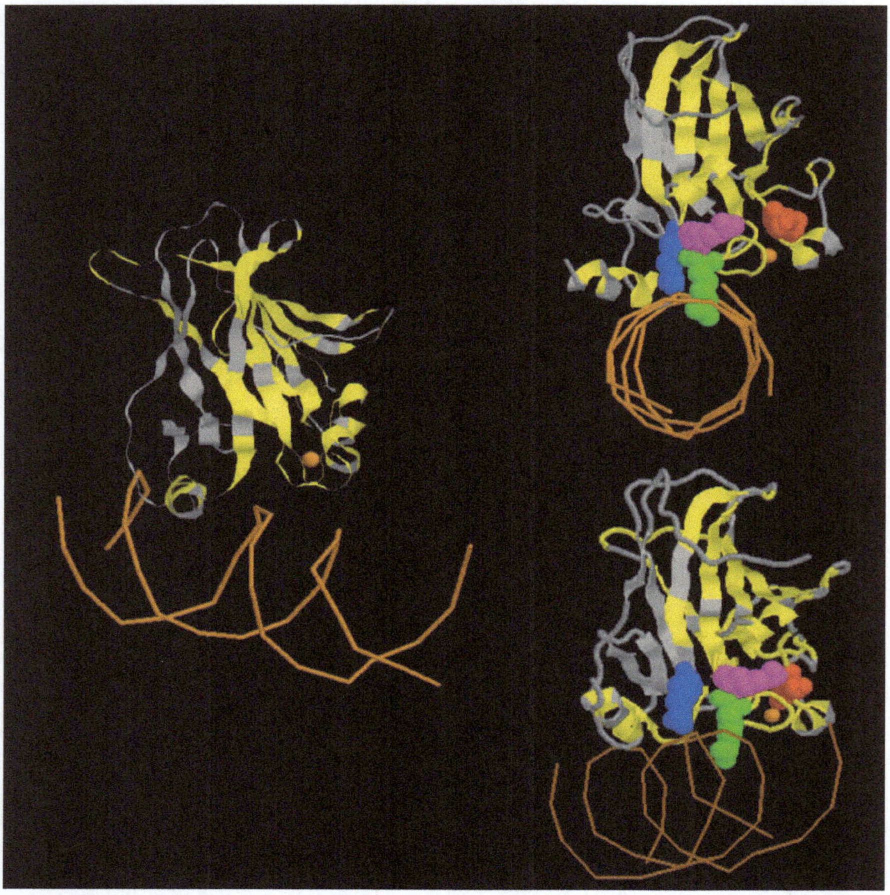

Figure 8. "Hot spot" codons on the three-dimensional structure of p53. Two views of the 3-D structure of the central region of p53 in contact with DNA. Zinc atom: orange. Yellow: residues most conserved between p53, p63 and p73. The position corresponding to residues often mutated in p53 or in p63 are highlighted on the right panel. Red: Arg175; green: Arg248, magenta: Arg249; blue: Arg273. Arg248 and 273 interact with DNA, whereas Arg175 and 249 are implicated in the maintenance of protein folding. Figures are generated with RasMol 2.7 using a modified version of the PDB data file 1TUP.

Expression of p63 and p73

The expression of the p63 and p73 isoforms is tissue and cell-specific. The distribution pattern of TA and ΔN isoforms depends on the differentiation status of cells.[35-37] In squamous epithelia, ΔN isoforms of p63 are mainly expressed in proliferating cells of the basal layers, whereas TA isoforms are preferentially detected in differentiated cells of the upper layers (Fig. 9). This distribution is consistent with the notion that p63 plays an active role in generating the "molecular gradient" that drives the maturation of squamous epithelia. The switch from ΔN to TA isoforms is thought to result from differential use of P1 (TA forms) and P2 (ΔN forms) promoters. However, the nature and regulation of the transcription factors that control these promoters remain to be analyzed.

The p73 protein plays a role in the development and survival of neuronal cells. In developing neurons, p73 is constitutively present as a truncated isoform whose levels are dramatically

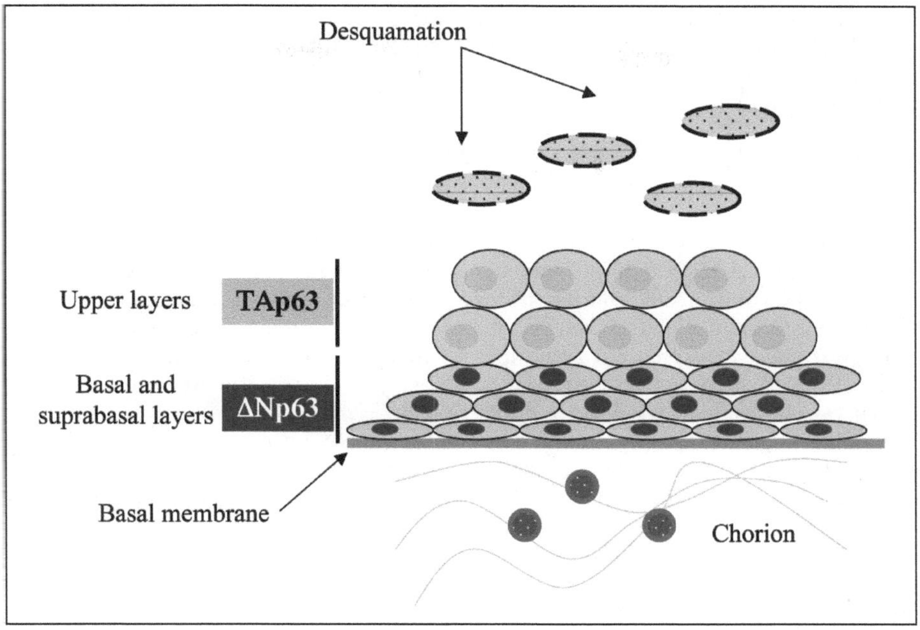

Figure 9. Expression of TA and ΔNp63 isoforms in normal squamous epithelium. Transversal representation of a squamous epithelium. The expression of ΔNp63 isoform is restricted to basal and suprabasal layers (including proliferating stem cells). TAp63 isoform is expressed in upper layers (cells engaged in differentiation). From P. Tanière, PhD, Thesis, 2002, with permission.

decreased when sympathetic neurons undergo apoptosis after withdrawal of nerve growth factor (NGF). Overexpression of the ΔNp73α protein protects these neurons from apoptosis induced by NGF withdrawal as well as by overexpression of p53. These results indicate that p73 expression is essential for the survival of neuronal cells.[38]

Regulation of p63 and p73 Stability

In the case of p53, post-translational stabilization through escape from proteasome degradation is the major mechanism for regulating protein level. There is evidence that p63 and p73 are also regulated through proteasome-dependent degradation. However, the effectors of this regulation are not clearly identified. It has been reported that p53 could trigger ΔNp63 degradation by the proteasome, suggesting that there are cross-talks between the degradation pathways of the various family members.[39]

Degradation of p53 by the proteasome involves two main cellular effectors, Mdm2 (and the parent protein Mdmx) and the inactive form of the Jun-N (amino)-terminal kinase (JNK).[40,41] JNK, in its inactive form, is thought to bind in the proline-rich domain of p53 and to phosphorylate a threonine residue at position 81.[42] It is interesting to note that the proline-rich domains of p63 and p73 both contain a conserved Threonine (see Fig. 3), raising the possibility that a form of regulation by JNK may also affect these proteins. However, there is no evidence so far that inactive JNK binds p63 or p73.

The inter-relations between Mdm2-Mdmx and p63/p73 are better documented but the data remain controversial. Protein interactions between p63 and Mdm2 or Mdmx, and their resulting effects on transactivation and protein degradation, are yet to be demonstrated.[43-45] In contrast, Mdm2 or Mdmx have been shown to interact with p73, resulting in an inhibition of transactivation but not in protein degradation. Quite the opposite, there is evidence that binding of Mdm2 increases the half-life of both proteins (p73 and Mdm2), thus creating a positive feedback loop in which p73 enhances Mdm2 accumulation.[46-50] In turn, accumulation of Mdm2 triggers a reduction of p53 levels, suggesting a role for high levels of p73 in controlling p53 protein levels.[46] However, the molecular effectors that control the destabilization of p73 are not yet identified.

Common and Specific Target Genes of p63 and p73

The structural similarity of the DNA binding domain of p63, p73 and p53 led several groups to investigate the capacity of p63 and p73 to regulate the same target genes as p53. The TA isoforms of p63 and p73 were found to transactivate some p53-target genes, such as *WAF1/CIP1*, *MDM2*, *GADD45* and *PIG3*, but with a lower efficiency.[16,51,52] This suggests that TA isoforms of p63 or p73 could compensate for p53 deficiency. However, this capacity may just illustrate the fact that all three family members have inherited from their common ancestor a capacity to induce the expression of genes involved in cell cycle arrest, DNA repair or apoptosis. In fact, many of the genes identified so far as "p53-targets" after in vitro overexpression of the protein, might be physiological targets of p63 or p73.

The new target *REDD1* (regulated in development and DNA damage responses, also known as *RTP801*), recently isolated by Ellisen and collaborators, provides an example of the difficulty in identifying which transcription factor is physiologically responsible of expression control.[53] Initially identified as a "p53-target", *REDD1* was found to have a developmentally-regulated expression pattern that overlaps with the known localization of p63 expression. The same gene is up-regulated in adult cells in response to ionizing radiation in a p53-dependent manner. Ectopic expression of *REDD1* enhances intracellular levels of reactive oxygen species (ROS) and modulates the sensitivity of cells to oxidative stress-induced apoptosis. Furthermore, in $p63^{-/-}$ fibroblasts, there is a marked decrease in ROS levels, suggesting that p63 plays a physiological role in maintaining the normal intracellular balance of reactive oxygen species. Modification of ROS levels may be important in the signaling of differentiation as well as of the cellular responses to genotoxic agents. It is therefore interesting that the same molecular mechanism can be triggered in distinct circumstances by either p63 or p53.

Recent studies have reported that p63 and p73 are able to specifically transactivate promoters of some genes involved in the terminal differentiation of epidermal cells. For example, p63 and p73 specifically regulate the expression of *IVL* (Involucrin) and *LOR* (Loricrin).[54] So far, there is no evidence that these promoters contain a responsive element that binds p53. Thus, the mechanism by which these two promoters are transactivated remains to be identified.

Regulation of p63 and p73 in Response to Stress

As described above, p63 and p73 are able to transactivate many of the reported p53-target genes, the products of which are involved in cell cycle arrest or apoptosis. The p53 protein responds to a broad range of genotoxic or nongenotoxic forms of stress, through complex post-translational modifications that have been extensively described.[55] For p63, however, very little is known regarding the possible regulation in response to stress. Upon DNA damage, ΔN isoforms are down-regulated, whereas TA isoforms are up-regulated.[56,57] Using a phosphatase inhibitor, Okada and collaborators have obtained evidence that accumulation of p63 in response to DNA-damage requires phosphorylation at at Ser/Thr residues.[58]

Regarding p73, up-regulation has been observed in response to various DNA-damaging agents, with inconsistencies, however, that may result from heterogeneity in the cellular models and the dose and the nature of DNA damage. In response to cisplatin and ionizing radiation, p73 is regulated by the nonreceptor tyrosine kinase c-Abl. This direct protein interaction occurs through the binding of the SH3 domain of c-Abl to the PXXP motif of p73, and results in the phophorylation by c-Abl at Tyrosine 99 of p73.[59-61] This interaction requires intact ATM function to phosphorylate and activate c-Abl. Recently, it was reported that the interaction of p73 with c-Abl also induces the phosphorylation of threonine residues which are adjacent to proline, and that the p38 MAP kinase pathway mediates this

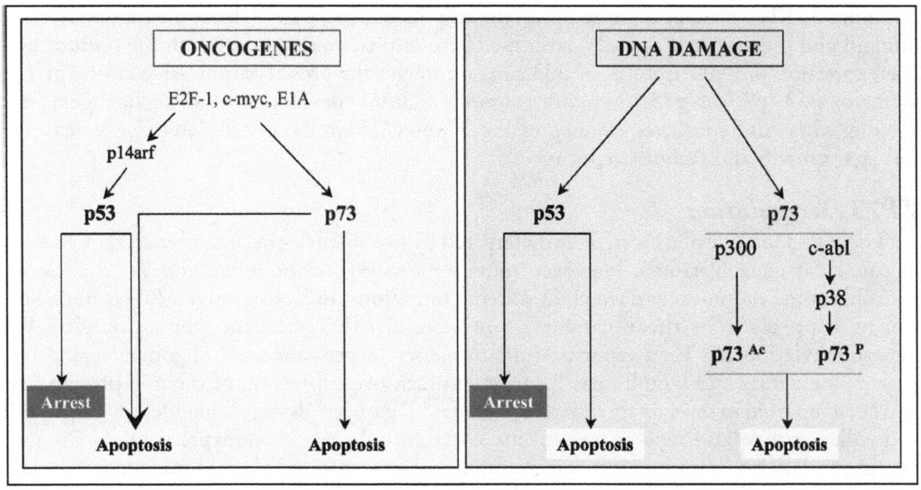

Figure 10. Pathways involving p73 in response to stress signals. Upon genotoxic stress, p73 is post-translationally modified by acetylations or phosphorylations, and induces apoptosis in a p53-independent manner. In this respect, p73 can work in parallel with p53. Upon oncogenic stress, however, the two proteins can either work in parallel or in synergy. p73 alone can also trigger apoptosis, or can modulate the p53 responses by tilting the balance from cell-cycle arrest-specific target-genes towards pro-apototic genes. See text for references.

response.[62] Other documented post-translational changes of p73 include acetylation by p300, which potentiates the pro-apoptotic functions of p73.[63] Sumoylation, another post-translational modification that affects p53,[64] has been described for p73α, but not for β, in response to stress.[65] However, the biological significance of this post-translational modification is unclear.

The p73 protein has been reported to mediate apoptosis in a p53-independent manner, in response to nongenotoxic stimuli. Oncogenic signaling by E2F-1 directly up-regulates TP73 expression, leading to the activation of p53-target genes and to apoptosis.[66-68] Moreover, ΔNp73 isoforms have recently been shown to inactivate the *RB* tumor suppressor gene.[69] Thus, p73 activation by deregulated E2F-1 activity might constitute a p53-independent, anti-tumorigenic safeguard mechanism. To add further complexity to these interactions, Flores and collaborators reported that p63 and p73 are required for p53-dependent apoptosis, in response to ectopic overexpression of oncogenes (e.g., adenovirus early region 1A: E1A) in combination with genotoxic agents.[70] These results suggest that there might be two classes of p53-family target genes: a first class of genes regulated by p53 alone, in the absence of p63 or p73 (*P21WAF1*, *MDM2*) and a second class, for which genetic and biochemical data indicate that p63 or p73 are required for p53 to be recruited and to function properly (*PERP*, *BAX* and *NOXA)*. The different pathways involving p53 and p73 in response to genotoxic or nongenotoxic stimuli are summarized in Figure 10.

Involvement of *TP63* and *TP73* in Cancer Development

TP63 Amplification

TP63 is rarely mutated in cancer. However, fluorescent in situ hybridization (FISH) analysis of the *TP63* locus revealed frequent amplification in primary squamous cell carcinoma of the lung, head and neck, and esophagus.[71,72,73] This amplification was initially identified by Sidransky and his collaborators, who named this locus AIS (amplified in squamous cell carcinoma).[71] The protein expressed from the amplified locus, p40^{AIS}, lacks the transactivation domain and is equivalent to a ΔN isoform. These results are consistent with the notion that overexpression of a ΔN isoform of p63 can counteract the growth suppressive effects of TA forms of p63, p73 or p53, therefore promoting tumor development as an oncogene. In keeping with this hypothesis, overexpression of p40 (AIS) in Rat1a cells led to an increase in soft agar growth and tumor size in mice.[71]

TP73 Deregulation

The *TP73* locus (1p36) is frequently targeted by loss of alleles in many cancers, including in particular neuroblastoma. However, mutation analysis of the remaining *TP73* allele in neuroblastoma did not reveal frequent somatic mutations, indicating that *TP73* is not a key tumor suppressor in these cancers.[32] In several other cancers, gene silencing by hypermethylation has been reported (inflammatory breast cancers;[74] oligodendroglial tumors;[75] leukemias and lymphomas[76,77]). In contrast, overexpression of the p73 protein has also been reported in several types of cancer (lung;[78,79] gastric;[80] breast;[81] bladder[82]). Tannapfel and collaborators have reported an inverse correlation between overexpression of p73 and mutation of *TP53* in liver cancer, suggesting a compensatory role.[83] It is important to note that the studies listed above have not determined the relative levels of expression of ΔN versus TA isoforms of p73. Since ΔNp73 has the potential to behave as an oncogene, it is possible that differential expression of ΔN versus TA isoforms may account for the apparently contradictory observations reported above.

Inactivation by Protein Interactions

Inactivation of p63 and/or p73 by interaction with mutant p53 has been recently put forward as one of the mechanisms by which mutant p53 may exert a dominant, gain-of-function effect in cancer development. Most mutations in *TP53* are missense and affect residues of the DNA-binding domain, inhibiting sequence-specific DNA-binding. However, all mutations are not equivalent in their functional effects and at least some mutants have acquired the capacity to exert dominant effects. This new property is thought to result from specific changes in protein conformation induced by the mutation. There is evidence that a conformationally altered p53 DNA-binding domain can form stable complexes with p73 and p63. This interaction is mediated by the binding of the central domain of mutant p53 to the oligomerization domain of p63 or p73.[84-86] Recent functional studies indicate that binding of mutant p53 down-regulates transactivation by p63 and p73.[85,86] Whether this mechanism contributes to explain the persistence and accumulation of mutant p53 protein in cancer cells, remains to be elucidated. In future studies, it will be important to determine whether tumors may contain different types of *TP53* mutations depending upon their *TP63* or *TP73* expression status.

Conclusion

The *TP53* gene family provides a good illustration of how genes with very specialized functions may have evolved from a common blueprint. The three members show obvious family traits in their structure, architecture and basic biochemistry. However, they have completely different lifestyles and functions, due to the fact that their modes of regulation are extremely specialized. A closer look at the family suggests that the most divergent members, in terms of function, are *TP53* and *TP63*. Whereas *TP53* is entirely specialized in stress response, *TP63* plays its roles in development and differentiation. The functional divergence between these two genes is underlined by the consequence of their germline mutations (predisposing to cancer for *TP53*; inducing malformations for *TP63*) and of their somatic alterations in cancer (inactivating mutations for *TP53*, activating amplifications for *TP63*). *TP73* occupies a somewhat intermediate position between the two "extremist" brothers. Although it plays basic roles in several aspects of normal differentiation (in particular in epithelial and neuronal cells), it has been implicated in response to genotoxic agents that also activate p53.

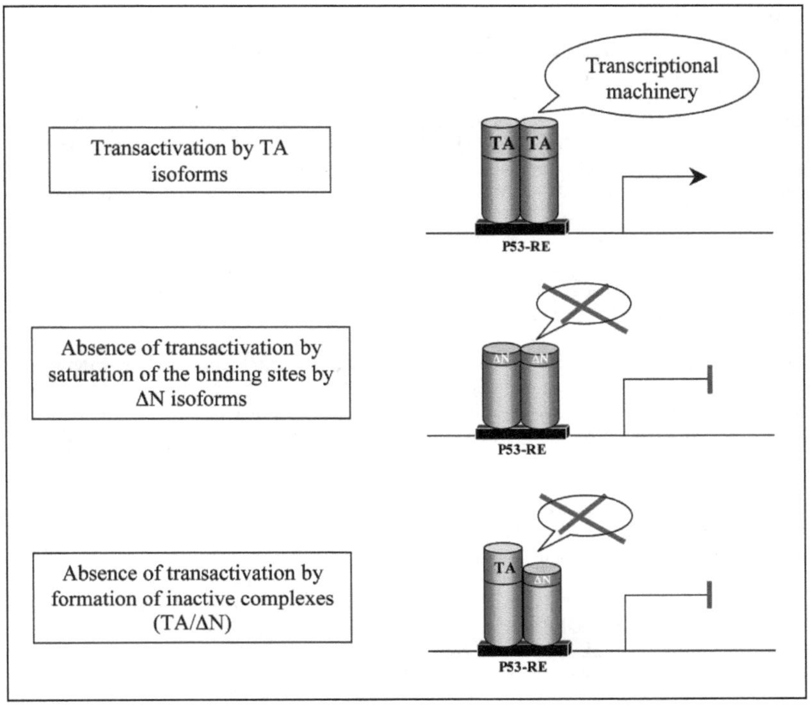

Figure 11. Dominant negative effects of ΔN isoforms on TA isoforms. Upper panel) TA isoforms (TAp63 or TAp73) bind p53-specific DNA sequences (RE for Responsive element) and recruit the transcription machinery. Middle panel) ΔN isoforms could exert a dominant-negative effect by saturating the p53-responsive elements. Lower panel) TA and ΔN isoforms could form hetero-oligomers, deficient in their capacity to recruit the transcription machinery.

It is very likely that the current picture of the role of *TP63* and *TP73* in cancer is, at best, very fragmentary. Further studies will be necessary to uncover the full extent of their contribution to carcinogenesis. In this respect, a very important characteristic of these two genes is their expression as both TA and ΔN isoforms, which confers to p63 and p73 the capacity to exert opposite effects on transcriptional regulation, depending on which type of isoform may predominate in the cell. As discussed above, TA isoforms exert primarily suppressive effects, whereas ΔN isoforms can counteract this suppression and therefore hold the potential to behave as oncogenes. Deregulation of the balance between the levels of the two types of isoforms should therefore be considered as a potential mechanism of carcinogenesis (Fig. 11), the demonstration of which awaits the development of knock out mice specifically lacking either TA or ΔN isoforms. Since the three family members may control a common subset of essential genes involved in growth regulation, overexpression of either ΔNp63 or ΔNp73 may block the suppressive effects of not only its TA counterpart, but also of other family members, including p53. The existence of amplifications of *TP63*, with concomitant overexpression of a ΔN-like protein in some squamous cancers, provides evidence that this mechanism may work in a subset of epithelial tumors. It is likely that, in cancer cells, other mechanisms than amplification may contribute to upset the balance between levels of TA and ΔN isoforms. These mechanisms may include differential promoter regulation, transcript stability, protein nuclear accumulation, and protein degradation. Further studies need to be developed to address these possibilities.

References

1. Lane DP. Cancer. p53, guardian of the genome. Nature 1992; 358(6381):15-16.
2. Koshland Jr DE. Molecule of the year. Science 1993; 262(5142):1953.
3. Kaghad M, Bonnet H, Yang A et al. Monoallelically expressed gene related to p53 at 1p36, a region frequently deleted in neuroblastoma and other human cancers.
4. Osada M, Ohba M, Kawahara C et al. Cloning and functional analysis of human p51, which structurally and functionally resembles p53. Nat Med 1998; 4(7):839-843.
5. Schmale H, Bamberger C. A novel protein with strong homology to the tumor suppressor p53. Oncogene 1997; 15(11):1363-1367.
6. Trink B, Okami K, Wu L et al. A new human p53 homologue. Nat Med 1998; 4(7):747-748.
7. Yang A, Kaghad M, Wang Y et al. p63, a p53 homolog at 3q27-29, encodes multiple products with transactivating, death-inducing, and dominant-negative activities. Mol Cell 1998; 2(3):305-316.
8. Hainaut P, Hollstein M. p53 and human cancer: The first ten thousand mutations. Adv Cancer Res 2000; 77:81-137.
9. Donehower LA, Harvey M, Slagle BL et al. Mice deficient for p53 are developmentally normal but susceptible to spontaneous tumors. Nature 1992; 356(6366):215-221.
10. Lavigueur A, Maltby V, Mock D et al. High incidence of lung, bone, and lymphoid tumors in transgenic mice overexpressing mutant alleles of the p53 oncogene. Mol Cell Biol 1989; 9(9):3982-3991.
11. Conseiller E, Debussche L, Landais D et al. CTS1: A p53-derived chimeric tumor suppressor gene with enhanced in vitro apoptotic properties. J Clin Invest 1998; 101(1):120-127.
12. De Laurenzi VD, Catani MV, Terrinoni A et al. Additional complexity in p73: Induction by mitogens in lymphoid cells and identification of two new splicing variants epsilon and zeta. Cell Death Differ 1999; 6(5):389-390.
13. De Laurenzi V, Costanzo A, Barcaroli D et al. Two new p73 splice variants, gamma and delta, with different transcriptional activity. J Exp Med 1998; 188(9):1763-1768.
14. Ueda Y, Hijikata M, Takagi S et al. New p73 variants with altered C-terminal structures have varied transcriptional activities. Oncogene 1999; 18(35):4993-4998.
15. Thanos CD, Bowie JU. p53 Family members p63 and p73 are SAM domain-containing proteins. Protein Sci 1999; 8(8):1708-1710.
16. Dohn M, Zhang S, Chen X. p63alpha and DeltaNp63alpha can induce cell cycle arrest and apoptosis and differentially regulate p53 target genes. Oncogene 2001; 20(25):3193-3205.
17. Ghioni P, Bolognese F, Duijf PH et al. Complex transcriptional effects of p63 isoforms: Identification of novel activation and repression domains. Mol Cell Biol 2002; 22(24):8659-8668.
18. Melino G, de Laurenzi V, Vousden KH. p73: Friend or foe in tumorigenesis. Nat Rev Cancer 2002; 2(8):605-615.
19. Stiewe T, Zimmermann S, Frilling A et al. Transactivation-deficient DeltaTA-p73 acts as an oncogene. Cancer Res 2002; 62(13):3598-3602.
20. Walker KK, Levine AJ. Identification of a novel p53 functional domain that is necessary for efficient growth suppression. Proc Natl Acad Sci USA 1996; 93(26):15335-15340.
21. Ruaro EM, Collavin L, Del Sal G et al. A proline-rich motif in p53 is required for transactivation-independent growth arrest as induced by Gas1. Proc Natl Acad Sci USA 1997; 94(9):4675-4680.
22. Sakamuro D, Sabbatini P, White E et al. The polyproline region of p53 is required to activate apoptosis but not growth arrest. Oncogene 1997; 15(8):887-898.
23. Van Beneden RJ, Walker CW, Laughner ES. Characterization of gene expression of a p53 homologue in the soft-shell clam (Mya arenaria). Mol Mar Biol Biotechnol 1997; 6(2):116-122.
24. Jacks T, Remington L, Williams BO et al. Tumor spectrum analysis in p53-mutant mice. Curr Biol 1994; 4(1):1-7.
25. Mills AA, Zheng B, Wang XJ et al. p63 is a p53 homologue required for limb and epidermal morphogenesis. Nature 1999; 398(6729):708-713.
26. Yang A, Schweitzer R, Sun D et al. p63 is essential for regenerative proliferation in limb, craniofacial and epithelial development. Nature 1999; 398(6729):714-718.
27. Yang A, Walker N, Bronson R et al. p73-deficient mice have neurological, pheromonal and inflammatory defects but lack spontaneous tumors. Nature 2000; 404(6773):99-103.
28. Van Bokhoven H, McKeon F. Mutations in the p53 homolog p63: Allele-specific developmental syndromes in humans. Trends Mol Med 2002; 8(3):133-139.
29. Van Bokhoven H, Brunner HG. Splitting p63. Am J Hum Genet 2002; 71(1):1-13.
30. Versteeg R, Caron H, Cheng NC et al. 1p36: Every subband a suppressor? Eur J Cancer 1995; 31A(4):538-541.

31. Herath NI, Kew MC, Whitehall VL et al. p73 is up-regulated in a subset of hepatocellular carcinomas. Hepatology 2000; 31(3):601-605.
32. Ichimiya S, Nimura Y, Kageyama H et al. p73 at chromosome 1p36.3 is lost in advanced stage neuroblastoma but its mutation is infrequent. Oncogene 1999; 18(4):1061-1066.
33. Imyanitov EN, Birrell GW, Filippovich I et al. Frequent loss of heterozygosity at 1p36 in ovarian adenocarcinomas but the gene encoding p73 is unlikely to be the target. Oncogene 1999; 18(32):4640-4642.
34. Chompret A. The Li-Fraumeni syndrome. Biochimie 2002; 84(1):75-82.
35. Hall PA, Campbell SJ, O'neill M et al. Expression of the p53 homologue p63alpha and deltaNp63alpha in normal and neoplastic cells. Carcinogenesis 2000; 21(2):153-160.
36. Nylander K, Coates PJ, Hall PA. Characterization of the expression pattern of p63 alpha and delta Np63 alpha in benign and malignant oral epithelial lesions. Int J Cancer 2000; 87(3):368-372.
37. Nylander K, Vojtesek B, Nenutil R et al. Differential expression of p63 isoforms in normal tissues and neoplastic cells. J Pathol 2002; 198(4):417-427.
38. Pozniak CD, Radinovic S, Yang A et al. An anti-apoptotic role for the p53 family member, p73, during developmental neuron death. Science 2000; 289(5477):304-306.
39. Ratovitski EA, Patturajan M, Hibi K et al. p53 associates with and targets Delta Np63 into a protein degradation pathway. Proc Natl Acad Sci USA 2001; 98(4):1817-1822.
40. Fuchs SY, Adler V, Buschmann T et al. JNK targets p53 ubiquitination and degradation in nonstressed cells. Genes Dev 1998; 12(17):2658-2663.
41. Michael D, Oren M. The p53-Mdm2 module and the ubiquitin system. Semin Cancer Biol 2003; 13(1):49-58.
42. Buschmann T, Potapova O, Bar-Shira A et al. Jun NH2-terminal kinase phosphorylation of p53 on Thr-81 is important for p53 stabilization and transcriptional activities in response to stress. Mol Cell Biol 2001; 21(8):2743-2754.
43. Kadakia M, Slader C, Berberich SJ. Regulation of p63 function by Mdm2 and MdmX. DNA Cell Biol 2001; 20(6):321-330.
44. Little NA, Jochemsen AG. Hdmx and Mdm2 can repress transcription activation by p53 but not by p63. Oncogene 2001; 20(33):4576-4580.
45. Wang X, Arooz T, Siu WY et al. MDM2 and MDMX can interact differently with ARF and members of the p53 family. FEBS Lett 2001; 490(3):202-208.
46. Wang XQ, Ongkeko WM, Lau AW et al. A possible role of p73 on the modulation of p53 level through MDM2. Cancer Res 2001; 61(4):1598-1603.
47. Balint E, Bates S, Vousden KH. Mdm2 binds p73 alpha without targeting degradation. Oncogene 1999; 18(27):3923-3929.
48. Dobbelstein M, Wienzek S, Konig C et al. Inactivation of the p53-homologue p73 by the mdm2-oncoprotein. Oncogene 1999; 18(12):2101-2106.
49. Ongkeko WM, Wang XQ, Siu WY et al. MDM2 and MDMX bind and stabilize the p53-related protein p73. Curr Biol 1999; 9(15):829-832.
50. Zeng X, Chen L, Jost CA et al. MDM2 suppresses p73 function without promoting p73 degradation. Mol Cell Biol 1999; 19(5):3257-3266.
51. Alarcon-Vargas D, Fuchs SY, Deb S et al. p73 transcriptional activity increases upon cooperation between its spliced forms. Oncogene 2000; 19(6):831-835.
52. Lee CW, La Thangue NB. Promoter specificity and stability control of the p53-related protein p73. Oncogene 1999; 18(29):4171-4181.
53. Ellisen LW, Ramsayer KD, Johannessen CM et al. REDD1, a developmentally regulated transcriptional target of p63 and p53, links p63 to regulation of reactive oxygen species. Mol Cell 2002; 10(5):995-1005.
54. De Laurenzi V, Rossi A, Terrinoni A et al. p63 and p73 transactivate differentiation gene promoters in human keratinocytes. Biochem Biophys Res Commun 2000; 273(1):342-346.
55. Pluquet O, Hainaut P. Genotoxic and nongenotoxic pathways of p53 induction. Cancer Lett 2001; 174(1):1-15.
56. Katoh I, Aisaki KI, Kurata SI et al. p51A (TAp63gamma), a p53 homolog, accumulates in response to DNA damage for cell regulation. Oncogene 2000; 19(27):3126-3130.
57. Liefer KM, Koster MI, Wang XJ et al. Down-regulation of p63 is required for epidermal UV-B-induced apoptosis. Cancer Res 2000; 60(15):4016-4020.
58. Okada Y, Osada M, Kurata S et al. p53 gene family p51(p63)-encoded, secondary transactivator p51B(TAp63alpha) occurs without forming an immunoprecipitable complex with MDM2, but responds to genotoxic stress by accumulation. Exp Cell Res 2002; 276(2):194-200.

59. Agami R, Blandino G, Oren M et al. Interaction of c-Abl and p73alpha and their collaboration to induce apoptosis. Nature 1999; 399(6738):809-813.
60. Gong JG, Costanzo A, Yang HQ et al. The tyrosine kinase c-Abl regulates p73 in apoptotic response to cisplatin-induced DNA damage. Nature 1999; 399(6738):806-809.
61. Yuan ZM, Shioya H, Ishiko T et al. p73 is regulated by tyrosine kinase c-Abl in the apoptotic response to DNA damage. Nature 1999; 399(6738):814-817.
62. Sanchez-Prieto R, Sanchez-Arevalo VJ, Servitja JM et al. Regulation of p73 by c-Abl through the p38 MAP kinase pathway. Oncogene 2002; 21(6):974-979.
63. Costanzo A, Merlo P, Pediconi N et al. DNA damage-dependent acetylation of p73 dictates the selective activation of apoptotic target genes. Mol Cell 2002; 9(1):175-186.
64. Rodriguez MS, Desterro JM, Lain S et al. SUMO-1 modification activates the transcriptional response of p53. EMBO J 1999; 18(22):6455-6461.
65. Minty A, Dumont X, Kaghad M et al. Covalent modification of p73alpha by SUMO-1. Two-hybrid screening with p73 identifies novel SUMO-1-interacting proteins and a SUMO-1 interaction motif. J Biol Chem 2000; 275(46):36316-36323.
66. Irwin M, Marin MC, Phillips AC et al. Role for the p53 homologue p73 in E2F-1-induced apoptosis. Nature 2000; 407(6804):645-648.
67. Stiewe T, Putzer BM. Role of the p53-homologue p73 in E2F1-induced apoptosis. Nat Genet 2000; 26(4):464-469.
68. Zaika A, Irwin M, Sansome C et al. Oncogenes induce and activate endogenous p73 protein. J Biol Chem 2001; 276(14):11310-11316.
69. Stiewe T, Stanelle J, Theseling CC et al. Inactivation of the RB tumor suppressor gene by oncogenic isoforms of the p53 family member p73. J Biol Chem 2003.
70. Flores ER, Tsai KY, Crowley D et al. p63 and p73 are required for p53-dependent apoptosis in response to DNA damage. Nature 2002; 416(6880):560-564.
71. Hibi K, Trink B, Patturajan M et al. AIS is an oncogene amplified in squamous cell carcinoma. Proc Natl Acad Sci USA 2000; 97(10):5462-5467.
72. Taniere P, Martel-Planche G, Saurin JC et al. TP53 mutations, amplification of P63 and expression of cell cycle proteins in squamous cell carcinoma of the oesophagus from a low incidence area in Western Europe. Br J Cancer 2001; 85(5):721-726.
73. Yamaguchi K, Wu L, Caballero OL et al. Frequent gain of the p40/p51/p63 gene locus in primary head and neck squamous cell carcinoma. Int J Cancer 2000; 86(5):684-689.
74. Ahomadegbe JC, Tourpin S, Kaghad M et al. Loss of heterozygosity, allele silencing and decreased expression of p73 gene in breast cancers: Prevalence of alterations in inflammatory breast cancers. Oncogene 2000; 19(47):5413-5418.
75. Dong S, Pang JC, Hu J et al. Transcriptional inactivation of TP73 expression in oligodendroglial tumors. Int J Cancer 2002; 98(3):370-375.
76. Kawano S, Miller CW, Gombart AF et al. Loss of p73 gene expression in leukemias/lymphomas due to hypermethylation. Blood 1999; 94(3):1113-1120.
77. Liu M, Taketani T, Li R et al. Loss of p73 gene expression in lymphoid leukemia cell lines is associated with hypermethylation. Leuk Res 2001; 25(6):441-447.
78. Tokuchi Y, Hashimoto T, Kobayashi Y et al. The expression of p73 is increased in lung cancer, independent of p53 gene alteration. Br J Cancer 1999; 80(10):1623-1629.
79. Mai M, Qian C, Yokomizo A et al. Loss of imprinting and allele switching of p73 in renal cell carcinoma. Oncogene 1998; 17(13):1739-1741.
80. Kang MJ, Park BJ, Byun DS et al. Loss of imprinting and elevated expression of wild-type p73 in human gastric adenocarcinoma. Clin Cancer Res 2000; 6(5):1767-1771.
81. Dominguez G, Silva J, Silva JM et al. Clinicopathological characteristics of breast carcinomas with allelic loss in the p73 region. Breast Cancer Res Treat 2000; 63(1):17-22.
82. Yokomizo A, Mai M, Tindall DJ et al. Overexpression of the wild-type p73 gene in human bladder cancer. Oncogene 1999; 18(8):1629-1633.
83. Tannapfel A, Engeland K, Weinans L et al. Expression of p73, a novel protein related to the p53 tumor suppressor p53, and apoptosis in cholangiocellular carcinoma of the liver. Br J Cancer 1999; 80(7):1069-1074.
84. Bensaad K, Le Bras M, Unsal K et al. Change of conformation of the DNA binding domain of p53 is the only key element for binding of and interference with p73. J Biol Chem 2003.
85. Gaiddon C, Lokshin M, Ahn J et al. A subset of tumor-derived mutant forms of p53 down-regulate p63 and p73 through a direct interaction with the p53 core domain. Mol Cell Biol 2001; 21(5):1874-1887.
86. Strano S, Fontemaggi G, Costanzo A et al. Physical interaction with human tumor-derived p53 mutants inhibits p63 activities. J Biol Chem 2002; 277(21):18817-18826.

Chapter 4

The Regulation of p53 Protein Function by Phosphorylation

Nicola J. Maclaine and Theodore Hupp*

Introduction

P53 is a sequence-specific DNA-binding protein and stress-activated transcription factor that controls the expression of hundreds of genes implicated in a variety of physiological responses to genome instability, virus infection and interferon production, DNA damage, and metabolic stresses.[1] The vast numbers of gene products mediating the p53 signal coordinately promote many repair processes, some of which include elimination of damaged proteins, DNA repair, ATP generation via oxidative phosphorylation, organellar functions that maintain autophagy signaling and mitochondrial function, the cell division cycle, and programmed cell death. The implications of this stress-induced transcription reprogramming by p53 is that cell and tissue integrity can be maintained, thereby contributing to organism health and viability.

Inactivating missense mutations in *p53* are very common in a wide range of human cancers, indicating a critical role for p53 as a cancer suppressor in very distinct tissue microenvironments.[2] These mutations reside predominantly in the core DNA-binding domain (Fig. 1), and result in a p53 protein with an altered conformation and reduced sequence-specific DNA-binding function.[3] These mutations suppress p53 transcription, reduce the cellular repair capacity, and stimulate cancer development. As p53 is a conformationally flexible and thermodynamically unstable protein, biophysical studies have suggested there is promise in drug developments aimed at stabilizing the mutant p53 conformation into a wild-type state, and reengaging the p53 dependent transcription.[4]

Transgenic technologies in mice have supported biochemical and clinical data showing a critical role for the DNA-binding function of p53 in cancer suppression. Animals null for *p53* strikingly develop cancer at an advanced rate.[5] By contrast, deletion of many of the p53-inducible genes do not give the same tumor incidence or tumor spectrum as *p53*-null animals,[6] further highlighting the role of p53 itself as a central hub in the integration of tissue repair triggers. There is one intriguing exception: animals double null for *ataxia telangiectasia mutated* (*ATM*) and the p53-inducible gene *p21* have a similar tumor spectrum and death incidence to the *p53*-null animals.[7] This suggests that ATM and p21 form a positive genetic circuit in the p53-dependent cancer suppression mechanism.

p53 Regulation

p53 protein function is regulated post-translationally by coordinated interaction with signaling proteins including protein kinases, acetyltransferases, methyltransferses, and ubiquitin-like modifying enzymes (Fig. 1). The majority of the sites of covalent modification occur at intrin-

*Corresponding Author: Theodore Hupp—University of Edinburgh, Institute of Genetics and Molecular Medicine, CRUK p53 Signal Transduction Laboratories, Crewe Road South, Edinburgh, EH4 2XR, Scotland, UK. Email: ted.hupp@ed.ac.uk

p53, edited by Ayeda Ayed and Theodore Hupp. ©2010 Landes Bioscience and Springer Science+Business Media.

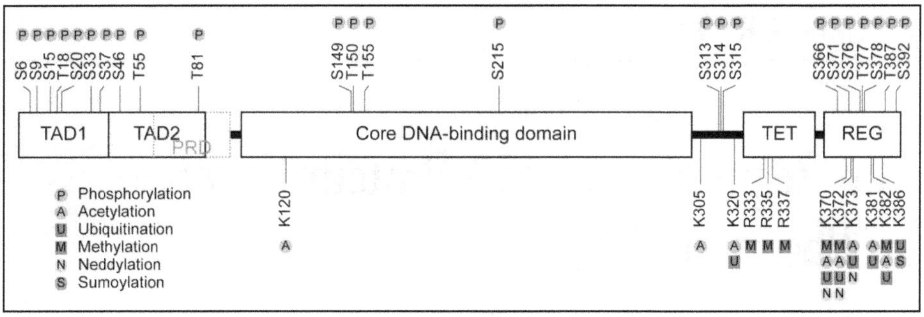

Figure 1. Sites of post-translational modifications on p53. The 393 amino acid domain structure of human p53 is depicted showing the sites of post-translational modification including phosphorylation, acetylation, ubiquitination, methylation, neddylation, and sumoylation. Abbreviations: N-terminal transactivation domain (TAD); proline-rich domain (PRD); tetramerization domain (TET); C-terminal regulatory domain (REG); arginine (R); lysine (K); serine (S); threonine (T).

sically unstructured linear peptide docking motifs that flank the DNA-binding domain of p53 which play a role in anchoring or in allosterically activating the enzymes that mediate covalent modification of p53 (Fig. 2A). Such unstructured linear domains are proving to be important in signaling control.[8-11] In undamaged cells, p53 protein has a relatively short half-life and is degraded by a ubiquitin-proteasome dependent pathway through the action of E3 ubiquitin ligases including MDM2, PirH2, COP-1, and CHIP.[12]

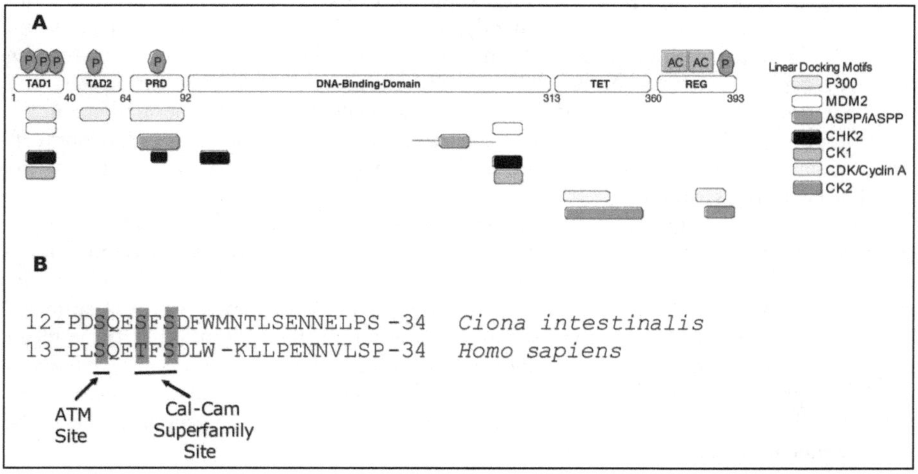

Figure 2. Linear peptide docking sites in p53. A) Linear peptide docking sites for enzymes that regulate p53 function. The N-terminus is composed of three transactivation motifs, TAD1, TAD2, and Proline-repeat domain (PRD). A key regulatory domain in the C-terminus (REG) contains the acetylation motifs and phosphorylation site and flanks the Tetramerization domain (TET). The overlapping, but distinct, linear polypeptide docking motifs for p53 regulators include the acetyltransferase p300, the E3 ubiquitin ligase MDM2, iASPP, and the protein kinases including CDK, CK2, CK1, and CHK2 are highlighted. B) Conservation of key phospho-acceptor sites between urochordate and human. The panel highlights the conservation of amino acids and phospho-acceptor sites in the *BOX-I* transactivation domain of p53 (TAD1 in (A)) between human and urochordate (*Ciona intestinalis*). The ATM phospho-acceptor site at Ser15 and the Calcium Calmodulin kinase/CK1 phospho-acceptors sites at Thr18 and Ser20 are highlighted as indicated.

Following stress, p53 is modified enzymatically producing a plethora covalent adducts that mediate many changes in its function (Fig. 1). Phosphorylation at multiple residues is the most well-studied covalent modification and will be the subject of this chapter. The key biochemical functions of p53, whose regulation can be accessed after phosphorylation, include sequence-specific DNA binding and specific protein-protein interactions. Of the dozens of phospho-acceptor sites reported on p53 only three (Ser15, Thr18, Ser20) are highly conserved between humans and urochordates (Fig. 2B), the latter being where a bona-fide p53-MDM2 axis has appeared in evolution (unpublished data). Especially striking is the conservation of primary amino acid homology in the p53 transactivation domain between the invertebrate sea squirt and humans, indicating that as yet undefined evolutionary selection pressures have maintained this amino acid sequence at least since this urochordate lineage. The only other highly conserved phosphorylation site in p53 is within the C-terminus of p53 and is conserved only amongst vertebrates; the CK2 site at Ser392 (Fig. 1). As such, we have focused our research and this chapter on studying two of these highly conserved phosphorylation sites in p53: the Ser20 site and the Ser392 site, as they form a paradigm to facilitate our understanding of how phosphorylation controls p53 function as a transcription factor. The many other sites of covalent modification on p53 (Fig. 1) also likely play important roles in p53 function or regulation, but there are relatively smaller amounts of genetic and biochemical data describing the effects of these modifications on p53 function.

Paradigm I: Kinase Regulation of the Specific DNA Binding Function of p53 (Fig. 3)

The Ser392 phospho-acceptor site is located in the C-terminal regulatory domain (REG) in a flexible and unstructured motif (Figs. 1 and 2) whose phosphorylation by casein kinase 2 (CK2) stimulates the sequence-specific DNA-binding function of p53[13] (Fig. 3A,B). This activation of p53 presumably occurs by two inter-related mechanisms: (i) changes in the conformation of the DNA binding domain that increases p53 thermostability as defined with biophysical studies using a phospho-mimetic S392D mutant p53 protein;[14] and (ii) reductions in p53 nonspecific DNA binding related to diffusion of p53 on chromatin.[15] Phosphorylation at p53 Ser392 also increases after either UV or ionizing radiation in cell lines and in mice spleenocytes in vivo[16,17] (Fig. 3C). These data are consistent with an activating rather than inhibitory role for phosphorylation of this site on p53 function. Another kinase of the CDK2 family can also stimulate the DNA-function of p53 via phosphorylation at Ser315[18] and this phosphorylation is elevated after DNA damage in cells, as defined using phospho-specific monoclonal antibodies generated to this site.[19] However, Ser315 and Ser376 phosphorylation induced by GSK3 under ER stress can promote nuclear export of p53 indicating that phosphorylation at a specific site(s) can have opposing effects depending upon context and stress.[20] There are many such examples of p53 phosphorylation regulating a range of biological outcomes which can be inferred from the vast arrays of covalent sites on p53 (Fig. 1).

In addition to the cellular and biochemical studies described above, immunochemical analysis in clinical biopsies has shown that UV irradiation of normal human skin results in p53 phosphorylation at the CK2 site in basal cell population which are thought to reflect stem cell populations (Fig. 3D).[21] These data suggest that cells have specifically evolved p53 activation pathways in the critical stems cells required to maintain tissue integrity. The pathways that selectively trigger high levels of p53 phosphorylation in these basal cells and/or why high levels of p53 phosphorylation are suppressed in suprabasal skin stem cells are still not known. However, current models suggest a novel type of cross-talk between ΔN-p63 and p53 activating kinases in the basal stem cell populations.[22]

The in vivo significance of the Ser392 (CK2 site) phosphorylation was relatively difficult to study physiologically, as the cell biology approaches involving mutant p53 gene replacement in cells lines has not always given unambiguous phenotypes that support an important role for p53 phosphorylation at the CK2 site in cells.[23] Critically, using animal transgenic models, a clarification of the role of the CK2 site on p53 biology has been obtained. Substitution mutation of the

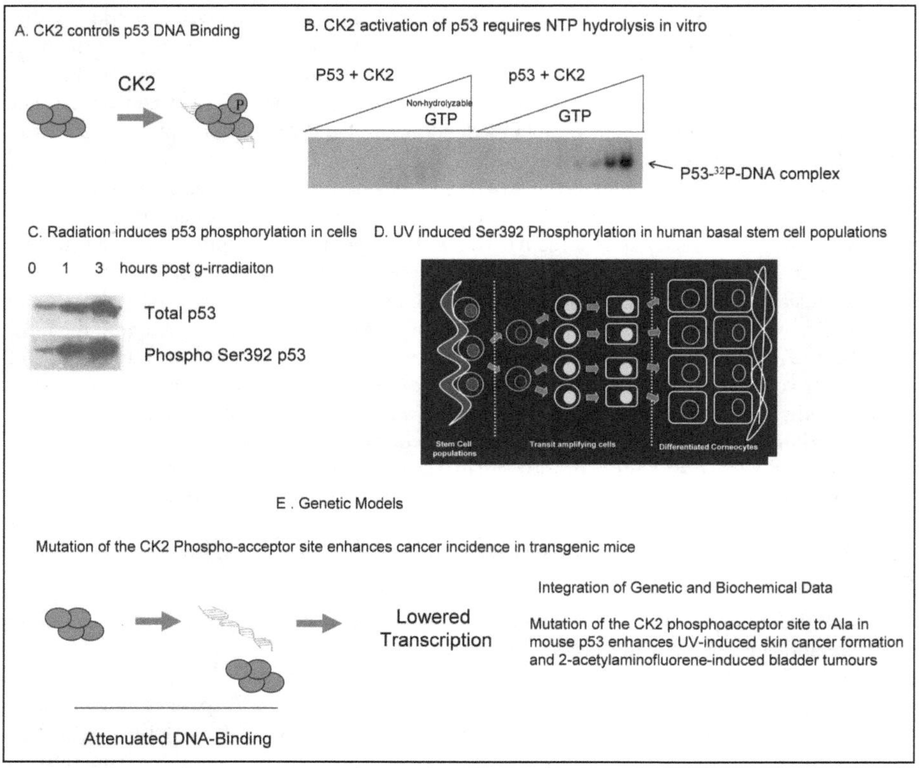

Figure 3. Paradigm I-CK2 phosphorylation in p53 control. A) Diagram highlighting the ability of CK2 phosphorylation of p53 to stabilize p53 :DNA interactions.[13] B) in vitro DNA-binding assay measures the ability of p53 to radioactive consensus site DNA (highlighted by arrows) with CK2 using (left panel) nonhydrolyzable GTP or (right panel) nonhydrolyzable GTP. C) Radiation of human cells results in stabilization of p53 protein levels (top panel) or elevated phosphorylation at the CK2 site (bottom panel) as defined using a phospho-specific monoclonal antibody that can detect Ser392-phospho-p53.[59] D) Models of UV irradiated normal human skin have been used to demonstrate the types of pathways that trigger p53 activation in skin.[60] Using this system, Ser392-site phosphorylation was detected in basal cell populations but not in transit amplifying or differentiated cell types despite the latter two showing stabilization of p53 protein. These data highlight a cell specific role for Ser392-site phosphorylation in UV irradiated human skin.[21, 22, 61] E) Genetic studies were developed using the Ser to Ala Ck2 site mutant on murine p53. Cells from these mice show reduction in p53 DNA binding and show enhanced cancer development after carcinogen exposure.[24, 25]

murine equivalent of Ser392 to Ala392 results in enhanced UV-induced skin cancer and elevated carcinogen-induced bladder cancer in transgenic mice[24, 25] (Fig. 3E). These data identify a p53-activating kinase pathway whose attenuation could modify tissue repair rates and cancer development in squamous tissue like skin and bladder. These data also together give an explanation, in part, as to why the CK2 site is one of the most highly conserved phospho-acceptor sites on p53 in vertebrates, as there must be some mechanism operating through natural selection that maintained conservation of this site during vertebrate evolution. Whether phosphorylation of p53 at the Ser392 site plays tumor suppressing role in other cancer types remains to be determined, but these approaches highlight the insights that can be acquired, by focusing on one phospho-acceptor site, using protein science, clinical science, and genetics to develop a paradigm in p53 control.

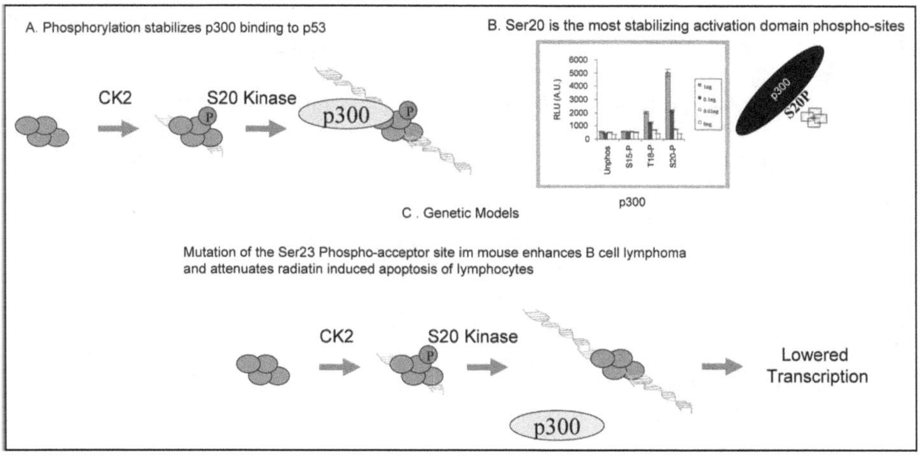

Figure 4. Kinase Paradigm II- Phosphorylation in regulation of p53 binding to p300. A) Diagram highlighting the ability of Ser20 phosphorylation of p53 to stabilize p53 :p300 interactions.[26] B) In vitro p300-peptide binding assay measures the ability of p300 to bind to biotinylated version of the indicated peptide (unphosphorylated, Ser15-P, Thr18-P, and Ser20-P). The data are depicted as p300 binding (in relative light units) as a function of p53 peptide Type. C) Genetic studies were developed using the Ser to Ala Ser20 site equivalent mutant on murine p53. Cells from these mice show reduction in radiation induced apoptosis in B-cells and enhanced spontaneous B-cell lymphome.[36]

Paradigm II: Kinase Regulation of the Interaction of p53 with the Acetyltransferase p300 (Fig. 4)

The second highly conserved phospho-acceptor site model studied, Ser20, is located in the N-terminal transactivation domain (TAD) in an unstructured linear motif (Fig. 1). This is one of three sites in the activation domain that are the most highly conserved in evolution; especially striking is that the Ser15, Thr/Ser18, and Ser20 phospho-acceptor sites and surrounding amino acids are nearly identical between the urochordate and human (Fig. 2B). The selection pressures that maintained this degree of conservation presumably relate to p53 transactivation potential, as phosphorylation of these three sites are linked to a stimulation rather than an inhibition of p53 function. In a neutral screen of these three phospho-acceptor sites, only the Ser20 phosphorylation can have a striking effect on stabilizing the p300 binding[26] (Fig. 4B). Subsequent studies fine mapping the phospho-peptide binding domain of p300 using peptide combinatorial libraries have confirmed that ability of at least two p300 minidomains to bind to peptides with this consensus: LSQXTFSXLXXLL.[27] Biophysical studies have also shown at least four mini-domains from p300 can bind to varying degrees to phospho-peptides derived from the activation domain of p53.[28] Together, these data provide biochemical evidence that the role of the highly conserved phospho-acceptor site in the p53 activation domain are stabilizing with respect to the binding of the coactivator p300 (Fig. 4A). There are other important roles of these phosphorylation sites including MDM2 binding; Thr18 phosphorylation has a striking effect on inhibiting MDM2 binding to p53,[29, 30] but MDM2 biophysics will not be discussed in this chapter.

As with the Ser392 site on p53, physiological significance of Ser20 phosphorylation on p53 function has been relatively ambiguous due to the difficulty of designing physiological models of p53 function when phospho-acceptor sites are mutated. However, in vitro cancer cell models have shown that mutation of Ser20 to Asp20, thereby mimicking constitutive phosphorylation of p53 Ser20, results in a p53 with enhanced transcription function in cell lines.[31,32] Further, as Ser20 site phosphorylation is elevated after DNA damage[33-35] and S20D phospho-mimetic peptides can inhibit p300 function in cells,[26] these data suggest that phosphorylation at p53 Ser20 forms a stimulatory rather than an inhibitory signal for p53 activity.

As with the Ser392 site, the transgenic approaches examining the role of activation domain phosphorylation have been illuminating. Transgenic mice with a phospho-acceptor site mutation at the Ser20 equivalent in murine p53 have been shown to develop spontaneous B-cell lymphoma,[36] providing evidence of the first spontaneous cancer-prone phenotype for a p53 regulatory site. Further, as B-cells from these transgenic mice exhibit attenuated ionizing radiation-induced apoptosis in vitro,[36] these data highlight a central role for Ser20 site phosphorylation in p53-dependent apoptotic activation in this cell type. Transgenic approaches have also shown a role for the Ser15 phospho-acceptor site in p53 biology in mouse models. Transgenic mice with phospho-acceptor site mutations at the murine equivalent of the Ser15 ATM target site have been shown to exhibit an accelerated aging-associated phenotype, along with an enhanced spontaneous development of late-onset lymphomas.[37] This indicates that the Ser15 phospho-acceptor site is important for the tumor suppression and anti-aging activity of p53.[37] The biochemical effects of Ser15 phosphorylation on p300 protein-protein interactions are not well-validated and for example no effect is observed using the assay described in Figure 4B. However, one study has shown that p53 acetylation can be stimulated by phosphorylation of p53 at Ser15 in vitro by the p300 homologue CBP.[38]

Together, these biochemical and genetic studies show that phosphorylation at two highly conserved sites can activate p53 function, although these studies do not necessarily explain what selection pressures have maintained the integrity of the Ser20 and Ser392 phospho-acceptor sites during evolution in the urochordate-chordate lineage. Nevertheless, the apparent cell- and damage-type specificity observed in post-translational modification signaling pathways highlights the need to develop tissue-specific experimental cancer models that reflect the physiological switches that can activate p53. These include changes in cytokines like transforming growth factor β (TGF-β)[39] or interferons, metabolic stresses like hypoxia, glucose starvation or acidification, external stresses including carcinogen damage to DNA, and internal signals such as oncogene activation (Fig. 8).

Paradigm III: DNA-Dependent Acetylation of p53 and the Proline-Repeat Transactivation Domain (Fig. 5)

In addition to phosphorylation of p53 regulating its ability of function as a transcription factor, acetylation also plays a fundamental role in p53 activity.[40] Acetylation of p53 protein itself is thought to anchor the protein to other protein components of the transcription machinery including holo-TFIID recruitment.[41] Since phosphorylation at Ser20 stabilizes the binding of p300, these data suggest there maybe a relationship between p53 phosphorylation and p53 acetylation induced by p300. In analysis of this reaction, a DNA-dependence in p53 acetylation was observed[42] (Fig. 5B) highlighting an intrinsic conformational restraint to p53 acetylation until bound to DNA (Fig. 5C). This acetylation at promoters then in turns allows an ordering of the covalent modifications as in Figure 5A. The intrinsic conformational restraint to p53 acetylation implies that DNA-binding "opens" p53 and allows the protein to function as a substrate in acetyltransferase reactions. In fact, NMR was used to localize sites in the DNA-binding domain of p53 that change conformation when p53 is DNA-bound (Fig. 6), this affirming p53 can change conformation in this DNA-bound state.[43, 44] he structure of the p53 tetramer is being assembled and will likely advance our knowledge of these large multi-protein interactions.[45]

Many protein-protein interactions are regulated by small linear peptide docking motifs.[10] Indeed p53 has a range of small linear domains that anchor or allosterically activate enzyme functions (Fig. 2A). In order to identify the mechanism of DNA-dependent acetylation of p53, we reasoned that p300 would have novel linear docking sites for DNA-bound p53 and as such p300 was subjected to combinatorial peptide screen during which a family of proline-repeat peptides were discovered as having high affinity for p300 (Fig. 7A). Subsequent examination demonstrated that the proline repeat domain was essential for p300 binding to p53 and for mediating DNA-dependent acetylation[42] thus indicating that the proline-repeat motif of p53

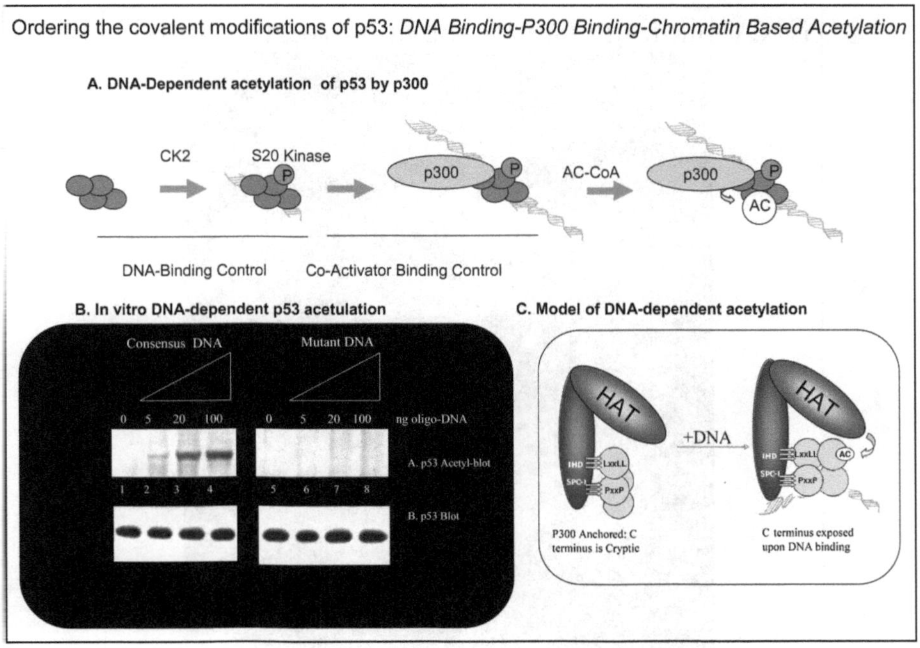

Figure 5. Paradigm III-DNA dependent p53 acetylation. A) The diagram highlights the various stages in which biochemical data show regulated effects of p53 colavant modification on p53 functions including DNA binding, p300 binding, and acetylation. B) The diagram shows the effects of p53 acetylation as a function of increasing consensus site DNA (left panel) or mutant consensus site DNA (right panel). The top immunoblot reflects p53 acetylation and the bottom total p53 protein. C) Model of p53 acetylation describing the intrinsic restraint to p53 acetylation (left) and how in the presence of DNA p53 undergoes some undefined conformational change that allows p300 to induce acetylation of the substrate.

is rate-limiting for this reaction. Structural studies on this effect are still lacking. The effects of the proline-repeat domain on p53 function was further investigated and shown to be regulated by proline isomerases which can convert phospho-proline motifs via isomerization to a conformation more stabilizing for p300 binding.[46] Together, these three paradigms highlight an orchestration of a series of enzymes including kinases and acetyltransferases which coordinate the p53 transcriptional response (Fig. 5A). As there are many sites of covalent modification onp53 (Fig. 1), it is likely that these other enzymatic steps also play critical roles in regulating p53 function and further genetic plus biochemical data will shed light on these pathways.

Paradigm IV: How Cells Integrate the p53 Response through Distinct Stresses

A fundamental paradigm in p53 function is that p53 "integrates" diverse stress signals towards a biological outcome. The integration mechanism is undefined but presumably involves both inhibition of p53's degradation pathway and activation of its transcription function. p53 is controlled by a variety of post-translational mechanisms (Fig. 1). Of the many types of activating covalent modifications observed on p53, phosphorylation has been the most well-studied both biochemically and genetically. A chemical biology screen was previously undertaken to determine the mechanisms underlying the integration of stress signals to p53 activation. The fundamental question that set out to answer is whether one common kinase pathway is able to target the Ser20 site within the transactivation domain of p53 in

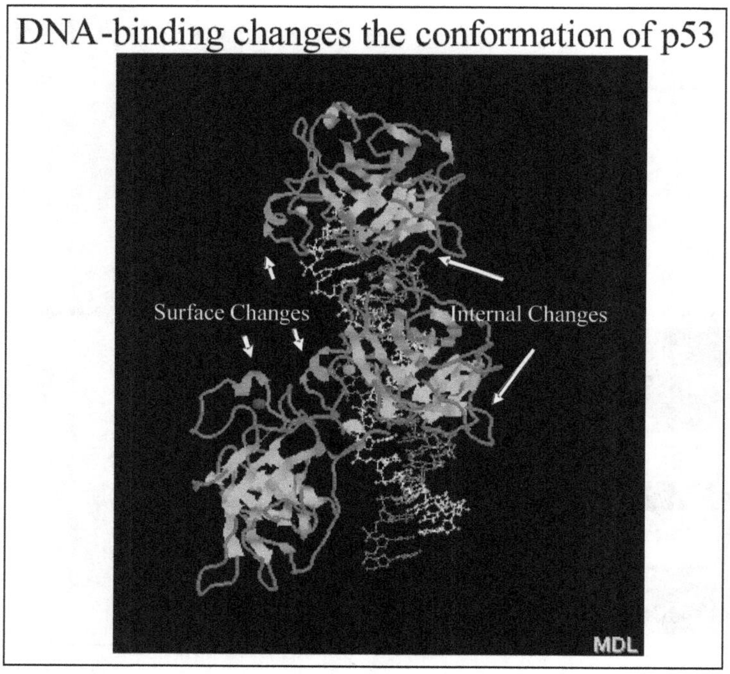

Figure 6. Conformational changes in p53 upon DNA binding. The diagram highlights the trimeric p53:DNA complex[62] and the green areas highlight the regions that change conformation as defined using NMR.[43,44] A color version of this image is available online at www.landesbioscience.com/curie.

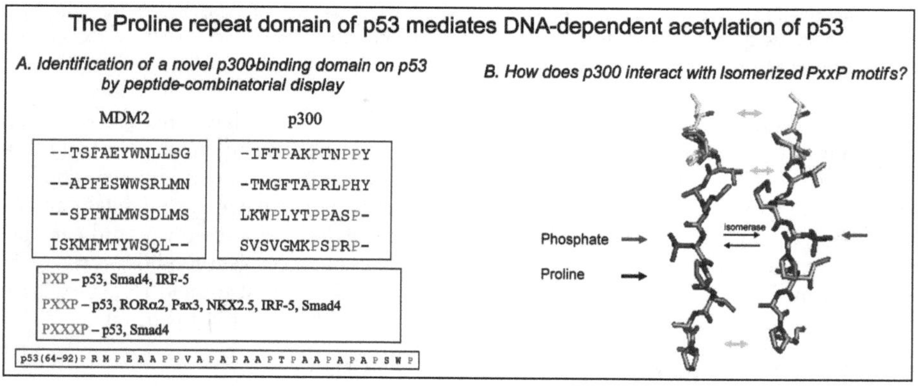

Figure 7. Proline repeat domain drives DNA-dependent p53 acetylation. A) In order to define novel contacts on p53 that mediate p300 acetylation, peptide combinatorial libraries were used to screen for novel p300 linear peptide binding domains. The left panel as a control highlights library integrity as defined by the ability of MDM2 to be used to pull out peptides that match the MDM2 docking site in the BOX-I domain of p53 : FxxxWxxL.[63, 64] The right panel shows peptides pulled out that interact with p300 ; PxxP repeat motifs predominate, as indicated. When such motifs are screened using peptide motif search engines against the human proteome, then these motifs can be found in transcription factors including p53 and SMAD4.[47] Deletion or mutation of the proline repeat motif of p53 prevents DNA-dependent acetylation of p53. Further, there are phosphorylation sites in the PxxP motif of p53 whose phosphorylation produces a proline isomerase (PIN) consensus site that results in proline isomerization[65] and stabilizes p300 binding by as yet undefined mechanism.[46]

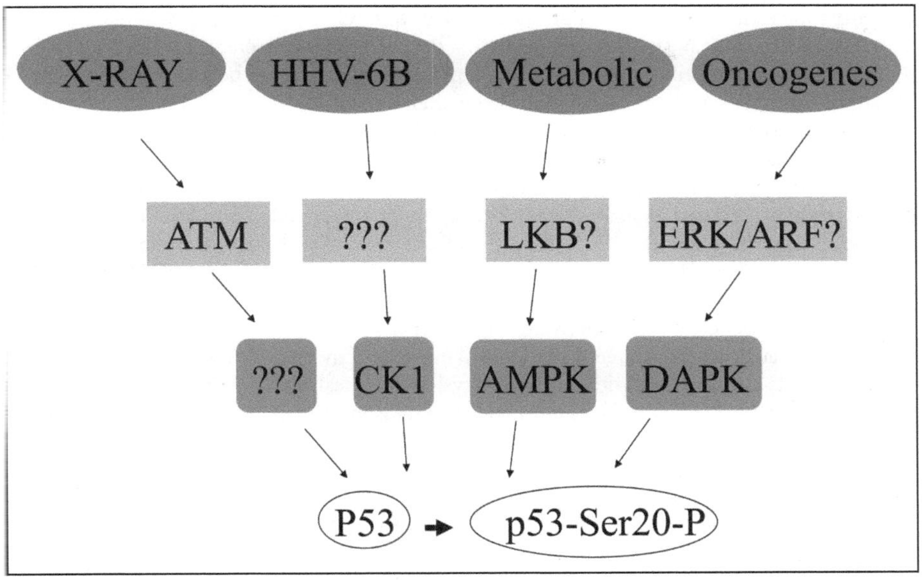

Figure 8. A model to explain how cells integrate distinct stresses to p53 activation via Ser20-site phosphorylation. p53 is activated by distinct stresses, some of which include as indicated, ionizing radiation, viral infection, metabolic stress induced by an altered AMP/ATP ratio, and oncogene activation. The X-ray-induced Ser20 site kinase is ATM-dependent, but its identity is unknown (highlighted by "?"). CK1 is the DNA virus HHV-6B-induced p53 Ser20 kinase, but the upstream sensor is currently undefined (highlighted by "?"). The Ser20 site kinase induced by an elevated AMP/ATP ratio is AMPK, and LKB is the likely upstream sensor. DAPK-1 is the p53 Ser20 kinase induced by inappropriate oncogene activation, and ERK or ARF are the likely upstream sensors. These data support the formation of a model suggesting that the phosphorylation of p53 at Ser20 is triggered by distinct stress-responsive signaling cascades. Future analysis will be required to determine the identity of all the enzymes that mediate stress-induced phosphorylation at this site and "integrate" the p53 response and developing disease models that deregulate these signaling cascades.

response to various stresses, or whether distinct kinases induced by different stresses are required to drive the same mechanism. We have focussed on the Ser20 site since it is the most highly conserved phospho-acceptor site between urochordates and humans (Fig. 2B) with well-documented genetic and biochemical effects. Phosphorylation at Ser20 has the most striking effect on stabilizing the p300:p53 transcription complex through interactions with multiple LxxLL peptide binding domains on p300.[42, 47] In that study,[48] it was shown that phosphorylation at the Ser20 site of p53 increases in response to distinct stresses, including ionizing radiation, virus infection or metabolic stress, and we investigated the kinase signaling pathways involved in this phosphorylation using small molecule kinase inhibitors. Thus, using a combination or biochemical fractionation, kinase inhibitors, and distinct stresses in cell lines, a model can be developed that begins to explain how cells activate p53 via Ser20 phosphorylation in response to distinct stresses. The model involves the recruitment of a family of enzymes that dock to the DNA-binding domain of p53 and catalyze phosphorylation.[49, 50] These enzymes in turn are activated by specific stresses including DNA damage, virus infection, AMP/ATP stress, and oncogene activation.

Other cellular stresses, including aberrant oncogene activation and subsequent induction of ARF[51, 52] o extracellular signal-regulated kinases (ERKs)[53] and death-associated protein kinase 1 (DAPK-1; Fig. 3)[54-56] have not been evaluated as of yet due to the lack of a common cell model that has an active ARF pathway (Fig. 8). However, given the role of ARF-p53 axis in regulating longevity and cancer suppression (reviewed in ref. 57), this signal will be important to evaluate. In fact, recent studies have shown that oncogene-induced senescence does not change p53 levels but increases its specific activity,[58] a phenomenon that can be accomplished by p53 phosphorylation at specific regulatory sites. Together, these data provide a paradigm explaining how distinct stresses can activate p53 (summarized in Fig. 8). In a biochemical approach to identify candidate kinases, we had previously identified many members of the calcium-calmodulin kinase superfamily, including CHK1/2, DAPK-1 and AMPK as p53 Ser20 site kinases.[50] The identification of CK1 as a major Ser20 site kinase was the first member outwith this superfamily that could target this site on p53.[49] However, all these enzymes have a common biochemical requirement for a high affinity docking site in the core DNA-binding domain of p53 to catalyze Ser20 site phosphorylation in the transactivation domain.[49, 50]

Studying the genetics of p53 activation by distinct stresses is relatively difficult given that there are no classic genetic screens for identifying p53 activating pathways. However, these models provide a framework with which to develop chemical biology screens to identify p53 activating enzymes that target the highly conserved phosphorylation sites in the p53 activation domain.

References

1. Riley T, Sontag E, Chen P et al. Transcriptional control of human p53-regulated genes. Nat Rev Mol Cell Biol 2008; 9:402-12.
2. Olivier M, Hussain SP, Caron de Fromentel C et al. TP53 mutation spectra and load: a tool for generating hypotheses on the etiology of cancer. IARC Sci Publ 2004; 247-70.
3. Vousden KH, Lu X. Live or let die: the cell's response to p53. Nat Rev Cancer 2002; 2:594-604.
4. Joerger AC, Fersht AR. Structural biology of the tumor suppressor p53. Annu Rev Biochem 2008; 77:557-82.
5. Donehower LA, Harvey M, Slagle BL et al. Mice deficient for p53 are developmentally normal but susceptible to spontaneous tumors. Nature 1992; 356:215-21.
6. Martin-Caballero J, Flores JM, Garcia-Palencia P et al. Tumor susceptibility of p21(Waf1/Cip1)-deficient mice. Cancer Res 2001; 61:6234-8.
7. Shen KC, Heng H, Wang Y et al. ATM and p21 cooperate to suppress aneuploidy and subsequent tumor development. Cancer Res 2005; 65:8747-53.
8. Neduva V, Russell RB. Peptides mediating interaction networks: new leads at last. Curr Opin Biotechnol 2006; 17:465-71.
9. Neduva V, Russell RB. DILIMOT: discovery of linear motifs in proteins. Nucleic Acids Res 2006; 34:W350-5.
10. Neduva V, Linding R, Su-Angrand I et al. Systematic discovery of new recognition peptides mediating protein interaction networks. PLoS Biol 2005; 3:e405.
11. Neduva V, Russell RB. Linear motifs: evolutionary interaction switches. FEBS Lett 2005; 579: 3342-5.
12. Carter S, Vousden KH. p53-Ubl fusions as models of ubiquitination, sumoylation and neddylation of p53. Cell Cycle 2008; 7:2519-28.
13. Hupp TR, Sparks A, Lane DP. Small peptides activate the latent sequence-specific DNA binding function of p53. Cell 1995; 83:237-45.
14. Nichols NM, Matthews KS. Human p53 phosphorylation mimic, S392E, increases nonspecific DNA affinity and thermal stability. Biochemistry 2002; 41:170-8.
15. McKinney K, Mattia M, Gottifredi V et al. p53 linear diffusion along DNA requires its C terminus. Mol Cell 2004; 16:413-24.
16. Blaydes JP, Hupp TR. DNA damage triggers DRB-resistant phosphorylation of human p53 at the CK2 site. Oncogene 1998; 17:1045-52.
17. Wallace M, Coates PJ, Wright EG et al. Differential post-translational modification of the tumor suppressor proteins Rb and p53 modulate the rates of radiation-induced apoptosis in vivo. Oncogene 2001; 20:3597-608.
18. Wang Y, Prives C. Increased and altered DNA binding of human p53 by S and G2/M but not G1 cyclin-dependent kinases. Nature 1995; 376:88-91.
19. Blaydes JP, Luciani MG, Pospisilova S et al. Stoichiometric phosphorylation of human p53 at Ser315 stimulates p53-dependent transcription. J Biol Chem 2001; 276:4699-708.

20. Qu L, Huang S, Baltzis D et al. Endoplasmic reticulum stress induces p53 cytoplasmic localization and prevents p53-dependent apoptosis by a pathway involving glycogen synthase kinase-3beta. Genes Dev 2004; 18:261-77.
21. Finlan LE, Nenutil R, Ibbotson SH et al. CK2-site phosphorylation of p53 is induced in DeltaNp63 expressing basal stem cells in UVB irradiated human skin. Cell Cycle 2006; 5:2489-94.
22. Finlan LE, Hupp TR. p63: the phantom of the tumor suppressor. Cell Cycle 2007; 6:1062-71.
23. Ashcroft M, Kubbutat MH, Vousden KH. Regulation of p53 function and stability by phosphorylation. Mol Cell Biol 1999; 19:1751-8.
24. Bruins W, Zwart E, Attardi LD et al. Increased sensitivity to UV radiation in mice with a p53 point mutation at Ser389. Mol Cell Biol 2004; 24:8884-94.
25. Bruins W, Jonker MJ, Bruning O et al. Delayed expression of apoptotic and cell-cycle control genes in carcinogen-exposed bladders of mice lacking p53. S389 phosphorylation. Carcinogenesis 2007; 28:1814-23.
26. Dornan D, Hupp TR. Inhibition of p53-dependent transcription by BOX-I phospho-peptide mimetics that bind to p300. EMBO Rep 2001; 2:139-44.
27. Finlan L, Hupp TR. The N-terminal interferon-binding domain (IBiD) homology domain of p300 binds to peptides with homology to the p53 transactivation domain. J Biol Chem 2004; 279:49395-405.
28. Teufel DP, Freund SM, Bycroft M et al. Four domains of p300 each bind tightly to a sequence spanning both transactivation subdomains of p53. Proc Natl Acad Sci U S A 2007; 104:7009-14.
29. Craig AL, Burch L, Vojtesek B et al. Novel phosphorylation sites of human tumor suppressor protein p53 at Ser20 and Thr18 that disrupt the binding of mdm2 (mouse double minute 2) protein are modified in human cancers. Biochem J 1999; 342(Pt 1):133-41.
30. Lee HJ, Srinivasan D, Coomber D et al. Modulation of the p53-MDM2 interaction by phosphorylation of Thr18: a computational study. Cell Cycle 2007; 6:2604-11.
31. Jabbur JR, Huang P, Zhang W. Enhancement of the antiproliferative function of p53 by phosphorylation at serine 20: an inference from site-directed mutagenesis studies. Int J Mol Med 2001; 7:163-8.
32. Jabbur JR, Zhang W. p53 Antiproliferative function is enhanced by aspartate substitution at threonine 18 and serine 20. Cancer Biol Ther 2002; 1:277-83.
33. Shieh SY, Taya Y, Prives C. DNA damage-inducible phosphorylation of p53 at N-terminal sites including a novel site, Ser20, requires tetramerization. EMBO J 1999; 18:1815-23.
34. Craig AL, Bray SE, Finlan LE et al. Signaling to p53: the use of phospho-specific antibodies to probe for in vivo kinase activation. Methods Mol Biol 2003; 234:171-202.
35. Craig A, Scott M, Burch L et al. Allosteric effects mediate CHK2 phosphorylation of the p53 transactivation domain. EMBO Rep 2003; 4:787-92.
36. MacPherson D, Kim J, Kim T et al. Defective apoptosis and B-cell lymphomas in mice with p53 point mutation at Ser 23. EMBO J 2004; 23:3689-99.
37. Armata HL, Garlick DS, Sluss HK. The ataxia telangiectasia-mutated target site Ser18 is required for p53-mediated tumor suppression. Cancer Res 2007; 67:11696-703.
38. Lambert PF, Kashanchi F, Radonovich MF et al. Phosphorylation of p53 serine 15 increases interaction with CBP. J Biol Chem 1998; 273:33048-53.
39. Cordenonsi M, Montagner M, Adorno M et al. Integration of TGF-beta and Ras/MAPK signaling through p53 phosphorylation. Science 2007; 315:840-3.
40. Tang Y, Zhao W, Chen Y et al. Acetylation is indispensable for p53 activation. Cell 2008; 133:612-26.
41. Li AG, Piluso LG, Cai X et al. An acetylation switch in p53 mediates holo-TFIID recruitment. Mol Cell 2007; 28:408-21.
42. Dornan D, Shimizu H, Perkins ND et al. DNA-dependent acetylation of p53 by the transcription coactivator p300. J Biol Chem 2003; 278:13431-41.
43. Weinberg RL, Veprintsev DB, Fersht AR. Cooperative binding of tetrameric p53 to DNA. J Mol Biol 2004; 341:1145-59.
44. Rippin TM, Freund SM, Veprintsev DB et al. Recognition of DNA by p53 core domain and location of intermolecular contacts of cooperative binding. J Mol Biol 2002; 319:351-8.
45. Tidow H, Melero R, Mylonas E et al. Quaternary structures of tumor suppressor p53 and a specific p53 DNA complex. Proc Natl Acad Sci U S A 2007; 104:12324-9.
46. Mantovani F, Tocco F, Girardini J et al. The prolyl isomerase Pin1 orchestrates p53 acetylation and dissociation from the apoptosis inhibitor iASPP. Nat Struct Mol Biol 2007; 14:912-20.

47. Dornan D, Shimizu H, Burch L et al. The proline repeat domain of p53 binds directly to the transcriptional coactivator p300 and allosterically controls DNA-dependent acetylation of p53. Mol Cell Biol 2003; 23:8846-61.
48. MacLaine NJ, Hupp T. Phosphorylation control of p53: A model that integrates distinct stresses to a common outcome. submmited. 2009
49. MacLaine NJ, Oster B, Bundgaard B et al. A central role for CK1 in catalyzing phosphorylation of the p53 transactivation domain at serine 20 after HHV-6B viral infection. J Biol Chem 2008; 283:28563-73.
50. Craig AL, Chrystal JA, Fraser JA et al. The MDM2 ubiquitination signal in the DNA-binding domain of p53 forms a docking site for calcium calmodulin kinase superfamily members. Mol Cell Biol 2007; 27:3542-55.
51. Palmero I, Pantoja C, Serrano M. p19ARF links the tumor suppressor p53 to Ras. Nature 1998; 395:125-6.
52. Zindy F, Eischen CM, Randle DH et al. Myc signaling via the ARF tumor suppressor regulates p53-dependent apoptosis and immortalization. Genes Dev 1998; 12:2424-33.
53. Mallette FA, Gaumont-Leclerc MF, Ferbeyre G. The DNA damage signaling pathway is a critical mediator of oncogene-induced senescence. Genes Dev 2007; 21:43-8.
54. Raveh T, Droguett G, Horwitz MS et al. DAP kinase activates a p19ARF/p53-mediated apoptotic checkpoint to suppress oncogenic transformation. Nat Cell Biol 2001; 3:1-7.
55. Anjum R, Roux PP, Ballif BA et al. The tumor suppressor DAP kinase is a target of RSK-mediated survival signaling. Curr Biol 2005; 15:1762-7.
56 Chen CH, Wang WJ, Kuo JC et al. Bidirectional signals transduced by DAPK-ERK interaction promote the apoptotic effect of DAPK. EMBO J 2005; 24:294-304.
57. Matheu A, Maraver A, Serrano M. The Arf/p53 pathway in cancer and aging. Cancer Res 2008; 68:6031-4.
58. Ruiz L, Traskine M, Ferrer I et al. Characterization of the p53 response to oncogene-induced senescence. PLoS ONE 2008; 3:e3230.
59. Blaydes JP, Craig AL, Wallace M et al. Synergistic activation of p53-dependent transcription by two cooperating damage recognition pathways. Oncogene 2000; 19:3829-39.
60. Finlan LE, Kernohan NM, Thomson G et al. Differential effects of 5-aminolaevulinic acid photodynamic therapy and psoralen + ultraviolet A therapy on p53 phosphorylation in normal human skin in vivo. Br J Dermatol 2005; 153:1001-10.
61. Finlan LE, Hupp TR. Epidermal stem cells and cancer stem cells: insights into cancer and potential therapeutic strategies. Eur J Cancer 2006; 42:1283-92.
62. Cho Y, Gorina S, Jeffrey PD, Pavletich NP. Crystal structure of a p53 tumor suppressor-DNA complex: understanding tumorigenic mutations. Science 1994; 265:346-55.
63. Bottger A, Bottger V, Garcia-Echeverria C et al. Molecular characterization of the hdm2-p53 interaction. J Mol Biol 1997; 269:744-56.
64. Bottger V, Bottger A, Howard SF et al. Identification of novel mdm2 binding peptides by phage display. Oncogene 1996; 13:2141-7.
65. Hupp TR, Walkinshaw M. Multienzyme assembly of a p53 transcription complex. Nat Struct Mol Biol 2007; 14:885-7.

Chapter 5

The p53-Mdm2 Loop:
A Critical Juncture of Stress Response

Yaara Levav-Cohen, Zehavit Goldberg, Osnat Alsheich-Bartok, Valentina Zuckerman, Sue Haupt and Ygal Haupt*

Abstract

The presence of a functional p53 protein is a key factor for the proper suppression of cancer development. A loss of p53 activity, by mutations or inhibition, is often associated with human malignancies. The p53 protein integrates various stress signals into a growth restrictive cellular response. In this way, p53 eliminates cells with a potential to become cancerous. Being a powerful decision maker, it is imperative that p53 be activated properly, efficiently and temporarily in response to stress. Equally important is that p53 activation will be extinguished upon recovery from stress, and that improper activation of p53 will be avoided. Failure to achieve these aims is likely to have catastrophic consequences for the organism. The machinery that governs this tight regulation is largely based on the major inhibitor of p53, Mdm2, which both blocks p53 activities and promotes its destabilization. The interplay between p53 and Mdm2 involves a complex network of positive and negative feedback loops. Relief from Mdm2 suppression is required for p53 to be stabilized and activated in response to stress. Protection from Mdm2 entails a concerted action of modifying enzymes and partner proteins. The association of p53 with the PML-nuclear bodies may provide an infrastructure in which this complex regulatory network can be orchestrated. In this chapter we use examples to illustrate the regulatory machinery that drives this network.

Introduction

The tumor suppressor p53 protein is pivotal in the prevention of cancer development. P53 determines cell fate through its activities as a transcription factor, and by engagement in critical protein interactions at the mitochondria (reviewed in ref. 1). P53 is normally labile, but in response to external and internal stress signals, it is triggered to become stable and active within the nucleus. As a transcription factor it controls the expression of genes that regulate cell growth and cell death (reviewed in refs. 2,3). Stabilized p53 induces either cell growth arrest (temporary, or permanent "senescence"), or programmed cell death (apoptosis). The growth restrictive activities of p53 prevent the proliferation of cells with damaged DNA or with a potential for neoplastic transformation; while p53-mediated permanent cell growth inhibition (apoptosis or senescence) drives tumor suppression. Given these functions of p53, it is not surprising that p53 serves as a serious obstacle to the step-by-step progression of cancer development. This barrier is very frequently removed at one of the steps, either by direct mutation of the *p53* gene, or by indirect mechanisms, such as an elevation in the expression levels of p53 inhibitors, or by down-regulation of p53 co-activators, such as ARF.[4]

*Correponding Author: Ygal Haupt—Research Division, The Peter MacCallum Cancer Centre, St. Andrew's Place, East Melbourne 3002, Victoria, Australia. Email: ygal.haupt@petermac.org

p53, edited by Ayeda Ayed and Theodore Hupp. ©2010 Landes Bioscience and Springer Science+Business Media.

The lability of p53 under normal cellular conditions is largely attributable to its inhibitor Mdm2 (Hdm2 in humans), which ensures that p53 has a short half-life and consequent low basal activity. Stresses that dramatically elicit a change in p53 status include: DNA damage, deregulated oncogenes, hypoxia, and nucleotide depletion among others (reviewed in ref. 5). The activation of p53 involves stabilization of the protein, which is mediated by extensive post-translational modifications, and protein-protein interactions with cooperating factors. Once stable, p53 engages in enhanced DNA binding and transcriptional activity. The summation of the incoming signals and the cellular context, dictates whether activated p53 will direct cells to growth arrest, senescence or apoptosis (reviewed in ref. 6). This chapter focuses on the regulation of p53 by Mdm2. Particular emphasis will be given to current models explaining how the p53/Mdm2 auto-regulatory loop is modulated or interrupted in response to stress. The different mechanisms involved will be illustrated by specific examples. The intention of this chapter is to explain how such a busy network of regulation may be coordinated within a cell in a spatial and temporal manner in response to a given stress signal.

The p53-Mdm2 Feedback Loop

Almost two decades of research have passed since the identification of *mdm2* as a p53 target gene[7,8] (also reviewed in refs. 9,10). The revelation that p53 induces Mdm2 expression, which then inhibits the biochemical and biological activities of p53, defined the first and the most important auto-regulatory loop that governs p53 regulation. This loop proves to be even more powerful than initially thought, as additional multiple regulatory loops are being found to interweave with it.[4,11-14] Several of these loops will be described in this chapter. Mdm2 binds p53 in the transactivation domain and blocks its ability to induce or suppress transcription (reviewed in refs. 9,10). The major and most efficient inhibitory effect of Mdm2 is to destroy the p53 protein via the ubiquitin-proteasome pathway[15,16] (also reviewed in ref. 17). Thus, through this negative feedback loop, Mdm2 shuts off its own expression (Fig. 1). The physiological significance of this auto-regulatory feedback loop was demonstrated by the clinical observation that amplification of Hdm2 in human cancers often correlates with wild type p53 status, supporting the notion that high expression of Hdm2 is sufficient for relieving a cell from p53 regulation, in the absence of p53 mutation (reviewed in refs. 9,18). Further, a single nucleotide polymorphism (SNP) in the *hdm2* gene that leads to increased Hdm2 expression,

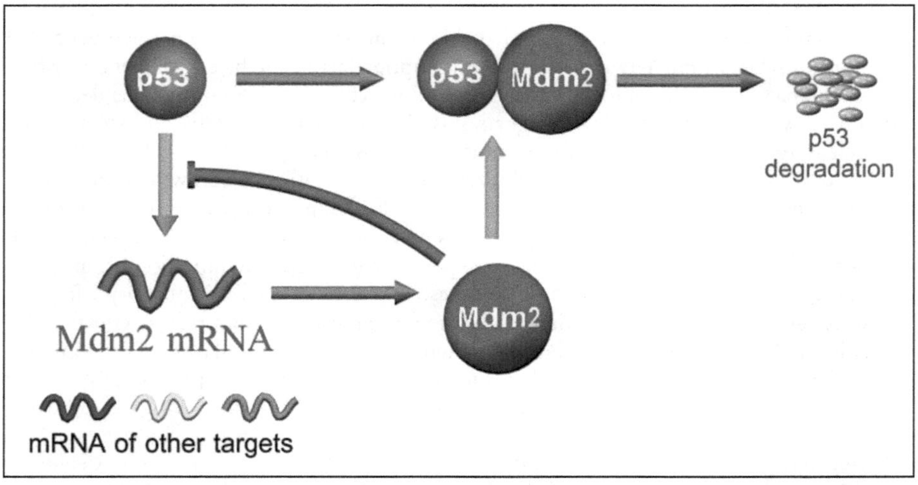

Figure 1. The p53/Mdm2 autoregulatory loop. Activated p53 induces the expression of multiple target genes. One of the genes is Mdm2 which binds p53, inhibits its transcriptional activity and promotes it for proteasomal degradation.

results in reduced levels of p53-dependent apoptosis and correlates with accelerated tumor formation[19] (for review see refs. 20,21). Consistently, reduced expression of Mdm2 protein in mice carrying a hypomorphic *mdm2* allele, results in increased frequency of p53 dependent apoptosis and leads to many defects in mice[22] (or reviewed in ref. 20). Even more convincing was the finding that mouse embryos lacking *mdm2* die early during development, but if crossed with p53 null mice this phenotype is rescued[9,23-25] (also reviewed in ref. 10).

Intriguingly, the presence of the Mdm2 analogue, Mdmx (also called Mdm4 or Hdmx in humans), is critical for proper mouse development. The embryonic lethality of *mdmx* null mice is rescued by the removal of p53, as in the case of *mdm2* deficient mice, supporting critical and non-overlapping roles for Mdm2 and Mdmx in the inhibition of p53, at least during early mouse development[26-28] (reviewed in ref. 29). Moreover, analogous to Hdm2, Hdmx expression is elevated in a considerable fraction of human cancers carrying wild-type p53, implying that high level of Mdmx expression is sufficient to inactivate p53, without concomitant p53 mutations.[18,30] Interestingly, in sharp contrast to Mdm2, the inhibition of p53 by Mdmx does not directly involve protein degradation and there is no evidence for existence of a regulatory loop between Mdmx and p53.[31] Mdmx binds p53 in its transactivation domain and inhibits p53 transcriptional activity.[32] Importantly, Mdmx forms heterodimers with Mdm2 and, although Mdmx lacks its own E3 ubiquitin ligase activity, it can assist Mdm2 in p53 degradation (reviewed in ref. 33).

Mdm2-Mediated p53 Ubiquitination and Degradation

The majority of studies investigating the inhibition of p53 by Mdm2 support the current model that Mdm2 promotes the ubiquitination and subsequent degradation of p53 through the proteasomal machinery (Fig. 1). This promotion of p53 degradation requires the E3 ligase activity of Mdm2, which is mediated by the RING-finger domain.[34,35] Although, a number of studies identified Mdm2 as the principal endogenous E3-ligase that promotes the efficient degradation of p53 (e.g., refs. 6,15,16,34,35), p53 is also degraded in *mdm2* deficient cells,[36] suggesting that other E3 ligase/s can promote p53 degradation in an Mdm2-independent manner in vivo. Indeed, a number of other E3 ligases were shown to regulate p53 protein levels in tissue culture and in biochemical studies in vitro; these include: Pirh2,[37] Cop1,[38] and Arf-BP1.[39] However, the exact cellular contexts and level of contribution of these ligases remain to be defined.

Confirmation that the primary physiological function of Mdm2 is to promote p53 degradation via its E3 ligase activity was recently shown by Itahana et al.[40] In this study, the loss of p53 rescued the early lethality of mice bearing a C462A mutation in Mdm2, a mutation that causes the abolishment of the E3 ligase activity of Mdm2, without affecting its interaction with p53. Nonetheless, accumulating evidence over the last decade indicates that Mdm2 can repress p53 activity by additional mechanisms, besides its direct interaction with p53. For example, Mdm2 can inhibit p53 acetylation mediated by p300/CBP[41,42] as well as by inhibiting and degrading PCAF.[43] In addition, Mdm2 can recruit the histone deacetylase HDAC1[44] and the nuclear corepressor KAP1,[45] providing alternative ways by which Mdm2 can repress the acetylation of either p53 or histones surrounding the p53 binding sites. Moreover, Mdm2 was also reported to promote NEDD8 conjugation of p53, a modification that inhibits its transcriptional activity.[46] Finally, Mdm2 induces monoubiquitination of histones in the vicinity of the p53 response elements, resulting in the transcriptional repression of p53.[47]

Mdm2 can also regulate the levels of nuclear p53 by a degradation-independent ubiquitination. Using an in vitro ubiquitination reconstitution assay, it has been shown that Mdm2 mediates the monomeric ubiquitination of p53 on multiple lysine residues, rather than adding ubiquitin chains onto one or few lysine residues (Fig. 2).[48] It was found, that when expressed in low levels, Mdm2 can mediate p53-monoubiquitination, which signals for p53 nuclear export and leads to its accumulation in the cytoplasm, thereby inhibiting its role as transcription factor.[49,50] However, proteasome-dependent degradation of nuclear p53 can also

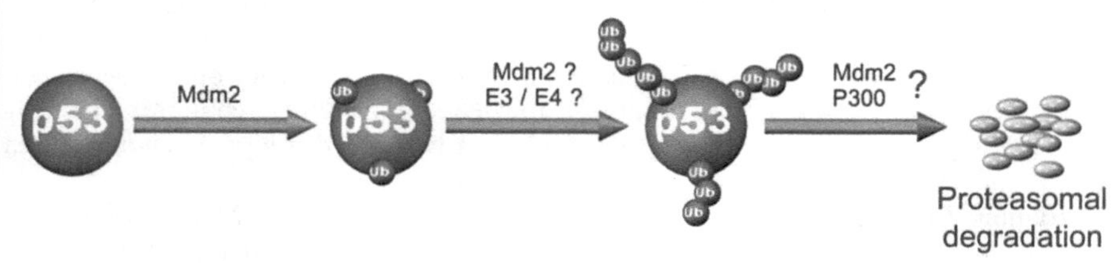

Figure 2. A model for the proteasomal degradation of p53 by Mdm2. Mdm2 is important for the mono-ubiquitination of p53. The possible involvement of Mdm2 and other factors in the subsequent steps leading to p53 degradation is shown. For simplicity the sub-cellular compartmentalization of this process is not shown.

effectively preventing its transactivation function,[50,51] suggesting monoubiquitination of cytoplasmic p53 is required for other processes. Indeed, the role that p53 plays in the cytoplasm, including direct signaling at the mitochondria and the induction of transcription-independent apoptosis,[52] may involve p53 modified in this manner.

P53 is subject to ubiquitination on at least 6 C-terminal lysines.[6] The six C-terminal lysines of p53 are the predominant sites for mdm2-mediated ubiquitination.[53] Although the in vitro data certainly demonstrate the importance of the six C-terminal lysines of p53 for Mdm2-mediated ubiquitination, knock-in studies in which the equivalent lysines have been mutated did not dramatically alter p53 protein levels.[54,55] These studies suggests that the C-terminal lysines are not essential for efficient p53 degradation in vivo and that additional E3 ligases, as well as ubiquitination of additional p53 lysines, are required for effective p53 regulation. Recent in vitro data suggest that lysine residues located in the DNA-binding domain and in the N-terminus may be ubiquitinated.[56]

An additional mechanism by which Mdm2 may regulate p53 has been suggested by Yin et al[57] who have shown that Mdm2, by virtue of its binding to p53, induces p53 translation from an internal initiation site, generating a smaller product of 47Kd (termed p53/47). This product lacks the Mdm2 binding site and hence is more stable than wt p53 and has altered specificity towards the apoptotic target gene, *bax*. The p47 form is still subject to degradation, presumably through oligomerization with wt p53, as was previously shown for another Mdm2-binding deficient mutant of p53.[58] Surprisingly, the overall apoptotic activity of the p53/47 does not differ from that of wt p53. Hence, the physiological relevance of this mode of regulation has yet to be defined.

It is difficult to weigh the relative contribution of p53 degradation to the overall down-regulation of p53 by Mdm2 (reviewed in ref. 17). The inhibition of p53 transcriptional and apoptotic activities by Mdm2 without promoting p53 degradation supports dual inhibitory mechanisms by Mdm2 (e.g., refs. 59,60). However, recent genetic studies support the notion that the major physiological importance of Mdm2 relates to its ability to regulate p53 protein abundance and that Mdm2 has little effect on p53 transcriptional activity on a "per molecule" basis (reviewed in ref. 61). Moreover, in vivo, Mdm2 that binds p53 but lacks its ubiquitination ligase activity is unable to efficiently suppress p53 functions (reviewed in ref. 62).

Breaking the p53/Mdm2 Regulatory Loop

The p53/Mdm2 auto-regulatory loop provides explanations for how the low basal level of p53 is maintained and how p53 returns to its basal level during the recovery from stress. At the same time, however, it raises the question: how does the p53 protein temporarily escape these intensive restrictive activities of Mdm2 (as well as of Mdmx) when cells are exposed to stress? Considerable effort has been devoted over the past decade to unravel the mechanisms by which the p53/Mdm2 loop is regulated and is interrupted under stress conditions (e.g., refs. 12,13,17,63,64). Three major mechanisms for this loop interruption have been identified; first, stress induces post-translational modifications of both p53 and Mdm2; second, interacting proteins act to disrupt p53/Mdm2 binding; and third, transportation of the two proteins leads to their spatial separation. A variety of proteins participate in these levels of regulation, where the choice of the particular regulator is largely dictated by the type of the incoming stress signal (e.g., ref. 12). It is not our intention to cover every known case, but rather to use selected examples to illustrate the principles of three major levels of regulation.

Stress Induced Phosphorylation

P53 Activation by the ATM/ATR-Chk Pathways

The p53 protein is subject to extensive post-translational modifications (reviewed in refs. 65-68), some of which affect its sensitivity to inhibition by Mdm2. Here we will focus on the role of phosphorylation in this regulation. Phosphorylation of p53, in particular within the N-terminal side, has critical impact on its functional interaction with Mdm2. Phosphorylation of threonine 18 (Thr18) by casein kinase I reduces the interaction between p53 and Mdm2.[69,70] This phosphorylation appears to require a preceding phosphorylation on serine 15 (Ser15) by several protein kinases including ATM and ATR (reviewed in refs. 18,66), which activate p53, but do not stabilize it.[71-74] Moreover, phosphorylation of threonine-proline motifs enables the binding of the prolyl isomerase PIN1 to induce cis-trans prolyl isomerizations. Recently the phosphorylation of p53 of three sites, Ser33, Thr81 and Ser315, was shown to affect the sensitivity of p53 to Mdm2 through the involvement of Pin1 (see below). Although several PIN1 sites exist in human p53, the PIN1 site in the proline-rich domain (threonine 81-proline 82) seems to be essential because proline 82 is isomerized by PIN1, enabling the recruitment of Chk2 to phosphorylate serine 20 and consequently reduce Mdm2 binding.[75]

Much attention has been focused on the phosphorylation of human p53 at Ser20 (Ser23 in the mouse), as it resides within the Mdm2 binding site. This phosphorylation is mediated by the checkpoint kinase (Chk) 1, MAPKAPK2 or JNK in response to UV-light and by Chk2 in response to γ-irradiation (IR) (reviewed in refs. 18,65,66). The phosphorylation of Ser20 reduces the binding affinity between p53 and Mdm2, and consequently p53 is activated and stabilized (reviewed in ref. 76). Thus, one mechanism by which p53 is protected from Mdm2 in response to DNA damage involves Ser20 phosphorylation (Fig. 3). This temporal protection would last as long as the DNA damage signal persists. How this signal is terminated is not clear. For instance does it involve dephosphorylation of Ser20 and other sites?

Another mechanism which allows p53 induction involves stimulation of ATM and ATR in response to DNA damage, leading to phosphorylation of Mdm2 and MdmX, mediating their rapid degradation.[77,78]

Biochemical analysis and studies using cultured cells indicate that phosphorylation of these p53 sites stimulates the recruitment of key transcriptional proteins, such as p300 and CBP,[79-85] leading to the acetylation of several key lysine residues in the carboxy-terminus of p53 that are normally targets for ubiquitination; this process is thought to help stabilize p53.[44,86]

The physiological role of Ser20 in human p53 was also tested in the transgenic mice bearing a substitution from Ser23 to alanine 23 (Ala23). Introduction of this substitution mutant p53 into ES cells or MEFs had no significant effect on the activation or accumulation of p53 in response to DNA damage in contrast to the case in the *Chk2* null mice.[87,88] Furthermore, Ser23 is phosphorylated in *Chk2* null mice in response to IR.[89] Can this apparent controversy be reconciled? Several possible explanations may be suggested. First, the regulatory role of Ser20 in human p53 may differ from that of Ser23 in the mouse. The lack of conservation of the Ser46 phosphorylation site in mouse p53 supports this notion (see below). Second, Ser20 (and Ser23) is not the only Chk2 phosphorylation site in p53 (reviewed in refs. 18,90). Third, Chk2 may also activate p53 by phosphorylation-independent mechanisms, for instance by

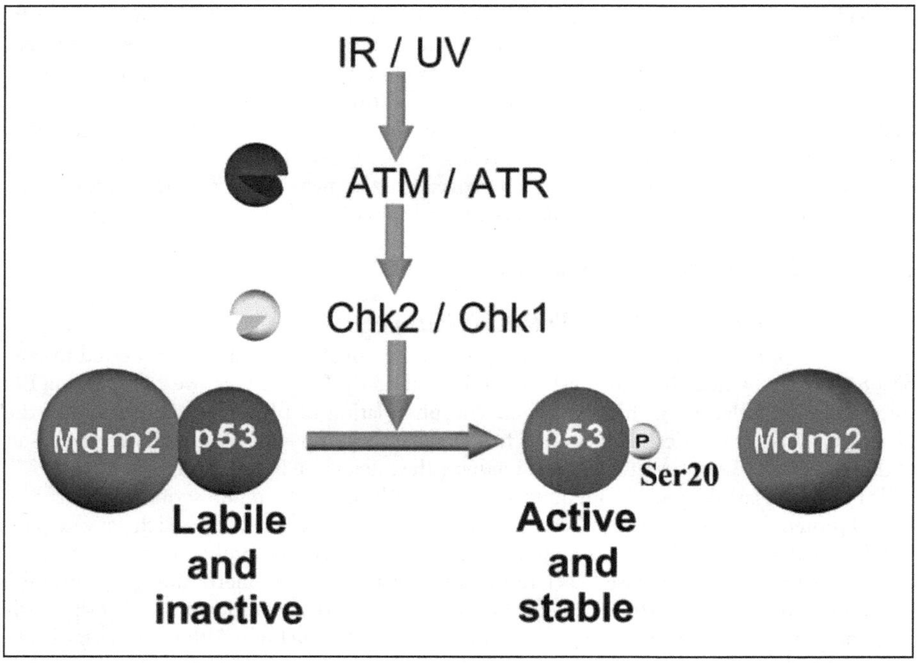

Figure 3. Interruption of the p53/Mdm2 loop by DNA damage-mediated phosphorylation. A major pathway by which p53 is protected from Mdm2 in response to DNA damage. A role for Ser20 phosphorylation is shown.

direct interaction with p53.[91] Whereas the precise mechanistic explanation for p53 activation by Chk2 needs further exploration, the evidence for the physiological relevance of Chk2 appears solid. The activation of p53 in response to DNA damage is severely impaired in *Chk2* null mice, and p53 mutations are not or rarely found in cancers bearing germ-line or somatic mutations in the Chk2 gene (reviewed in ref. 76). Despite the subtle phenotype, one study reported a partial defect in p53 accumulation and apoptosis in irradiated thymocytes.[87] Mutant mice developed B-cell lymphomas, but with a long period of latency (around 18 months, compared with 6-10 months in $Trp53^{-/-}$ mice).

It was also surprising that the mutation of Ser18 in mouse p53 (human Ser15) led to a modest phenotype[92,93] that had normal p53 stability in unstressed and DNA-damaged cells, normal cell-cycle control, cell-type-specific partial defects in apoptosis and normal tumor suppression. Recently, a targeted double mutation in S18A/S23A was analyzed in-vivo.[94] The double mutant knock-in mice, display reduced apoptosis in thymocytes and develop some malignancies, lending support to the physiological importance of these two key phosphorylation sites. Overall, these mice provide evidence to support the idea that DNA damage pathways can, at least partially, influence tumor suppressor function.

Mdm2 Inactivation by the ATM-c-Abl Pathway

Mdm2 is a phosphoprotein, that is subject to both phosphorylation and de-phosphorylation of specific sites in response to DNA damage[95,96] (also reviewed in refs. 97,98). A search for phosphorylation sites in Mdm2 relevant to p53 regulation revealed multiple sites that are targeted by several protein kinases (reviewed in refs. 97,98). The interaction between Mdm2 and p53 is impaired upon phosphorylation of Mdm2 by DNA-dependent protein kinase.[99]

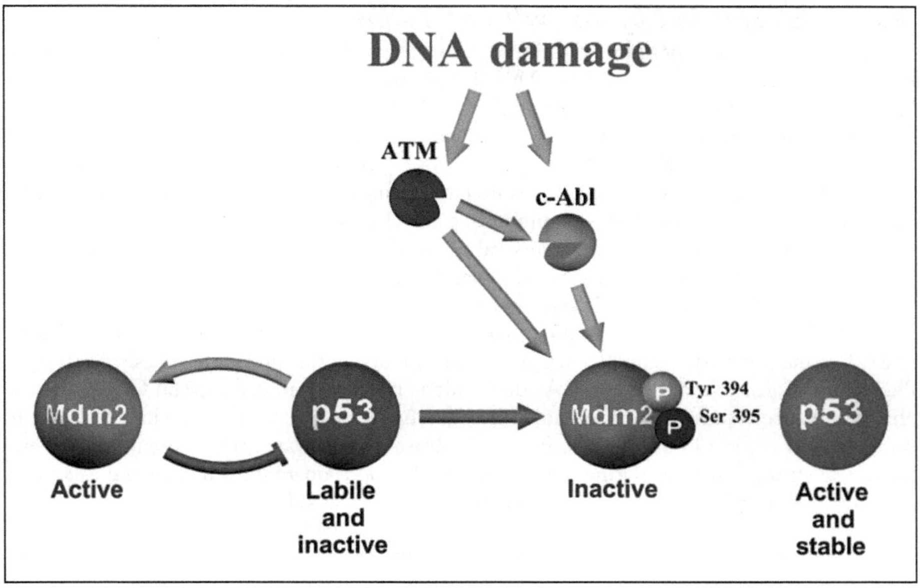

Figure 4. A role for c-Abl and ATM in the protection of p53 from Mdm2 in response to DNA damage. The role for the phosphorylation of Hdm2 on Tyr394 by c-Abl and Ser395 by ATM is shown. Whether these adjacent phosphorylations have a synergistic effect is not known yet.

A similar effect is observed when cyclin A-Cdk2 phosphorylates Mdm2 on threonine 216, which also augments the Mdm2/ARF binding.[100]

An important antagonist of Mdm2 is the c-Abl tyrosine kinase. C-Abl is essential for the efficient accumulation of p53 in response to DNA damage,[101] (reviewed in ref. 102). This is achieved by protecting p53 from Mdm2-induced nuclear export, ubiquitination, and degradation.[60,101,102] The kinase activity of c-Abl is essential for the neutralization of Mdm2. Indeed, c-Abl phosphorylates Mdm2 at tyrosine 394[103] and at tyrosine 276.[104] The latter modification enhances interaction of Mdm2 with ARF and leads to decreased p53 turnover.[104] Prevention of phosphorylation on tyrosine 394 enhances the ability of Mdm2 to promote p53 degradation and to inhibit p53 transcriptional and apoptotic activities.[103] Intriguingly, the adjacent amino acid Ser395 is phosphorylated by ATM in response to DNA damage.[105] This phosphorylation impairs the nuclear export and degradation of p53.[106] Interestingly, c-Abl is activated by ATM in response to DNA damage (reviewed in refs. 107,108), raising the possible scenario that ATM and c-Abl may work in concert to neutralize Mdm2 under certain stress conditions (**Fig.** 4). Although Tyr394 phosphorylation occurs independently of the phosphorylation of Ser395,[103] the effect of these two kinases on the neutralization of Mdm2 may be synergistic. Additional studies show that the complex phosphorylation of Mdm2 upon DNA damage by PI-3 kinases including ATM, ATR and DNA-PK may increase auto-ubiquitination of Mdm2 and enhance its degradation, resulting in augmented p53 abundance and transcription.[78] Additionally, it was recently proposed that phosphorylation of Mdm2 by ATM on multiple sites near its RING domain may disrupt Mdm2 oligomerization thus specifically suppressing p53 poly-ubiquitination.[109,110]

It is important to note that the inhibition of p53 by Mdmx is also restrained by ATM- and c-Abl-mediated phosphorylations that accelerate Mdmx degradation by Mdm2 and inhibit Mdmx binding with p53, respectively. (Reviewed in refs. 1,32,63.)

Stimulation of Mdm2 by the Akt/PTEN Pathway

The phosphorylation of Mdm2 can also be stimulatory, as is the case with the mitogen-activated kinase, Akt. Upon growth stimulation, Mdm2 is phosphorylated by Akt at Ser166, Ser186 and Ser188, which enhance the nuclear accumulation of Mdm2 and its ability to inhibit p53.[111-113] These phosphorylations also augment Mdm2 interaction with p300, reduce the affinity of Mdm2 for p19ARF[114] and inhibit Mdm2 self-ubiquitination.[115] (Reviewed in refs. 97,116,117.) Consequently, Akt stimulates the inhibition and destabilization of p53 via Mdm2. Interestingly p53 can counteract this inhibitory axis by promoting the cleavage and degradation of the Akt protein.[118] This feedback loop generates a survival signal when Akt is activated, whereas under death inducing conditions, p53 opposes survival signals by eliminating Akt[119] (Fig. 5). The regulatory loop involving p53, Mdm2 and Akt is further regulated by additional feedback loops (reviewed in ref. 11,119). The first involves the p53 target gene, Cyclin G, which recruits the phosphatase PP2AB' to the Mdm2/p53 complex where it de-phosphorylates Mdm2 at the Akt site Ser166.[120] Phosphorylation of Mdm2 by cyclin A/cdk2 inhibits its activity, thus the cyclin G-PP2A phosphatase enhances Mdm2 activity and inhibits p53. Mice with the cyclin G gene knocked out are viable,[121] and cyclin G null mouse embryo fibroblasts have elevated p53 protein levels in the absence of stress,[120] demonstrating that this feedback loop is operational in vivo and acts upon the basal levels of p53 in a cell, not only the higher p53 activated levels after stress.

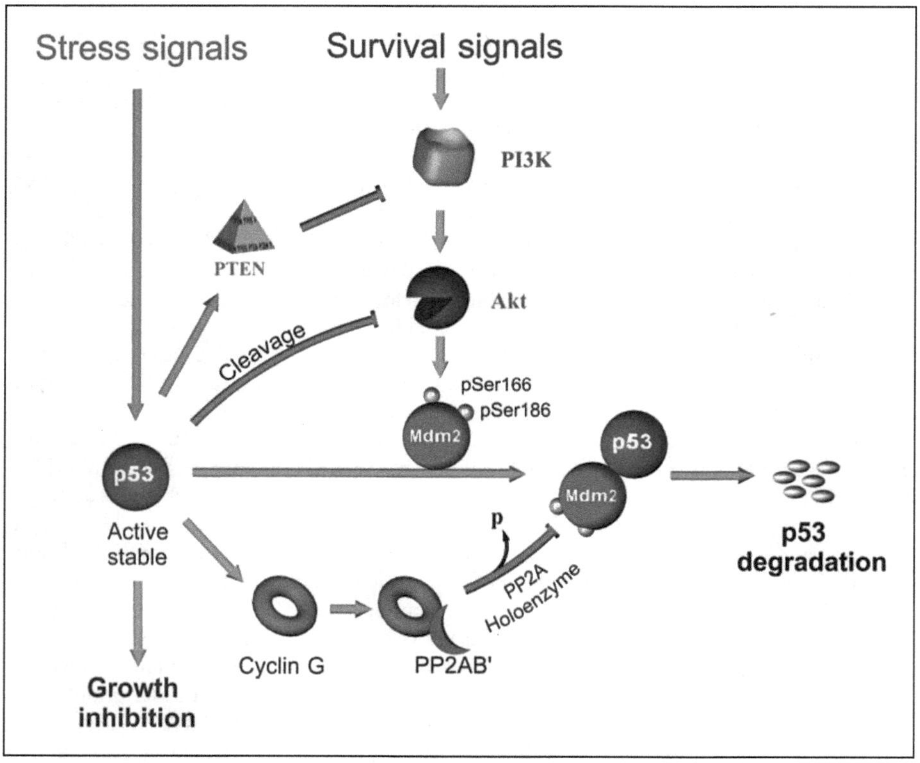

Figure 5. A model for the p53/Mdm2/Akt regulatory loop. The negative regulation of p53 by Akt is induced in response to survival signals and involves the activation of Mdm2. p53 counteracts this pathway through at least 3 different loops: the cleavage of Akt, the inhibition of PI3K through PTEN, and the dephosphorylation of Mdm2 at Akt sites through the induction of cyclin G and the subsequent recruitment of PP2AB′. In this pathway, survival is achieved by inhibition of p53 by Akt, whereas apoptosis is achieved by counteracting Akt by p53.

The kinase activity of Akt is constitutively activated in human cancer as a result of dysregulation of its regulators, including loss of the tumor suppressor PTEN,[122] which is involved in the second loop. PTEN is a phosphatase that de-phosphorylates the Akt activating kinase, PI3-Kinase (reviewed in refs. 11,123,124). PTEN can inhibit Akt, thereby affecting the sub-cellular localization of Mdm2, PTEN can also downregulate Mdm2 and increase p53 stability.[125] And finally, PTEN is a transcriptional target of p53[126] generating another positive feedback loop. Thus, activation of p53, with the subsequent induction of effector target genes, counteracts the survival action of Akt at multiple levels, shifting the balance towards growth inhibition (Fig. 5).

Recently, it was shown that the PTEN–Akt pathway participates in checkpoint control in response to DNA damage. $PTEN^{-/-}$ cells have high Akt activity, and are defective in checkpoint control in response to DNA damage.[127] Importantly, activated Akt is able to phosphorylate Chk1 at Ser 280, thereby reducing Chk1 nuclear localization.[128] In addition, Chk1 phosphorylated at Ser 280 is located in the cytoplasm. Together, these observations indicate that loss of PTEN and subsequent activation of Akt can lead to Chk1 phosphorylation and reduced nuclear localization of Chk1, which in turn will compromise the DNA damage response.

Overall, different stress or mitogenic signals dictates the patterns of Mdm2 phosphorylation. The summation of these phosphorylation events determines the extent to which p53 exerts its biological activities or is being suppressed by Mdm2.

A Role for the Proline Rich Region of p53

The polyproline region of p53 (PPR; also referred to as the proline-rich-doman, PRD; residues 62-91) consists of 5 PXXP motifs, which are partially conserved in evolution. The regulatory role of the PPR in tumor suppression by p53 was initially noted by A. Levine and colleagues.[129] This followed a series of studies attributing various regulatory functions to the PPR. These include the induction of p53-mediated apoptosis, but notably not growth arrest,[130] reflecting altered specificity of p53 for apoptotic target genes versus growth arrest promoting genes.[131] Subsequent attempts to clarify the genes that are activated both in humans and mice have led to some contradictory findings.[132] Since there was no clear functional distinction among the affected target genes, it has been suggested that the impact of the PPR on the transcriptional specificity of p53 does not adequately explain the impaired apoptotic activity of p53ΔPro.[133] Furthermore, since the lack of the PPR has little effect on transcription in a gene reporter assay, it is possible that the PPR may affect gene expression at the chromatin level.[133] Interestingly, the PPR is required for p53-induced apoptosis in response to chemotherapeutic treatment, and this cell death is transcriptionally-independent.[134]

Searching for an additional explanation for the impaired activities of p53ΔPro, Berger et al,[135] demonstrated a link between the PPR of p53 and p53 regulation by Mdm2. p53ΔPro mutant was shown to be excessively sensitive to Mdm2-mediated ubiquitination and degradation, as well as to Mdm2-mediated inhibition of transcriptional and apoptotic activities.[135] This sensitivity of p53Δpro to inhibition by Mdm2 results from the enhanced affinity of p53 to Mdm2 relative to that of wt p53.[135] Further studies in mice confirmed the importance of the PPR of p53 for Mdm2 interactions and p53 stability.[132]

This suggests that the PPR may serve as an anchor for p53 stabilizing proteins. Indeed, a region within the PPR serves as a binding site for the corepressor mSin3A. This interaction is important for transcriptional repression by p53 and for the stabilization of p53.[136] However, this stabilization appears to be independent of Mdm2, suggesting an additional mechanism for the PPR in p53 degradation.[136]

An interesting insight into this story is based on the demonstrated role for Pin1 in the regulation of p53. Pin1 is a peptidylprolyl isomerase, which converts a cis-trans configuration of a peptide bond between proline and an adjacent residue. A special case is when the preceding residue serves as a phosphorylation site, such as serine or threonine. Interestingly, one of the residues that govern the binding of Pin1 to p53 resides within the PPR, proline 82, is affected by the phosphorylation of threonine 81 (in addition to two other residues) in response to UV light.[137,138] Importantly, in response to Pin1 isomerization p53 is relieved from Mdm2.[137] This may represent an additional mechanism for how the PPR of p53 influences the p53/Mdm2 loop.

Additional studies have emphasized the importance of the human p53 PPR for interactions with the ASPP family of p53 regulators, including the p53 inhibitor iASPP and its relatives ASPP, which are activators of p53. Importantly, these proteins preferentially bind to p53 codon-72 in the PPR and modulate p53 apoptotic function, through selective enhancement of pro-apoptotic gene-target activation (i.e., *PIG3*). Intriguingly, the evolutionarily conserved p53 inhibitor iASPP, binds more efficiently to the p53Pro72 polymorphism, which is a poorer activator of apoptosis than p53 bearing the Arg codon.[139]

Modulation by Protein-Protein Interactions

Both p53 and Mdm2 form complexes with various modulators, which can be classified into two large groups. Proteins from the first group enhance specific biochemical activities of p53. For instance, Ref1 increases the DNA binding and repair activities of p53 in the presence of selenomethionine.[140] In most cases, however, the mechanisms by which members of this group activate p53 are not clear. Members of the other groups confer protection for p53 from the inhibitory effects of Mdm2. It is likely that contribution from both groups is required for maximal activation of p53. For example, the mere stabilization of p53, by proteasomal degradation, is insufficient for p53 activation. Likewise, the activation of a labile p53 protein may not provide sufficient signal for triggering growth inhibition. In accordance with the topic of this chapter we shall focus here on the second group of proteins, using examples to illustrate the major mechanisms employed.

In principle, prevention of p53-Mdm2 interaction ought to be sufficient for the protection of p53 from Mdm2. The proof of principle was demonstrated by introducing to cells antibodies or peptides directed to the interaction site in p53, and observing a reduction at the levels and activities of the p53 protein (e.g., ref. 141). Surprisingly, this mechanism is employed by a minority of p53 regulators or co-factors. For instance, the TAF(II)31 transcriptional co-activator of p53 competes with Mdm2 for p53 binding.[142] Other proteins impair their physical separation by imposing spatial separation (see below). Perhaps the reason why this mechanism is not widely used is because it involves a considerable risk of unscheduled activation of p53 with severe consequences. In fact, the majority of p53 co-activators that protect p53 from Mdm2 do so without interrupting their physical interaction. These include pRb, ZBP-89, ARF, Werner's syndrome protein (WRN), β-catenin, and c-Abl (reviewed in ref. 6,143). The type of stress signals dictates, at least in part, which of the co-activators will come to action. DNA damage triggers proteins such as c-Abl to protect p53 from Mdm2, whereas deregulation of oncogenes, such as Myc and beta-catenin, trigger the ARF pathway (reviewed in ref. 12). The fact that Mdm2 masks the transactivation domain of p53 makes it difficult to explain how proteins activate p53 when bound to Mdm2. It is possible that the nature of the interaction between p53 and Mdm2 is altered in a manner that alleviates the restrictive effect of Mdm2 from p53 N-terminal domain. A more trivial explanation is that upon binding of the co-activator to p53, a small pool of p53 is relieved from Mdm2 and is sufficient to perform its biological role.

The ARF Oncogenic Pathway

An important regulatory loop of p53 involves the tumor suppressor product of the *CDKN2A* locus, ARF (alternative reading frame) (also called p19^{ARF} in mouse and p14^{ARF} in human), that is normally expressed at low levels in cells (reviewed in ref. 11). ARF serves as the prime nodal point integrating oncogenic signals into growth inhibition through the activation of p53.[144] This is the case when the expression of oncogenes such as c-Myc, Ras, or β-Catenin are deregulated.[145,146] Activation of ARF leads to growth arrest, senescence or even apoptosis under certain conditions. These effects are achieved by activation of p53 through several mechanisms. ARF promotes the nucleolar localization of Mdm2, thereby spatially separating Mdm2 from nucleoplasmic p53.[144] Also, a nucleolar-independent mechanisms has also been identified[147,148] in which ARF directly inhibits the E3 ubiquitin ligase activity of Mdm2. Moreover, in addition to suppressing Mdm2-mediated effects on p53, ARF modulates the activity of other E3 ligases such as ARF-BP1, where the ARF/ARF-BP1 interaction was found to be involved in both p53-dependent and p53-independent functions of ARF.[39] ARF also enhances p53 function by promoting the phosphorylation and inhibiting the transcriptional activity of the RelA NF-κB subunit. The NF-κB family of transcription factors display anti-apoptotic activity and antagonize the p53 pathway through induction of Mdm2 and repression of p53. Thus, by counteracting the functions of Rel A, ARF increases the effectiveness of the p53 pathway.[149]

Because ARF is also activated by mitogenic signals it is imperative that unscheduled activation of p53 under growth promoting conditions be avoided. One compensatory mechanism was demonstrated for the Ras-Raf growth promoting pathway. Activation of p53 through the Ras-Raf-ARF axis is counteracted by the parallel induction of Mdm2.[150] Failure of the p53-ARF axis is a common event in most human cancers. Generally, mutations in both genes within the same tumor are not common events.[4] However, at least in certain tissues or cell types the p53 and ARF pathways may function independently (reviewed in ref. 151). Mice lacking ARF are highly prone to tumor development,[152] underscoring the role of ARF in tumor suppression in mice. In humans, however, mutations at the CDK2A locus (which encodes INK4A, also known as p16, and ARF in overlapping reading frames) target mainly INK4A (p16) and rarely target ARF[153,154] suggesting that ARF may be less crucial to tumor suppression in humans. Two recent studies concluded that p53 did not have a tumor suppressor function in *ARF*-null mice.[155,156] Using a knock-in mouse that expresses a wild-type p53-oestrogen receptor fusion protein ($p53ER^{TAM}$) which is dependent upon 4-hydroxytamoxifen (4-OHT) for activity, Christophorou and colleagues[155] showed that the restoration of p53 function six days before administering a single whole-body dose of ionizing radiation led to widespread p53-dependent cell death in radiosensitive tissues in a manner similar to that observed in wild-type mice. However, although there was a substantial p53 response, it provided no protection against the subsequent onset of lymphoma development. By contrast, when p53 function was absent during irradiation but was restored for a six-day period eight days after administering the radiation, when precancrerous cells are presumbly present, a significant level of protection from tumor formation was observed. Notably, this acquired protection was lost when the mice were crossed onto an *ARF*-null background. A similar conclusion has been reached by Serrano's group,[156] who studied the role of ARF in tumor suppression in transgenic mice that expressed an additional copy of *Trp53* (known as $p53^{super}$ mice), which is known to provide added protection against the development of cancer.[157] They showed that wild-type and $p53^{super}$ mice, regardless of whether they are in an *ARF*-competent or *ARF*-null background, respond normally to DNA damage as measured by the number of apoptotic thymocytes detected following a high dose of ionizing radiation. However, although the $p53^{super}$ mice have extra protection against spontaneous and drug-induced tumor development, they are not protected in the absence of ARF. Moreover, when MEFs from these animals were used in a two-oncogene focus assay, focus formation was detectable only in the absence of either p53 or ARF, suggesting that ARF is required to suppress the transformed phenotype arising from oncogene expression.

The Spatial Distribution Mode of Regulation: The Nuclear Cytoplasmic Boundary

It has long been shown that p53 is a dynamic protein that is shuttled between the nucleus and the cytoplasm in a cell-cycle-dependent manner (reviewed in refs. 158,159). Deregulation of p53 by mutations or by elevation of inhibitory proteins such as Mdm2 and HPV-E6, can bias the shuttle towards the cytoplasm. Stress conditions, on the other hand, promote the nuclear accumulation of p53.

Many recent studies demonstrated that the nuclear versus cytoplasmic effects of p53 are determined by multiple post-translational modifications that affect its interaction with other proteins, its shuttling between the cytoplasm and the nucleus and its biological activities. Poly(ADP)ribosylation of p53 leads to its nuclear accumulation.[160] In contrast, monoubiquitylation by Mdm2 stimulates the nuclear export of p53, which on arrival at the mitochondria is deubiquitylated by mitochondrial HAUSP, thus generating the apoptotically active non-ubiquitylated p53.[52] Other post-translational modifications of p53 (such as phosphorylation and sumoylation of carboxy-terminal serines and lysines) can stimulate nuclear export and/or mitochondrial association. Moreover, the transcription factor Foxo3a (Foxo3) promotes p53 cytoplasmic accumulation by increasing its nuclear export.[161] This indicates that the entire context of post-transcriptional p53 modifications and protein interactions can affect the precise subcellular localization and function of p53.

Many studies have indicated that p53 plays active roles in the cytoplasm, such as direct signaling at the mitochondria and the induction of apoptosis.[52] Cytoplasmic p53 can localize to the mitochondria, and induce apoptosis via interactions with antiapooptotic members of the Bcl family such as Bcl-XL and Bcl2, resulting in the permeabilization of the outer mitochondrial membrane, the release of cytochrome c and other apoptotic activators from the mitochondria.[162-164] In addition, p53 can interact with the proapoptotic factor Bak, releasing it from the negative inhibition of the anti-apoptotic Bcl2-family member Mcl1.[165] The pro apoptotic effects of cytoplasmic p53 are not dependent on transcription, in principle. However, the control of transcription by nuclear p53 contributes to the function of cytoplasmic p53. For instance, the p53 target protein PUMA controls the sequestration of cytoplasmic p53 by the anti apoptotic Bcl-XL protein, releasing p53 to activate Bax.[166] Understanding the extra-nuclear activities of p53 will likewise furnish new opportunities to pharmacologically modulate the p53 system.

PML Nuclear Bodies as a Regulatory Junction

A dynamic redistribution of p53 within the nucleus may provide a means by which p53 regulation is coordinated in response to a given stress signal. This transportation of p53 within the nucleus is mediated by the promyelocytic leukemia protein (PML), which forms small structures termed PML nuclear bodies (PML-NB). These nuclear structures increase in numbers and size when cells are exposed to stress, such as γIR or deregulation of oncogene expression (reviewed in ref. 167). Several lines of evidence strongly link p53 with the PML-NBs. First, p53 is recruited to these structures in response to IR or UV light, and Ras activation.[168-171] Second, p53 interacts with PML (isoform IV), an interaction that is required for the activation of p53 by PML.[169,172] Third, PML is critical for the activation of p53 in response to stress, and for p53-dependent apoptosis.[172] Indeed, p53 transcriptional activity is impaired in PML null primary cells, and these mice are radio-resistant, even to lethal doses of γIR.[172] Furthermore, p53 activity is compromised in acute promyelocytic leukaemia, which explains the low frequency of p53 mutations in this form of cancer.[167] Fourth, *PML* itself is a p53 target gene,[173] suggesting a positive feedback loop between the two proteins.

Fifth, a growing list of p53 regulators has been demonstrated to be localized to the PML-NBs. In response to oncogenic Ras activation, p53 is co-localized along with the CBP acetyltransferase into the PML-NBs, inducing the formation of a trimeric p53-PML-CBP complex and the acetylation of p53 at Lys382, leading to p53-induced senescence.[171] Interestingly, the histone deacetylase SIRT1 can also localize to the PML-NBs and reduce the acetylation of p53 at Lys382 and thereby antagonize the induction of p53 activity.[174,175] In addition to CBP co-localization to the PML-NBs, it was recently shown that UV irradiation also induces the accumulation of another acetyl transferase into the PML-NBs, TIP60,[176] which induces p53-dependent apoptosis by acetylating p53 at Lys120.[177,178] Apart from regulating p53 acetylation, PML was shown to promote the Ser46 phosphorylation of p53 by HIPK2,[179] a modification that increases p53-mediated apoptosis in response to UV light exposure.[180,181] All

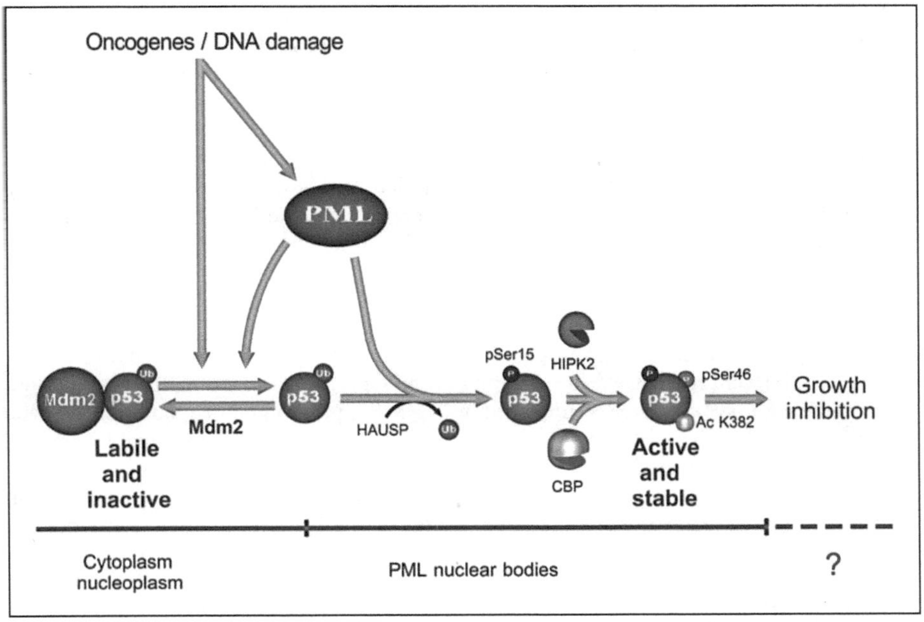

Figure 6. A model for the role of PML in the integrated regulation of p53 in the PML-NBs. In response to oncogenic activation or DNA damage PML is activated by SUMOylation (s) and recruits p53 into the PML-NBs. Consequently, p53 undergoes series of modifications, including acetylation (Ac) by CBP, phosphorylation of S46 by HIPK2 and de-ubiquitination by HAUSP. In addition, PML protects p53 from Mdm2 mediated degradation. These effects of PML lead to the activation and stabilization of p53. It is not clear yet whether these various modifications occur within the PML-NBs.

together, these findings raise the attractive hypothesis that the PML-NBs may serve not only as a meeting junction for p53 and its regulators in response to DNA damage, but that they could also block the physical interaction between p53 and Mdm2. Indeed, PML can promote the phosphorylation of p53 on Thr18 and Ser20, mediated by Chk2 and CK1, respectively.[182,183] These residues reside within the N-terminal transactivation domain of p53 and are known to be the most important in attenuating the p53-Mdm2 interaction.[76,184]

Interestingly, CK1 can directly phosphorylate Mdm2 at its acidic domain, a modification that can further weaken the p53-Mdm2 interaction.[185] Likewise, while recruited into the PML-NBs in response to stress, Chk2 can also phosphorylate PML itself and activate its apoptotic function.[186] In addition, it was recently published that PML can facilitate Chk2 autophosphorylation and activation,[187] revealing even more complex interaction between PML and Chk2.

Finally, PML can also regulate Mdm2. It was shown that PML can bind Mdm2,[188,189] protecting p53 from Mdm2 mediated degradation through the physical inhibition of their interaction by forming a trimeric PML-p53-Mdm2 complex,[190] or by sequestrating Mdm2 to the nucleoli upon DNA damage.[191] PML can also affect Mdm2 expression indirectly, for example, by inhibiting the eIF4E dependent mRNA export of Mdm2, leading to reduced levels of Mdm2 protein.[192] Interestingly, p53 overexpression results in transcriptional repression of eIF4E.[193] Consistently, Mdm2 overexpression, leads to reduced p53 and increased eIF4E levels.[193] This might suggest the existence of a feedback loop between eIF4E, Mdm2, p53 and PML. Another example involves the transcriptional repressor Daxx, which also associates with the PML-NBs, which may also play a role in the PML-NBs/p53- induced apoptosis.[194] Recently, it was shown that Daxx association with Axin can increase the phosphorylation of p53 at Ser46, mediated by HIPK2.[195] In addition, under unstressed conditions, Daxx can form a ternary complex with Mdm2 and the deubiquitinating enzyme HAUSP, enhancing Mdm2 stability. Upon DNA damage, this complex is dissociated, resulting in the degradation of Mdm2, which in turn stabilizes p53.[196,197]

Thus, several competing p53 regulation mechanisms converge at PML-NBs and it is becoming clear that much of this action involves PML directly and is also attributed to the recruitment of key p53 modulators. It is likely that many other modulators are involved and additional PML-dependent modifications of p53 are present.The identification of this link between p53 and the PML-NBs may shed new light on how the complex network of p53 regulation is coordinated in a timely manner within the nucleus (Fig. 6).

Acknowledgement

Due to space limitations many original important studies have not been cited directly but rather through recent reviews. We are grateful to Mati Goldberg for drawing the illustrations. Work in the author's laboratory is supported by NHMRC project grants (grant No. 509196, 509197), by the VESKI award, by the Prostate Cancer Foundation of Australia, by Cancer Council Victoria, and by the EC FP6 funding of the European Commission (contract 503576). This publication reflects only the author's views. The European Commission is not liable for any use that may be made of the information herein.

References

1. Zuckerman V, Wolyniec K, Sionov RV et al. Tumor suppression by p53: the importance of apoptosis and cellular senescence. J Pathol 2009; 219(1):3-15.
2. Sionov RV, Haupt Y. The cellular response to p53: the decision between life and death. Oncogene 1999; 18(45):6145-57.
3. Wang J, Yang J. Interaction of tumor suppressor p53 with DNA and proteins. Curr Pharm Biotechnol 2009; 11(1):122-7.
4. Vogelstein B, Lane D, Levine AJ. Surfing the p53 network. Nature 2000; 408(6810):307-10.
5. Appella E, Anderson CW. Signaling to p53: breaking the posttranslational modification code. Pathol Biol (Paris) 2000; 48(3):227-45.
6. Woods DB, Vousden KH. Regulation of p53 function. Exp Cell Res 2001; 264(1):56-66.
7. Barak Y, Juven T, Haffner R, Oren M. mdm2 expression is induced by wild type p53 activity. EMBO J 1993; 12(2):461-8.
8. Perry ME, Levine AJ. Tumor-suppressor p53 and the cell cycle. Curr Opin Genet Dev 1993; 3(1):50-4.
9. Momand J, Wu HH, Dasgupta G. MDM2—master regulator of the p53 tumor suppressor protein. Gene 2000; 242(1-2):15-29.
10. Bond GL, Hu W, Levine AJ. MDM2 is a central node in the p53 pathway: 12 years and counting. Curr Cancer Drug Targets 2005; 5(1):3-8.
11. Harris SL, Levine AJ. The p53 pathway: positive and negative feedback loops. Oncogene 2005; 24(17):2899-908.
12. Horn HF, Vousden KH. Coping with stress: multiple ways to activate p53. Oncogene 2007; 26(9):1306-16.
13. Lavin MF, Gueven N. The complexity of p53 stabilization and activation. Cell Death Differ 2006; 13(6):941-50.
14. Haupt Y. p53 Regulation: a family affair. Cell Cycle 2004; 3(7):884-5.
15. Haupt Y, Maya R, Kazaz A, Oren M. Mdm2 promotes the rapid degradation of p53. Nature 1997; 387(6630):296-9.
16. Kubbutat MH, Jones SN, Vousden KH. Regulation of p53 stability by Mdm2. Nature 1997; 387(6630):299-303.
17. Marine JC, Lozano G. Mdm2-mediated ubiquitylation: p53 and beyond. Cell Death Differ 2010; 17(1):93-102.
18. Toledo F, Wahl GM. Regulating the p53 pathway: in vitro hypotheses, in vivo veritas. Nat Rev Cancer 2006; 6(12):909-23.
19. Bond GL, Hu W, Bond EE et al. A single nucleotide polymorphism in the MDM2 promoter attenuates the p53 tumor suppressor pathway and accelerates tumor formation in humans. Cell 2004; 119(5):591-602.
20. Bond GL, Hu W, Levine A. A single nucleotide polymorphism in the MDM2 gene: from a molecular and cellular explanation to clinical effect. Cancer Res 2005; 65(13):5481-4.

21. Levine AJ, Hu W, Feng Z. The P53 pathway: what questions remain to be explored? Cell Death Differ 2006; 13(6):1027-36.
22. Mendrysa SM, McElwee MK, Michalowski J et al. mdm2 Is critical for inhibition of p53 during lymphopoiesis and the response to ionizing irradiation. Mol Cell Biol 2003; 23(2):462-72.
23. Juven-Gershon T, Oren M. Mdm2: the ups and downs. Mol Med 1999; 5(2):71-83.
24. Jones SN, Roe AE, Donehower LA, Bradley A. Rescue of embryonic lethality in Mdm2-deficient mice by absence of p53. Nature 1995; 378(6553):206-8.
25. Léveillard T, Gorry P, Niederreither K, Wasylyk B. MDM2 expression during mouse embryogenesis and the requirement of p53. Mech Dev 1998; 74(1-2):189-93.
26. Parant J, Chavez-Reyes A, Little NA et al. Rescue of embryonic lethality in Mdm4-null mice by loss of Trp53 suggests a nonoverlapping pathway with MDM2 to regulate p53. Nat Genet 2001; 29(1):92-5.
27. Finch RA, Donoviel DB, Potter D et al. mdmx is a negative regulator of p53 activity in vivo. Cancer Res 2002; 62(11):3221-5.
28. Migliorini D, Lazzerini Denchi E et al. Mdm4 (Mdmx) regulates p53-induced growth arrest and neuronal cell death during early embryonic mouse development. Mol Cell Biol 2002; 22(15):5527-38.
29. Marine JC, Jochemsen AG. Mdmx and Mdm2: brothers in arms? Cell Cycle 2004; 3(7):900-4.
30. Laurie NA, Donovan SL, Shih CS et al. Inactivation of the p53 pathway in retinoblastoma. Nature 2006; 444(7115):61-6.
31. Marine JC, Jochemsen AG. Mdmx as an essential regulator of p53 activity. Biochem Biophys Res Commun 2005; 331(3):750-60.
32. Marine JC, Dyer MA, Jochemsen AG. MDMX: from bench to bedside. J Cell Sci 2007; 120(Pt 3):371-8.
33. Wade M, Wahl GM. Targeting Mdm2 and Mdmx in cancer therapy: better living through medicinal chemistry? Mol Cancer Res 2009; 7(1):1-11.
34. Honda R, Yasuda H. Activity of MDM2, a ubiquitin ligase, toward p53 or itself is dependent on the RING finger domain of the ligase. Oncogene 2000; 19(11):1473-6.
35. Fang S, Jensen JP, Ludwig RL et al. Mdm2 is a RING finger-dependent ubiquitin protein ligase for itself and p53. J Biol Chem 2000; 275(12):8945-51.
36. Ringshausen I, O'Shea CC, Finch AJ et al. Mdm2 is critically and continuously required to suppress lethal p53 activity in vivo. Cancer Cell 2006; 10(6):501-14.
37. Leng RP, Lin Y, Ma W et al. Pirh2, a p53-induced ubiquitin-protein ligase, promotes p53 degradation. Cell 2003; 112(6):779-91.
38. Dornan D, Wertz I, Shimizu H et al. The ubiquitin ligase COP1 is a critical negative regulator of p53. Nature 2004; 429(6987):86-92.
39. Chen D, Kon N, Li M et al. ARF-BP1/Mule is a critical mediator of the ARF tumor suppressor. Cell 2005; 121(7):1071-83.
40. Itahana K, Mao H, Jin A et al. Targeted inactivation of Mdm2 RING finger E3 ubiquitin ligase activity in the mouse reveals mechanistic insights into p53 regulation. Cancer Cell 2007; 12(4):355-66.
41. Kobet E, Zeng X, Zhu Y et al. MDM2 inhibits p300-mediated p53 acetylation and activation by forming a ternary complex with the two proteins. Proc Natl Acad Sci USA 2000; 97(23):12547-52.
42. Ito A, Lai CH, Zhao X et al. p300/CBP-mediated p53 acetylation is commonly induced by p53-activating agents and inhibited by MDM2. EMBO J 2001; 20(6):1331-40.
43. Jin Y, Zeng SX, Lee H, Lu H. MDM2 mediates p300/CREB-binding protein-associated factor ubiquitination and degradation. J Biol Chem 2004; 279(19):20035-43.
44. Ito A, Kawaguchi Y, Lai CH et al. MDM2-HDAC1-mediated deacetylation of p53 is required for its degradation. EMBO J 2002; 21(22):6236-45.
45. Wang C, Ivanov A, Chen L et al. MDM2 interaction with nuclear corepressor KAP1 contributes to p53 inactivation. EMBO J 2005; 24(18):3279-90.
46. Xirodimas DP, Saville MK, Bourdon JC et al. Mdm2-mediated NEDD8 conjugation of p53 inhibits its transcriptional activity. Cell 2004; 118(1):83-97.
47. Minsky N, Oren M. The RING domain of Mdm2 mediates histone ubiquitylation and transcriptional repression. Mol Cell 2004; 16(4):631-9.
48. Lai Z, Ferry KV, Diamond MA et al. Human mdm2 mediates multiple mono-ubiquitination of p53 by a mechanism requiring enzyme isomerization. J Biol Chem 2001; 276(33):31357-67.
49. Carter S, Bischof O, Dejean A, Vousden KH. C-terminal modifications regulate MDM2 dissociation and nuclear export of p53. Nat Cell Biol 2007; 9(4):428-35.
50. Li M, Brooks CL, Wu-Baer F et al. Mono- versus polyubiquitination: differential control of p53 fate by Mdm2. Science 2003; 302(5652):1972-5.
51. Joseph TW, Zaika A, Moll UM. Nuclear and cytoplasmic degradation of endogenous p53 and HDM2 occurs during down-regulation of the p53 response after multiple types of DNA damage. FASEB J 2003; 17(12):1622-30.

52. Marchenko ND, Wolff S, Erster S et al. Monoubiquitylation promotes mitochondrial p53 translocation. EMBO J 2007; 26(4):923-34.
53. Lohrum MA, Woods DB, Ludwig RL et al. C-terminal ubiquitination of p53 contributes to nuclear export. Mol Cell Biol 2001; 21(24):8521-32.
54. Feng L, Lin T, Uranishi H, Gu W, Xu Y. Functional analysis of the roles of posttranslational modifications at the p53 C terminus in regulating p53 stability and activity. Mol Cell Biol 2005; 25(13):5389-95.
55. Krummel KA, Lee CJ, Toledo F, Wahl GM. The C-terminal lysines fine-tune P53 stress responses in a mouse model but are not required for stability control or transactivation. Proc Natl Acad Sci USA 2005; 102(29):10188-93.
56. Chan WM, Mak MC, Fung TK et al. Ubiquitination of p53 at multiple sites in the DNA-binding domain. Mol Cancer Res 2006; 4(1):15-25.
57. Yin Y, Stephen CW, Luciani MG, Fåhraeus R. p53 Stability and activity is regulated by Mdm2-mediated induction of alternative p53 translation products. Nat Cell Biol 2002; 4(6):462-7.
58. Inoue T, Geyer RK, Howard D et al. MDM2 can promote the ubiquitination, nuclear export, and degradation of p53 in the absence of direct binding. J Biol Chem 2001; 276(48):45255-60.
59. Yap DB, Hsieh JK, Lu X. Mdm2 inhibits the apoptotic function of p53 mainly by targeting it for degradation. J Biol Chem 2000; 275(47):37296-302.
60. Unger T, Juven-Gershon T, Moallem E et al. Critical role for Ser20 of human p53 in the negative regulation of p53 by Mdm2. EMBO J 1999; 18(7):1805-14.
61. Marine JC, Francoz S, Maetens M et al. Keeping p53 in check: essential and synergistic functions of Mdm2 and Mdm4. Cell Death Differ 2006; 13(6):927-34.
62. Clegg HV, Itahana K, Zhang Y. Unlocking the Mdm2-p53 loop: ubiquitin is the key. Cell Cycle 2008; 7(3):287-92.
63. Wade M, Wang YV, Wahl GM. The p53 orchestra: Mdm2 and Mdmx set the tone. Trends Cell Biol 2010; [Epub ahead of print].
64. Stommel JM, Wahl GM. A new twist in the feedback loop: stress-activated MDM2 destabilization is required for p53 activation. Cell Cycle 2005; 4(3):411-7.
65. Appella E, Anderson CW. Post-translational modifications and activation of p53 by genotoxic stresses. Eur J Biochem 2001; 268(10):2764-72.
66. Bode AM, Dong Z. Post-translational modification of p53 in tumorigenesis. Nat Rev Cancer 2004; 4(10):793-805.
67. Kruse JP, Gu W. SnapShot: p53 posttranslational modifications. Cell 2008; 133(5):930-30.e1.
68. Olsson A, Manzl C, Strasser A, Villunger A. How important are post-translational modifications in p53 for selectivity in target-gene transcription and tumor suppression? Cell Death Differ 2007; 14(9):1561-75.
69. Bean LJ, Stark GR. Regulation of the accumulation and function of p53 by phosphorylation of two residues within the domain that binds to Mdm2. J Biol Chem 2002; 277(3):1864-71.
70. Sakaguchi K, Saito S, Higashimoto Y et al. Damage-mediated phosphorylation of human p53 threonine 18 through a cascade mediated by a casein 1-like kinase. Effect on Mdm2 binding. J Biol Chem 2000; 275(13):9278-83.
71. Banin S, Moyal L, Shieh S et al. Enhanced phosphorylation of p53 by ATM in response to DNA damage. Science 1998; 281(5383):1674-7.
72. Canman CE, Lim DS. The role of ATM in DNA damage responses and cancer. Oncogene 1998; 17(25):3301-8.
73. Khanna KK, Keating KE, Kozlov S et al. ATM associates with and phosphorylates p53: mapping the region of interaction. Nat Genet 1998; 20(4):398-400.
74. Lakin ND, Hann BC, Jackson SP. The ataxia-telangiectasia related protein ATR mediates DNA-dependent phosphorylation of p53. Oncogene 1999; 18(27):3989-95.
75. Berger M, Stahl N, Del Sal G, Haupt Y. Mutations in proline 82 of p53 impair its activation by Pin1 and Chk2 in response to DNA damage. Mol Cell Biol 2005; 25(13):5380-8.
76. Bartek J, Falck J, Lukas J. CHK2 kinase—a busy messenger. Nat Rev Mol Cell Biol 2001; 2(12):877-86.
77. Meulmeester E, Pereg Y, Shiloh Y, Jochemsen AG. ATM-mediated phosphorylations inhibit Mdmx/Mdm2 stabilization by HAUSP in favor of p53 activation. Cell Cycle 2005; 4(9):1166-70.
78. Stommel JM, Wahl GM. Accelerated MDM2 auto-degradation induced by DNA-damage kinases is required for p53 activation. EMBO J 2004; 23(7):1547-56.
79. Dornan D, Hupp TR. Inhibition of p53-dependent transcription by BOX-I phospho-peptide mimetics that bind to p300. EMBO J Rep 2001; 2(2):139-44.
80. Dumaz N, Meek DW. Serine15 phosphorylation stimulates p53 transactivation but does not directly influence interaction with HDM2. EMBO J 1999; 18(24):7002-10.

81. Feng H, Jenkins LM, Durell SR et al. Structural basis for p300 Taz2-p53 TAD1 binding and modulation by phosphorylation. Structure 2009; 17(2):202-10.
82. Jenkins LM, Yamaguchi H, Hayashi R et al. Two distinct motifs within the p53 transactivation domain bind to the Taz2 domain of p300 and are differentially affected by phosphorylation. Biochemistry 2009; 48(6):1244-55.
83. Lambert PF, Kashanchi F, Radonovich MF et al. Phosphorylation of p53 serine 15 increases interaction with CBP. J Biol Chem 1998; 273(49):33048-53.
84. Lee CW, Arai M, Martinez-Yamout MA et al. Mapping the interactions of the p53 transactivation domain with the KIX domain of CBP. Biochemistry 2009; 48(10):2115-24.
85. Teufel DP, Bycroft M, Fersht AR. Regulation by phosphorylation of the relative affinities of the N-terminal transactivation domains of p53 for p300 domains and Mdm2. Oncogene 2009; 28(20):2112-8.
86. Sakaguchi K, Herrera JE, Saito S et al. DNA damage activates p53 through a phosphorylation-acetylation cascade. Genes Dev 1998; 12(18):2831-41.
87. MacPherson D, Kim J, Kim T et al. Defective apoptosis and B-cell lymphomas in mice with p53 point mutation at Ser 23. EMBO J 2004; 23(18):3689-99.
88. Wu Z, Earle J, Saito S et al. Mutation of mouse p53 Ser23 and the response to DNA damage. Mol Cell Biol 2002; 22(8):2441-9.
89. Takai H, Naka K, Okada Y et al. Chk2-deficient mice exhibit radioresistance and defective p53-mediated transcription. EMBO J 2002; 21(19):5195-205.
90. Vousden KH, Prives C. Blinded by the Light: The Growing Complexity of p53. Cell 2009; 137(3):413-31.
91. Falck J, Lukas C, Protopopova M et al. Functional impact of concomitant versus alternative defects in the Chk2-p53 tumor suppressor pathway. Oncogene 2001; 20(39):5503-10.
92. Chao C, Hergenhahn M, Kaeser MD et al. Cell type- and promoter-specific roles of Ser18 phosphorylation in regulating p53 responses. J Biol Chem 2003; 278(42):41028-33.
93. Sluss HK, Armata H, Gallant J, Jones SN. Phosphorylation of serine 18 regulates distinct p53 functions in mice. Mol Cell Biol 2004; 24(3):976-84.
94. Chao C, Herr D, Chun J, Xu Y. Ser18 and 23 phosphorylation is required for p53-dependent apoptosis and tumor suppression. EMBO J 2006; 25(11):2615-22.
95. Hay TJ, Meek DW. Multiple sites of in vivo phosphorylation in the MDM2 oncoprotein cluster within two important functional domains. FEBS Lett 2000; 478(1-2):183-6.
96. Blattner C, Hay T, Meek DW, Lane DP. Hypophosphorylation of Mdm2 augments p53 stability. Mol Cell Biol 2002; 22(17):6170-82.
97. Meek DW, Knippschild U. Posttranslational modification of MDM2. Mol Cancer Res 2003; 1(14):1017-26.
98. Meek DW, Hupp TR. The regulation of MDM2 by multisite phosphorylation-Opportunities for molecular-based intervention to target tumors? Semin Cancer Biol 2009; [Epub ahead of print].
99. Mayo LD, Turchi JJ, Berberich SJ. Mdm-2 phosphorylation by DNA-dependent protein kinase prevents interaction with p53. Cancer Res 1997; 57(22):5013-6.
100. Zhang Y, Xiong Y. A p53 amino-terminal nuclear export signal inhibited by DNA damage-induced phosphorylation. Science 2001; 292(5523):1910-5.
101. Sionov RV, Coen S, Goldberg Z et al. c-Abl regulates p53 levels under normal and stress conditions by preventing its nuclear export and ubiquitination. Mol Cell Biol 2001; 21(17):5869-78.
102. Levav-Cohen Y, Goldberg Z, Zuckerman V et al. C-Abl as a modulator of p53. Biochem Biophys Res Commun 2005; 331(3):737-49.
103. Goldberg Z, Vogt Sionov R, Berger M et al. Tyrosine phosphorylation of Mdm2 by c-Abl: implications for p53 regulation. EMBO J 2002; 21(14):3715-27.
104. Dias SS, Milne DM, Meek DW. c-Abl phosphorylates Hdm2 at tyrosine 276 in response to DNA damage and regulates interaction with ARF. Oncogene 2006; 25(50):6666-71.
105. Khosravi R, Maya R, Gottlieb T et al. Rapid ATM-dependent phosphorylation of MDM2 precedes p53 accumulation in response to DNA damage. Proc Natl Acad Sci USA 1999; 96(26):14973-7.
106. Maya R, Balass M, Kim ST et al. ATM-dependent phosphorylation of Mdm2 on serine 395: role in p53 activation by DNA damage. Genes Dev 2001; 15(9):1067-77.
107. Kharbanda S, Yuan ZM, Weichselbaum R, Kufe D. Determination of cell fate by c-Abl activation in the response to DNA damage. Oncogene 1998; 17(25):3309-18.
108. Zhu J, Wang JY. Death by Abl: a matter of location. Curr Top Dev Biol 2004; 59:165-92.
109. Cheng Q, Chen L, Li Z et al. ATM activates p53 by regulating MDM2 oligomerization and E3 processivity. EMBO J 2009; 28(24):3857-67.
110. Cheng Q, Chen J. Mechanism of p53 stabilization by ATM after DNA damage. Cell Cycle 2010; 9(3):472-8.

111. Mayo LD, Donner DB. A phosphatidylinositol 3-kinase/Akt pathway promotes translocation of Mdm2 from the cytoplasm to the nucleus. Proc Natl Acad Sci USA 2001; 98(20):11598-603.
112. Mayo LD, Dixon JE, Durden DL et al. PTEN protects p53 from Mdm2 and sensitizes cancer cells to chemotherapy. J Biol Chem 2002; 277(7):5484-9.
113. Ogawara Y, Kishishita S, Obata T et al. Akt enhances Mdm2-mediated ubiquitination and degradation of p53. J Biol Chem 2002; 277(24):21843-50.
114. Zhou BB, Elledge SJ. The DNA damage response: putting checkpoints in perspective. Nature 2000; 408(6811):433-9.
115. Feng J, Tamaskovic R, Yang Z et al. Stabilization of Mdm2 via decreased ubiquitination is mediated by protein kinase B/Akt-dependent phosphorylation. J Biol Chem 2004; 279(34):35510-7.
116. Testa JR, Bellacosa A. AKT plays a central role in tumorigenesis. Proc Natl Acad Sci USA 2001; 98(20):10983-5.
117. Levav-Cohen Y, Haupt S, Haupt Y. Mdm2 in growth signaling and cancer. Growth Factors 2005; 23(3):183-92.
118. Gottlieb TM, Leal JF, Seger R et al. Cross-talk between Akt, p53 and Mdm2: possible implications for the regulation of apoptosis. Oncogene 2002; 21(8):1299-303.
119. Oren M, Damalas A, Gottlieb T et al. Regulation of p53: intricate loops and delicate balances. Ann NY Acad Sci 2002; 973:374-83.
120. Okamoto K, Li H, Jensen MR et al. Cyclin G recruits PP2A to dephosphorylate Mdm2. Mol Cell 2002; 9(4):761-71.
121. Kimura SH, Ikawa M, Ito A et al. Cyclin G1 is involved in G2/M arrest in response to DNA damage and in growth control after damage recovery. Oncogene 2001; 20(25):3290-300.
122. Wu X, Senechal K, Neshat MS et al. The PTEN/MMAC1 tumor suppressor phosphatase functions as a negative regulator of the phosphoinositide 3-kinase/Akt pathway. Proc Natl Acad Sci USA 1998; 95(26):15587-91.
123. Mayo LD, Donner DB. The PTEN, Mdm2, p53 tumor suppressor-oncoprotein network. Trends Biochem Sci 2002; 27(9):462-7.
124. Lee MH, Lozano G. Regulation of the p53-MDM2 pathway by 14-3-3 sigma and other proteins. Semin Cancer Biol 2006; 16(3):225-34.
125. Freeman DJ, Li AG, Wei G et al. PTEN tumor suppressor regulates p53 protein levels and activity through phosphatase-dependent and -independent mechanisms. Cancer Cell 2003; 3(2):117-30.
126. Stambolic V, MacPherson D, Sas D et al. Regulation of PTEN transcription by p53. Mol Cell 2001; 8(2):317-25.
127. Puc J, Keniry M, Li HS et al. Lack of PTEN sequesters CHK1 and initiates genetic instability. Cancer Cell 2005; 7(2):193-204.
128. Puc J, Parsons R. PTEN loss inhibits CHK1 to cause double stranded-DNA breaks in cells. Cell Cycle 2005; 4(7):927-9.
129. Walker KK, Levine AJ. Identification of a novel p53 functional domain that is necessary for efficient growth suppression. Proc Natl Acad Sci USA 1996; 93(26):15335-40.
130. Sakamuro D, Sabbatini P, White E, Prendergast GC. The polyproline region of p53 is required to activate apoptosis but not growth arrest. Oncogene 1997; 15(8):887-98.
131. Venot C, Maratrat M, Dureuil C et al. The requirement for the p53 proline-rich functional domain for mediation of apoptosis is correlated with specific PIG3 gene transactivation and with transcriptional repression. EMBO J 1998; 17(16):4668-79.
132. Toledo F, Krummel KA, Lee CJ et al. A mouse p53 mutant lacking the proline-rich domain rescues Mdm4 deficiency and provides insight into the Mdm2-Mdm4-p53 regulatory network. Cancer Cell 2006; 9(4):273-85.
133. Zhu J, Jiang J, Zhou W et al. Differential regulation of cellular target genes by p53 devoid of the PXXP motifs with impaired apoptotic activity. Oncogene 1999; 18(12):2149-55.
134. Baptiste N, Friedlander P, Chen X, Prives C. The proline-rich domain of p53 is required for cooperation with anti-neoplastic agents to promote apoptosis of tumor cells. Oncogene 2002; 21(1):9-21.
135. Berger M, Vogt Sionov R et al. A role for the polyproline domain of p53 in its regulation by Mdm2. J Biol Chem 2001; 276(6):3785-90.
136. Zilfou JT, Hoffman WH, Sank M et al. The corepressor mSin3a interacts with the proline-rich domain of p53 and protects p53 from proteasome-mediated degradation. Mol Cell Biol 2001; 21(12):3974-85.
137. Zacchi P, Gostissa M, Uchida T et al. The prolyl isomerase Pin1 reveals a mechanism to control p53 functions after genotoxic insults. Nature 2002; 419(6909):853-7.
138. Zheng H, You H, Zhou XZ et al. The prolyl isomerase Pin1 is a regulator of p53 in genotoxic response. Nature 2002; 419(6909):849-53.

139. Bergamaschi D, Samuels Y, Sullivan A et al. iASPP preferentially binds p53 proline-rich region and modulates apoptotic function of codon 72-polymorphic p53. Nat Genet 2006; 38(10):1133-41.
140. Seo YR, Kelley MR, Smith ML. Selenomethionine regulation of p53 by a ref1-dependent redox mechanism. Proc Natl Acad Sci USA 2002; 99(22):14548-53.
141. Böttger A, Böttger V, Sparks A et al. Design of a synthetic Mdm2-binding mini protein that activates the p53 response in vivo. Curr Biol 1997; 7(11):860-9.
142. Buschmann T, Lin Y, Aithmitti N et al. Stabilization and activation of p53 by the coactivator protein TAFII31. J Biol Chem 2001; 276(17):13852-7.
143. Bai L, Merchant JL. ZBP-89 promotes growth arrest through stabilization of p53. Mol Cell Biol 2001; 21(14):4670-83.
144. Sherr CJ. Parsing Ink4a/Arf: "pure" p16-null mice. Cell 2001; 106(5):531-4.
145. Sherr CJ, Weber JD. The ARF/p53 pathway. Curr Opin Genet Dev 2000; 10(1):94-9.
146. Damalas A, Kahan S, Shtutman M et al. Deregulated beta-catenin induces a p53- and ARF-dependent growth arrest and cooperates with Ras in transformation. EMBO J 2001; 20(17):4912-22.
147. Llanos S, Clark PA, Rowe J, Peters G. Stabilization of p53 by p14ARF without relocation of MDM2 to the nucleolus. Nat Cell Biol 2001; 3(5):445-52.
148. Honda R, Yasuda H. Association of p19(ARF) with Mdm2 inhibits ubiquitin ligase activity of Mdm2 for tumor suppressor p53. EMBO J 1999; 18(1):22-7.
149. Rocha S, Campbell KJ, Perkins ND. p53- and Mdm2-independent repression of NF-kappa B transactivation by the ARF tumor suppressor. Mol Cell 2003; 12(1):15-25.
150. Cell 2000; 103(2):321 et al. Opposing effects of Ras on p53: transcriptional activation of mdm2 and induction of p19ARF. Cell 2000; 103(2):321-30.
151. Hayon IL, Haupt Y. p53: an internal investigation. Cell Cycle 2002; 1(2):111-6.
152. Kamijo T, Zindy F, Roussel MF et al. Tumor suppression at the mouse INK4a locus mediated by the alternative reading frame product p19ARF. Cell 1997; 91(5):649-59.
153. Collins CJ, Sedivy JM. Involvement of the INK4a/Arf gene locus in senescence. Aging Cell 2003; 2(3):145-50.
154. Ruas M, Peters G. The p16INK4a/CDKN2A tumor suppressor and its relatives. Biochim Biophys Acta 1998; 1378(2):F115-77.
155. Christophorou MA, Ringshausen I, Finch AJ et al. The pathological response to DNA damage does not contribute to p53-mediated tumor suppression. Nature 2006; 443(7108):214-7.
156. Efeyan A, Garcia-Cao I, Herranz D et al. Tumor biology: Policing of oncogene activity by p53. Nature 2006; 443(7108):159.
157. Khan SH, Moritsugu J, Wahl GM. Differential requirement for p19ARF in the p53-dependent arrest induced by DNA damage, microtubule disruption, and ribonucleotide depletion. Proc Natl Acad Sci USA 2000; 97(7):3266-71.
158. Jimenez GS, Khan SH, Stommel JM, Wahl GM. p53 regulation by post-translational modification and nuclear retention in response to diverse stresses. Oncogene 1999; 18(53):7656-65.
159. Liang SH, Clarke MF. Regulation of p53 localization. Eur J Biochem 2001; 268(10):2779-83.
160. Murray-Zmijewski F, Slee EA, Lu X. A complex barcode underlies the heterogeneous response of p53 to stress. Nat Rev Mol Cell Biol 2008; 9(9):702-12.
161. You H, Yamamoto K, Mak TW. Regulation of transactivation-independent proapoptotic activity of p53 by FOXO3a. Proc Natl Acad Sci USA 2006; 103(24):9051-6.
162. Mihara M, Erster S, Zaika A et al. p53 has a direct apoptogenic role at the mitochondria. Mol Cell 2003; 11(3):577-90.
163. Chipuk JE, Kuwana T, Bouchier-Hayes L et al. Direct activation of Bax by p53 mediates mitochondrial membrane permeabilization and apoptosis. Science 2004; 303(5660):1010-4.
164. Tomita Y, Marchenko N, Erster S et al. WT p53, but not tumor-derived mutants, bind to Bcl2 via the DNA binding domain and induce mitochondrial permeabilization. J Biol Chem 2006; 281(13):8600-6.
165. Leu JI, Dumont P, Hafey M et al. Mitochondrial p53 activates Bak and causes disruption of a Bak-Mcl1 complex. Nat Cell Biol 2004; 6(5):443-50.
166. Chipuk JE, Bouchier-Hayes L, Kuwana T et al. PUMA couples the nuclear and cytoplasmic proapoptotic function of p53. Science 2005; 309(5741):1732-5.
167. Pearson M, Pelicci PG. PML interaction with p53 and its role in apoptosis and replicative senescence. Oncogene 2001; 20(49):7250-6.
168. Ferbeyre G, de Stanchina E, Querido E et al. PML is induced by oncogenic ras and promotes premature senescence. Genes Dev 2000; 14(16):2015-27.
169. Fogal V, Gostissa M, Sandy P et al. Regulation of p53 activity in nuclear bodies by a specific PML isoform. EMBO J 2000; 19(22):6185-95.

170. Carbone R, Pearson M, Minucci S, Pelicci PG. PML NBs associate with the hMre11 complex and p53 at sites of irradiation induced DNA damage. Oncogene 2002; 21(11):1633-40.
171. Pearson M, Carbone R, Sebastiani C et al. PML regulates p53 acetylation and premature senescence induced by oncogenic Ras. Nature 2000; 406(6792):207-10.
172. Guo A, Salomoni P, Luo J et al. The function of PML in p53-dependent apoptosis. Nat Cell Biol 2000; 2(10):730-6.
173. de Stanchina E, Querido E, Narita M et al. PML is a direct p53 target that modulates p53 effector functions. Mol Cell 2004; 13(4):523-35.
174. Langley E, Pearson M, Faretta M et al. Human SIR2 deacetylates p53 and antagonizes PML/p53-induced cellular senescence. EMBO J 2002; 21(10):2383-96.
175. Vaziri H, Dessain SK, Ng Eaton E et al. hSIR2(SIRT1) functions as an NAD-dependent p53 deacetylase. Cell 2001; 107(2):149-59.
176. Cheng Z, Ke Y, Ding X et al. Functional characterization of TIP60 sumoylation in UV-irradiated DNA damage response. Oncogene 2008; 27(7):931-41.
177. Sykes SM, Mellert HS, Holbert MA et al. Acetylation of the p53 DNA-binding domain regulates apoptosis induction. Mol Cell 2006; 24(6):841-51.
178. Tang Y, Luo J, Zhang W, Gu W et al. Tip60-dependent acetylation of p53 modulates the decision between cell-cycle arrest and apoptosis. Mol Cell 2006; 24(6):827-39.
179. Möller A, Sirma H, Hofmann TG et al. PML is required for homeodomain-interacting protein kinase 2 (HIPK2)-mediated p53 phosphorylation and cell cycle arrest but is dispensable for the formation of HIPK domains. Cancer Res 2003; 63(15):4310-4.
180. D'Orazi G, Cecchinelli B, Bruno T et al. Homeodomain-interacting protein kinase-2 phosphorylates p53 at Ser 46 and mediates apoptosis. Nat Cell Biol 2002; 4(1):11-9.
181. Hofmann TG, Möller A, Sirma H et al. Regulation of p53 activity by its interaction with homeodomain-interacting protein kinase-2. Nat Cell Biol 2002; 4(1):1-10.
182. Louria-Hayon I, Grossman T, Sionov RV et al. The promyelocytic leukemia protein protects p53 from Mdm2-mediated inhibition and degradation. J Biol Chem 2003; 278(35):33134-41.
183. Alsheich-Bartok O, Haupt S, Alkalay-Snir I et al. PML enhances the regulation of p53 by CK1 in response to DNA damage. Oncogene 2008; 27(26):3653-61.
184. Schon O, Friedler A, Bycroft M et al. Molecular mechanism of the interaction between MDM2 and p53. J Mol Biol 2002; 323(3):491-501.
185. Winter M, Milne D, Dias S et al. Protein kinase CK1delta phosphorylates key sites in the acidic domain of murine double-minute clone 2 protein (MDM2) that regulate p53 turnover. Biochemistry 2004; 43(51):16356-64.
186. Yang S, Kuo C, Bisi JE, Kim MK. PML-dependent apoptosis after DNA damage is regulated by the checkpoint kinase hCds1/Chk2. Nat Cell Biol 2002; 4(11):865-70.
187. Yang S, Jeong JH, Brown AL et al. Promyelocytic leukemia activates Chk2 by mediating Chk2 autophosphorylation. J Biol Chem 2006; 281(36):26645-54.
188. Wei X, Yu ZK, Ramalingam A et al. Physical and functional interactions between PML and MDM2. J Biol Chem 2003; 278(31):29288-97.
189. Zhu H, Wu L, Maki CG. MDM2 and promyelocytic leukemia antagonize each other through their direct interaction with p53. J Biol Chem 2003; 278(49):49286-92.
190. Kurki S, Latonen L, Laiho M. Cellular stress and DNA damage invoke temporally distinct Mdm2, p53 and PML complexes and damage-specific nuclear relocalization. J Cell Sci 2003; 116(Pt 19):3917-25.
191. Bernardi R, Scaglioni PP, Bergmann S et al. PML regulates p53 stability by sequestering Mdm2 to the nucleolus. Nat Cell Biol 2004; 6(7):665-72.
192. Culjkovic B, Topisirovic I, Skrabanek L et al. eIF4E is a central node of an RNA regulon that governs cellular proliferation. J Cell Biol 2006; 175(3):415-26.
193. Zhu N, Gu L, Findley HW, Zhou M. Transcriptional repression of the eukaryotic initiation factor 4E gene by wild type p53. Biochem Biophys Res Commun 2005; 335(4):1272-9.
194. Gostissa M, Morelli M, Mantovani F et al. The transcriptional repressor hDaxx potentiates p53-dependent apoptosis. J Biol Chem 2004; 279(46):48013-23.
195. Li Q, Wang X, Wu X et al. Daxx cooperates with the Axin/HIPK2/p53 complex to induce cell death. Cancer Res 2007; 67(1):66-74.
196. Cummins JM, Vogelstein B. HAUSP is required for p53 destabilization. Cell Cycle 2004; 3(6):689-92.
197. Tang J, Qu LK, Zhang J et al. Critical role for Daxx in regulating Mdm2. Nat Cell Biol 2006; 8(8):855-62.

Chapter 6

Cooperation between MDM2 and MDMX in the Regulation of p53

Jeremy Blaydes*

Abstract

Murine double minute-2 (MDM2) was first described as a p53-associated protein and potential oncogene in the early 1990s.[1,2] Its paralogue MDMX was subsequently identified in a screen for p53-binding proteins.[3] Extensive evidence now confirms both proteins to be oncogenic in both mice and humans, largely through their ability to negatively regulate the tumor-suppressor activities of p53. It is now clear that the two proteins form heterodimers, and act in concert to regulate p53 activity in proliferating and stressed cells. In this chapter I firstly review the several mechanisms whereby MDM2 and MDMX are potentially able to regulate p53 function independently of each other. I then examine how heterodimerization between the two molecules influences how they regulate the abundance and activity of both p53, and each other. I conclude by examining current models of how the dynamic equilibrium between p53 and its two negative regulators is maintained in proliferating cells, and is targeted by multiple signaling pathways to control the magnitude and duration of the p53-dependent transcriptional response to genotoxic stress.

Introduction

Both *mdm2* and *mdmx* genes and their protein products were originally identified in murine cells, and much of our knowledge on MDM2 and MDMX protein function is derived from studies of both mice and humans. I will thus use the terms MDM, MDM2 and MDMX to refer generically to the proteins of both species, and provide details on the human proteins unless specifically referring to murine Mdm2 and MdmX. The full length form of MDM2 has 491 amino acids and the unmodified protein has a predicted MW of 55 kDa. It has a mobility on SDS-PAGE gels equivalent to a protein of 90 kDa[4] and can be referred to as $p90^{MDM2}$ to distinguish it from $p75^{MDM2}$ proteins, which are translated from AUGs at codons 62 and 102,[5] as well as from MDM2 proteins derived from alternative mRNA splicing and caspase cleavage. Full length MDMX has 490 amino acids and a gel mobility equivalent to an 80 kDa protein. MDM2 and MDMX have a broadly comparable domain structure (Fig. 1), which includes a highly conserved p53-binding domain in their N-termini. Both proteins contain a RING domain which facilitates dimerization between MDM molecules.[6,7] The acidic domains and extreme C-termini of both molecules also contribute to dimerization.[8-10] Heterodimers form preferentially over MDM2 homodimers, whereas MDMX RING domains less readily homodimerize.[7,11-13] Structural studies show MDM2-MDM2 and MDM2-MDMX RING domains complexes to be symmetrical dimers.[11,14] The RING domain of MDM2 also

*Jeremy Blaydes—School of Medicine, University of Southampton. Somers Cancer Research Building, MP824 Southampton General Hospital, Southampton, SO16 6YD, UK. Email j.p.blaydes@soton.ac.uk

p53, edited by Ayeda Ayed and Theodore Hupp. ©2010 Landes Bioscience and Springer Science+Business Media.

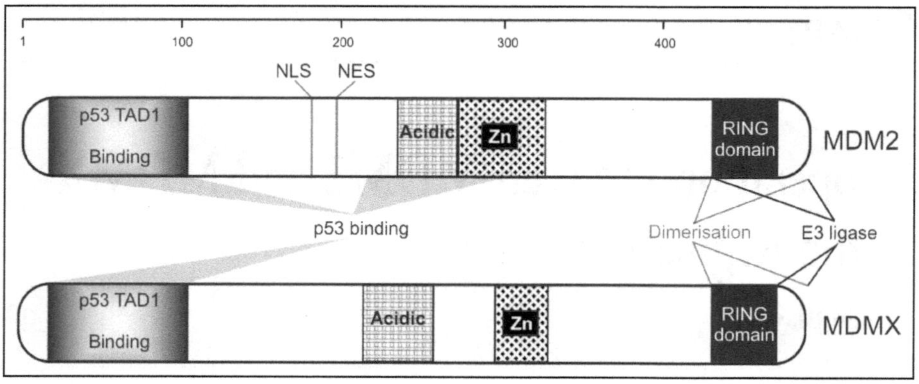

Figure 1. Domain structure of MDM2 and MDMX.

functions as an E3 ubiquitin-ligase, and MDM2, either as a homodimer or as heterodimer with MDMX, is able to facilitate the transfer of ubiquitin from the UbcH5 E2 ubiquitin conjugating enzyme to p53, MDMX or itself.[15-20] There is evidence from structural and mutagenesis studies that RING domain dimerization is required for the E3 ubiquitin-ligase activity of MDM2.[10-14,21] Monomeric MDM2 RING domains can be active in vitro, but only when concentrations of E2 are saturating.[9] In the absence of MDM2, MDMX has at most a very weak ability to promote p53 ubiquitination,[22-25] as the MDMX RING domain lacks E3 ubiquitin-ligase activity.[14] One study has shown that, in proliferating MCF-7 breast cancer cells, the majority of MDMX is associated with MDM2,[12] and it has been speculated that MDM2 concentrations in unstressed cells may be too low to favor MDM2 homodimer formation.[13] However, in general, neither the relative abundance nor multimerization status of endogenous MDM proteins are well defined. Multiple independent studies have demonstrated that both MDM proteins play a key role in the suppression of cellular p53 activity. What is clear, however, is that whereas studies in which either MDM2 or MDMX are examined independently are extremely informative, models of the role of MDM proteins in p53 regulation need to take into account their ability to function in concert.[26-28]

Both *mdm2* and *mdmx* are essential genes during murine development. The embryonic lethality caused by homozygous inactivating mutation of either gene is rescued in a p53-null background.[29-33] Mice in which the only form of Mdm2 expressed is an E3 ligase-incompetent mutant (C462A) also undergo p53-dependent death in utero.[34] Loss of *mdmx* can be rescued by over-expression of *mdm2* at approximately four-times normal levels ($mdm2^{Tr}$ mice).[35] Experiments using conditional alleles (reviewed in ref. 26) have confirmed the requirement for both *mdm2* and *mdmx* in several embryonic tissues, including post-mitotic neurons.[36] In general, the phenotypes caused by *mdm2* inactivation are more severe than from *mdmx* inactivation. For example *mdm2* inactivation in cardiomyocytes is lethal in embryogenesis, whereas mice with *mdmx* deletion in these cells are born normally, but die prematurely of heart failure; again, these effects are p53-dependent.[37,38] In adult tissues, an absolute requirement for *mdm2* but not *mdmx* is also observed in smooth muscle cells of the gut,[39] and mice with an hypomorphic allele of *mdm2,* in which the mRNA is expressed at 30% of normal levels in all adult tissues, have shown lymphoid progenitors and epithelial cells of the gastrointestinal tract to be dependent upon MDM2 for the suppression of spontaneous p53-dependent apoptosis.[40] There was markedly greater evidence of p53 activation in these animals than in conventional $mdm2^{+/-}$ heterozygotes, indicating a clear dose-dependency of *mdm2* gene expression for effective suppression of p53. As a consequence of reduced suppression of p53, insufficiency of Mdm2 can result in suppression of tumorigenesis in murine models.[41,42] Interventional studies of human cells are effectively limited to cell culture analysis, the interpretation of which is complicated by culture-associated stresses potentially enhancing cellular p53 activity. For the record, however, the proliferation of normal human fibroblasts in culture is dependent upon the negative regulation of p53 function by MDM2.[43]

Enforced over-expression of MDM2 in rodent fibroblast or haemopoetic stem cells can promote their immortalization or, when combined with additional oncogenic events, tumorigenic transformation.[1,44-46] MDMX over-expression has similar effects.[47,48] For both proteins, these effects are largely p53-dependent, though other p53-independent effects of MDM2 have been described.[49] *mdm2^{Tr}* mice, which express *mdm2* at approximately four times normal levels, are prone to increased rates of tumor formation, particularly of lymphoma and sarcoma.[50] Ectopic over-expression of *mdmx* in mouse retina cooperates with loss of *Rb* and *p107* to promote the clonal expansion of cells with morphological features resembling aggressive invasive retinoblastoma.[48] Retinoblastoma is also a good example of the involvement of *MDM* genes in human cancers, as *MDM2* and *MDMX* are amplified at a frequency of 10% and 65% of these tumors respectively, and human retinoblastoma cell lines with *MDMX* amplification are killed by a pharmacological inhibitor of p53-binding by MDM proteins.[48] Amplification of either *MDM2* or *MDMX* occurs at varying frequencies in a wide range of human tumors, and generally correlates with the absence of an inactivating mutation in p53, suggesting that the selection pressure for p53 mutation in these tumors has been overcome.[51,52] An inherited single nucleotide polymorphism in a promoter region of *MDM2* which correlates with a modest increase in MDM2 protein expression can also correlate with an earlier age of onset of cancer in some groups of patients,[53,54] though clearly not all.[55] There are also numerous examples of tumors in which MDM protein expression is abnormally upregulated through undetermined mechanisms.[51,52] Together these studies have identified MDM proteins as promising targets for cancer therapy, and several chemically and mechanistically distinct chemical inhibitors of MDM-dependent inhibition of p53 function have now been developed.[48,56-58]

Mechanisms of Inhibition of p53 by MDM2 and MDMX

p53 turns over very rapidly in proliferating cells, and is stabilized in response to many stresses, so clearly the targeted destruction of p53 is a critical aspect of its regulation.[59] However, cellular p53 activity can also be regulated at the level of p53 synthesis, activity, and sub-cellular distribution. It has been proposed that MDM2 acts primarily to promote p53 degradation, and MDMX to inhibit p53 activity.[28] However, perhaps this has been a helpful simplification, MDM proteins can act on p53 at all of the above points; and it remains a matter of debate how all these processes interconnect to regulate p53 in a cell during normal proliferation or under stress. With this in mind, herein follows a brief synopsis on the mechanisms of p53 regulation by MDM proteins, from p53 synthesis to destruction.

Synthesis of p53 Protein

At present there is little evidence as to whether MDM proteins regulate *TP53* mRNA transcription or splicing; p53 does up-regulate the transcription of its own gene following genotoxic stress [60] and this will presumably be inhibited by MDM2 and MDMX. Through binding to nascent p53 polypeptides, MDM2 promotes translation of *TP53* mRNA,[61] as well as the ATP-dependent folding of the p53 protein into a sequence-specific DNA-binding competent form.[62]

Function of p53 as an Activating Transcription Factor

Tetramers of p53 function primarily as sequence-specific DNA-binding transcriptional activators of promoters to which they bind. p53-dependent transcriptional activation of responsive promoters involves the recruitment of TBP-associated factors (TAF_{II}s) and p300/CBP histone acetyl transferases to transcriptional activation domain 1 (TAD1) of p53.[63,64] Amino acids 22 and 23 of p53 are critical for these interactions, and also bind the N-terminal hydrophobic pocket of MDM2. P53–TAF_{II} and p53–MDM2 interactions are mutually exclusive,[65-68] as are p53-p300/CBP and p53-MDM2 interactions.[69] Thus, while p53 in complex with MDM2 can still associate with DNA,[62,70,71] the complexes are unable to activate transcription.[71a] The p53-binding domain in MDMX is structurally and functionally conserved with that in MDM2[72] and MDMX inhibiting p53-dependent transcription independently of MDM2,[3,36] presumably also through coactivator exclusion, although this has not been formally proven. This mechanism of inhibition of p53 activity by MDM proteins is not dependent on their RING domain-mediated dimerization.[73,74]

In addition to the acetylation of histones at p53-bound promoters,[64] p300 and CBP also acetylate multiple lysines within the C-terminal regulatory domain of p53. This p53 acetylation stabilizes the p53-p300 complex,[75] enhances sequence-specific DNA-binding by p53, and inhibits the MDM2-dependent ubiquitination of p53 on these lysines.[69] MDM2 and MDMX inhibit this acetylation by preventing p300/CBP binding[76,77] and MDM2 also recruits HDAC1 to deacetylate p53.[78] Acetylation of p53 by p300/CBP is stimulated by its sequence-specific DNA-binding,[75] so p53 acetylation may be involved in sustaining p53-dependent transcription at promoters, with deacetylation providing a rapid mechanism to attenuate this activity.[69] Finally, MDM2 may also suppress p53-dependent transcription through an intrinsic transcriptional repressor activity,[79] the recruitment of transcriptional corepressors to p53-responsive promoters,[80] and the ubiquitination of histones at promoters to which it is recruited.[70]

Inhibition of p53 Function by MDM2-Mediated Ubiquitination of p53

The high affinity interaction between TAD1 of p53 and the N-terminal p53-binding domain of MDM2 results in a conformational shift in MDM2 that permits the formation of a second interaction between the central acidic and Zn domains of MDM2 (residues 221 to 302) and a ubiquitination signal within the DNA-binding domain of p53. This second interaction is essential for p53-ubiquitination, whereas the interaction between the N-termini is not, provided the conformational constraints of the MDM2 N-terminus are relieved either by its deletion, or by binding to p53-TAD1 mimetics in trans.[81-84] In contrast, while MDMX has an N-terminal p53-binding domain, its central domain cannot facilitate p53-ubiquitination.[85] Within an MDM dimer, the central domain can function in trans with the RING domain, to promote p53-ubiquitination,[85] although interaction in trans is presumably not obligatory given that MDM2-MDMX heterodimers are active as E3 ubiquitin-ligases for p53. Finally, in addition to a functional RING domain, the extreme C-terminal tail of either MDM2 or MDMX interacts with the RING domain in trans and is required for dimerization and E3 ubiquitin-ligase activity, apparently acting as an essential secondary docking site for the E2 ligase.[9,10,13,14]

MDM2 directs the ubiquitination of p53 at lysines 370, 372, 373, 381, 382 and 386[86,87] or NEDDyation at lysines 370, 372 and 373.[88] The E3 ubiquitin-ligase activity of MDM2 is, however, somewhat promiscuous, as when these C-terminal lysines in p53 are mutated, additional sites of ubiquitination in its DNA-binding domain are revealed.[89] When MDM2 is present at low abundance, it directs the mono-ubiquitination of one or more of these lysines on p53, whereas increased concentrations of MDM2,[90] or the action of an E4 ubiquitin-ligase such as p300, promotes the formation of polyubiquitin chains on p53.[91] Ubiquitination of p53 is reversible, and the p53-binding de-ubiquitinating enzyme HAUSP is a major determinant of overall p53 ubiquitination in a cell.[92]

DNA Binding and Nuclear Localization of p53

Mono-ubiquitination or NEDDylation of p53 inhibits its ability to activate transcription.[88,93] For mono-ubiquitination, recent experiments have clarified that this involves two mechanisms; inhibition of sequence-specific DNA-binding[93] and promotion of export from the nucleus to the cytoplasm. The latter is due both to the exposure of a C-terminal nuclear export sequence in p53 and the promotion of release of p53 from nuclear MDM2.[90,94]

Degradation of p53

Probably the most well recognized cellular role of MDM2 is the promotion of the degradation of p53 via the proteasome.[95,96] This is best understood mechanistically under conditions of high cellular MDM2 concentration, where MDM2 is able to direct the formation of polyubiquitin chains on C-terminal lysines of p53, which directs it to the proteasome. This process can occur entirely within the nucleus, as proteasome inhibitors result in the accumulation of nuclear p53.[90,92,97-99] Physiologically, it has been suggested that this mechanism of p53 degradation occurs primarily under conditions of elevated MDM2 abundance, for example during recovery from cellular stress, where MDM2 expression is increased due to enhanced p53-dependent transcription from the P2-promoter of the *MDM2* gene.[98,99] There is also very clear evidence that MDM2 plays an essential role in p53 degradation in early stages of mammalian development,[36,100] and also in cells in culture.[34,101,102]

What remains incompletely resolved however is whether, and if so how, MDM2 is involved in the degradation of p53 in nonstressed, nontransformed cells, particularly in adult tissues in vivo. This is clearly an important question, particularly when considering potential effects on nontumor cells of the MDM2-directed therapeutic strategies which are under development. A number of sophisticated animal studies have attempted to address this issue (reviewed in ref. 28); most notably conditional homozygous *mdm2* gene inactivation in terminally differentiated smooth muscle cells of 8-10 week old mice, which leads to accumulation of p53 protein in these cells.[39] Additionally, in mice engineered to express Mdm2 at 30% of normal levels, a modest elevation of p53 protein abundance is observed in the spleen,[40] though not in other tissues. In a separate study, mice treated with MI-219, a potent and highly selective small molecule inhibitor of the MDM2-p53 interaction, showed a modest accumulation of p53 in the thymus and small intestine.[58] In the thymus, where this was examined by immunohistochemistry, p53 accumulated only in a small proportion of the total cell population. This was in contrast to ionizing radiation which caused p53 to accumulate in the majority of the cells in the thymus. While the lack of p53 accumulation in the majority of cells in response to MDM2 inhibition could be interpreted as demonstrating that p53 is primarily degraded in an MDM2-independent manner, an alternative explanation, given that both the transcription and translation of p53 mRNA can be cell cycle dependent,[103] is simply that the p53 is not being actively synthesized in most of cells. In contrast, ionizing radiation can promote both the synthesis and stabilization of p53.[104] In summary, while key questions as to the status of the p53 pathway in the adult remains to be clarified, MDM2 is clearly involved in the normal process of p53 degradation in at least some unstressed adult tissues.

If it is indeed true that MDM2 homodimers only directly promote the formation of polyubiquitin chains on p53 at high cellular MDM2 concentrations, then how might MDM2 be involved in promoting p53 turnover when it is expressed at its normal levels in nonstressed tissues? Several mechanisms, which are not mutually exclusive, are possible; (i) MDM2-mediated mono-ubiquitination might precede the subsequent poly-ubiquitination of p53 by other ubiquitin-ligases; e.g., p300.[91] Several other p53-ubiquitin-ligases have been discovered recently, though any role they might have in cooperating with MDM2 to promote p53 degradation is unclear.[92] MDM2-induced mono-ubiquitination of p53 can also result in the recruitment of ubiquitin-binding E3 ligases, though to date only the PIASγ SUMO-ligase, which does not promote p53 degradation, is known to be recruited in this way;[94] (ii) MDM2-p53 complexes are recruited to the proteasome by MDM2 itself; this might occur through polyubiquitin chains on MDM2 (as discussed by Wahl and colleagues ref. 105) or through the interaction of MDM2 with proteasome-binding adapters such as hHR23A,[106] or components of the proteasome itself.[107] (iii) As discussed in the following section, MDM2-MDMX heterodimers may play a prominent role in the ubiquitination and degradation of p53 in unstressed cells.

Models of Cooperative Regulation of p53 by MDM2 and MDMX

There are two key levels at which we can to understand how MDM proteins cooperate with each other to regulate p53 activity in the cell; (1) what functional interactions between the proteins can potentially occur in experimental systems and (2) at physiological concentrations of the three proteins, how is p53 maintained inactive in proliferating cells and activated in response to stress?

Functional Interaction between MDM Proteins

Through their ability to form heterodimers, MDM2 and MDMX can impact each others' function through their effects on their respective sub-cellular localization, abundance, and, in the case of MDM2, E3 ligase activity.

Sub-Cellular Localization

While MDM2 can target the degradation of cytoplasmic p53,[108] in order to effectively inhibit p53's function as a nuclear transcription factor, MDM proteins must themselves be able to enter the nucleus.[25,109] MDM2 contains sequences for both nuclear localization (NLS; a.a. 181-185) and nuclear export (NES; a.a.191-202) and in most cells, localization of both endogenous and transiently over-expressed MDM2 is predominantly nuclear.[110] The MDM2 NLS is not conserved in MDMX, and the endogenous protein displays a mixed nuclear and cytoplasmic distribution.[74,111] When transfected into cells MDMX is predominantly cytoplasmic suggesting that it may be imported into the nucleus through saturable pathways.[74,112] MDMX can be recruited to the nucleus via RING domain-dependent heterodimerization with MDM2.[25,109,112] Thus, through this mechanism, MDM2 may be an important regulator of MDMX-mediated p53-inhibition, although the relative importance of this pathway in regulating MDMX localization, compared to interactions with other proteins such as p53[112] and 14-3-3,[113] has not been established.

MDM2 and MDMX Abundance

Rates of degradation are critical in determining the abundance of these two proteins in the cell. In normally proliferating cells, endogenous MDM2 is rapidly degraded by the proteasome and has a half-life of 30-120 min.[114] Endogenous MDMX is degraded with similar or somewhat slower kinetics.[19,115] MDM2 can catalyze the attachment of polyubiquitin chains to both itself[22,81] and MDMX.[116] In transient over-expression systems MDM2-dependent ubiquitination can clearly target both proteins for degradation; over-expressed wild-type MDM2 is rapidly degraded by the proteasome whereas MDM2 mutants which lack E3 ligase activity are not.[16] Over-expressed MDMX is relatively stable unless MDM2 is co-transfected.[116] It has been thought that MDM2-mediated ubiquitination is the primary mechanism by which both these proteins are degraded in cells.[27] However, this conclusion has been questioned by the recent finding that in mice which express only the E3-ligase incompetent C462A mutant of Mdm2, this mutant Mdm2 is degraded by the proteasome with similar kinetics to those of Mdm2 in wild-type animals.[34] This indicates that other E3-ubiquitin ligases for MDM2 exist, although whether these are somehow up-regulated in these animals due to loss of the normal mechanism of auto-E3 ubiquitin-ligase dependent degradation of MDM2 is not yet known. Finally, the rate of degradation of both proteins is also critically dependent upon HAUSP–mediated de-ubiquitination; in the absence of HAUSP the abundance of endogenous MDM2 and MDMX can be reduced to below the limit of detection.[117]

Based on results from transient over-expression studies it was proposed that an important role of MDMX, through the formation of heterodimers, was to inhibit auto-ubiquitination of MDM2, thus raising MDM2 concentration and promoting p53 degradation.[6,108,109] However, in in vitro experiments with recombinant proteins, MDMX can promote MDM2 auto-ubiquitination.[22,81] Resolution of these apparently contradictory findings will require a better understanding of the auto-E3 ligase activities of the MDM2-MDMX heterodimer. In vitro, the minimum domains of MDM2 required for auto-E3 ligase activity are the RING domain plus C-terminus (within residues 432-491) consistent with a role for the C-terminus interacting with the RING domain in trans to promote dimerization and E2-binding.[9,14] Heterodimers of MDM2(432-491) and MDMX(421-490) also have auto-E3 ligase activity, as the MDMX C-terminus is able to cooperate with the MDMX RING domain. However ubiquitin is only conjugated to one residue in the heterodimer; K442 in MDMX.[14] This is consistent with only the MDM2 RING domain having E3 ligase activity, and it being able to ubiquitinate the MDMX RING domain in trans, but not itself in cis. How then can full length MDMX promote the ubiquitination of full length MDM2?[22,81] First, if MDM2 is at low concentrations such that it is primarily monomeric, the addition of MDMX would promote the formation of heterodimers in which the MDMX C-terminus would cooperate with the MDM2 RING domain to promote its auto-E3 ligase activity. As well as ubiquitination of the MDMX RING domain, it could ubiquitinate other lysines in the heterodimer, including multiple lysines in MDM2, resulting in the high molecular weight smears or ladders of ubiquitinated MDM2 seen when full length proteins (e.g., ref. 81) or even GST-RING domain fusion proteins[14] are assayed. Further support for a role for the MDM2-MDMX heterodimer in promoting MDM2 degradation in cells potentially comes from experiments showing that MDMX could rescue the p53 ubiquitination activity of E3-ligase defective C-terminal domain mutant of MDM2 (Y489A), by providing a functional C-terminal domain to cooperate with the MDM2(Y489A) RING domain.[10] In these experiments, MDM2(Y489A) protein levels also appeared to be reduced by MDMX, suggesting its auto-E3 ubiquitin-ligase-dependent degradation was enhanced by heterodimer formation. Thus, if the auto-ubiquitination activity of MDM2 is important in its degradation, loss of MDMX should result in MDM2 stabilization. This can be demonstrated in the H1299 cell line following transient siRNA-mediated inhibition of MDMX synthesis.[81] However Mdm2 stability is not different in p53 null mouse embryo fibroblasts (MEFs) derived from either *mdm2*$^{+/-}$*mdmx*$^{+/+}$ or *mdm2*$^{+/-}$*mdmx*$^{-/-}$ animals.[36] Similar to the Mdm2(C462A) mice,[34] it is possible that E3-ligases other than MDM2 play a greater role in MDM2 degradation in the MEFs than in the H1299 cells.

In transient over-expression experiments, MDM2 would most likely be expressed at concentrations high enough for homodimers to form. In this situation, and potentially in some cancer cells where MDM2 is over-expressed, or in cells when MDM2 is transiently upregulated in response to stress, the effect of MDMX on MDM2 turnover would depend on the relative MDM2 auto-E3 ubiquitin-ligase activity of an MDM2 homodimer compared to an MDM2-MDMX heterodimer. This has yet to be determined. Many of the early transient transfection studies which demonstrated that MDMX stabilizes MDM2 employed a C-terminally tagged version of MDMX in the transfection experiments. We now know that the presence of such a tag inhibits MDMX degradation by MDM2,[116] and thus the effects on MDM2 turnover demonstrated in these experiments must also be questioned. Indeed, some studies in which MDMX without a C-terminal tag were used showed no stabilization of MDM2 by MDMX.[10,23] Thus this aspect of the interplay between MDM2 and MDMX probably requires some further clarification.

Activity of MDM2 as an E3 Ubiquitin-Ligase towards p53

It is now becoming apparent that how MDMX influences MDM2-dependent ubiquitination of p53 depends on both the absolute and relative concentrations of MDM2 and MDMX in the cell. Three distinct situations appear to be most relevant; (i) MDM2 low/MDMX not in excess; (ii) MDM2 high/MDMX not in excess; and (iii) MDMX in excess. (i) In vitro studies clearly demonstrate that MDMX is able to promote activity of MDM2 as an E3 ubiquitin-ligase for p53.[81] It seems probable that the primary mechanism underlying this is the formation of heterodimers under conditions where MDM2 concentrations are too low to form homodimers. Similar to its promotion of MDM2 auto-ubiquitination, the

MDMX C-terminus cooperates with the MDM2 RING domain to form a functional E3 ligase unit.[10,14] For p53 ubiquitination, the p53 docking site in the acidic domain of MDM2 is also required,[84] a feature which is lacking in MDMX. Both MDM2 and MDMX have a high affinity N-terminal p53-binding pocket; when this domain is present in MDM2, in the context of MDM2 homodimers, its binding to p53 is required in order to relieve its negative regulation of p53-binding to the MDM2 acidic domain.[84] Any effects of heterodimerization with MDMX on this requirement, or whether the N-terminal p53-binding site of MDMX needs to be occupied for heterodimers to have E3 ligase activity towards p53, remain to be determined. Evidence that MDM2-MDMX heterodimers are prominent mediators of p53 ubiquitination in proliferating cells comes from experiments employing the MCF-7 cell line. In these cells, the majority of MDMX is present in the form of heterodimers[12] and transient siRNA-mediated knockdown of MDMX results in an increase in the abundance of p53.[81] Notably, this also results in an increase in the abundance of MDM2, presumably through both increased p53-dependent transcription, and reduced auto-ubiquitination, which could potentially reduce p53 levels in any cells that do survive. As discussed by Francoz et al,[36] such an increase in p53-dependent MDM2 transcription following loss of MDMX could account for the lack of increased p53 protein abundance in cells from embryos of *mdmx*-null mice, compared to cells from normal embryos; indeed the fact that p53 abundance does not decrease in cells from these mice despite an increase Mdm2 is further evidence than MDMX is required to cooperate with MDM2 to promote p53 degradation. However, in this case, the concurrent increase in p53-dependent transcription of other genes is sufficient to induce cell cycle arrest and embryonic lethality.[31,32,36]

Under situation (ii), when functional MDM2 homodimers can form, the effects of MDMX on p53 ubiquitination are far less clear, as the relative p53 E3-ligase activity of MDM2 homodimers compared to the heterodimers is unknown. In vitro, MDMX only has substantial effects on MDM2 activity when MDM2 concentrations are low.[81] However in *mdm2*^{Tr} mice in which Mdm2 is expressed at sufficiently high levels to render MdmX unnecessary for p53 suppression, loss of MdmX results in increased cell proliferation in culture, and tumorigenesis in p53-expressing animals.[35] This latter finding is potential evidence that heterodimers can be less effective than MDM2 homodimers in suppressing cellular p53 activity, though further molecular analysis of the cells would be required to substantiate this conclusion. When MDMX is in excess (iii) p53 ubiquitination would be inhibited due to competition for p53-binding between heterodimers and E3 ligase-incompetent MDMX monomers.[109] This can result in the existence of a stable, functionally inactive pool of p53, and could occur in tumors where the *MDMX* gene is amplified. A comparable situation arises in tumors in which alternate splicing results in the expression of a C-terminally truncated form of MDMX, termed MDMX-S. MDMX-S which does not bind MDM2 and thus neither stimulates p53 ubiquitination, nor is itself degraded by MDM2; so both MDMX-S, and inactive p53 to which it is bound, accumulate in cells.[118] Finally, an interesting situation arises in some cell lines which are exposed to the MDM2 inhibitor, nutlin-3a. Nutlin-3a obscures the p53-binding pocket at the N-terminus of MDM2; it binds MDMX with much lower affinity. By preventing efficient binding of MDM2 to p53, it results in the activation and stabilization of p53.[102] This results in the p53-dependent transcription of MDM2 and subsequent MDM2-dependent degradation of MDMX. Cells in which this decrease in MDMX does not occur are more resistant to the pro-apoptotic activity of nutlin-3a.[119] Thus MDMX is clearly important in inhibiting p53 in these circumstances; this is presumably primarily through inhibition of p53 activity through binding of MDMX to its transactivation domain. However it would also be of interest to establish whether the elevated levelss of MDM2 protein plays a role in p53 inhibition in these circumstances; for example, through promoting the nuclear localization of MDMX, or by forming an MDMX-MDM2-nutlin-3a complex which potentially retains p53-binding and p53 E3 ubiquitin-ligase activity.

Models of p53 Regulation by MDM Proteins in Proliferating and Stressed Cells

Proliferating Cells

Through promoting transcription of a large numbers of genes, p53 can have diverse effects on the cell. However one of the most consistent consequences of p53 activation in many cell types is cell cycle arrest, in large part due to increased *WAF1* gene expression.[59] This is consistent with the widely recognized role of p53 as "guardian of the genome"; i.e., by inducing a transient inhibition of DNA synthesis in response to DNA damage, it permits DNA repair and thus prevents potentially deleterious mutations being fixed into the genome. While the fact that p53 and its regulators also have a role in post-mitotic cells such as neurons[36] clearly indicate that its activity is not solely restricted to proliferating cells, it is perhaps not surprising that there is a striking cross-talk between p53 regulation and proproliferative signaling pathways. In addition to promoting p53 protein synthesis,[103,120] signals that promote cell cycle entry also increase the post-translational activation of p53, most notably through enhanced transcription of p14^{ARF}, a potent inhibitor of the p53 E3 ubiquitin-ligase activity of MDM2.[121] The same signals also up-regulate negative regulators of p53; transcription from the P2 promoter of the *MDM2* gene is inducible by both activated p53 and the growth factor receptor-RAS-RAF-MEK-ERK-AP1/ETS signaling cascade.[122-124] MEK also promotes MDM2 synthesis by enhancing the nuclear export of *MDM2* mRNA through a MNK1-eIF4E dependent pathway.[125] AKT phosphorylates MDM2 at serines 166 and 188, resulting in the accumulation of MDM2 by inhibiting its degradation.[110,126] Regulation of MDMX is relatively poorly characterized; however *MDMX* transcription is also induced by mitogenic growth factors.[111] Thus, in cells proliferating in the presence of balanced mitogenic signals, net cellular p53 activity remains low due to the maintenance of a dynamic equilibrium with its negative regulators MDM2 and MDMX. However, the proliferating cells are highly dependent upon MDM proteins, most likely in the form of MDM2-MDMX heterodimers, for the suppression of p53 activity, and are thus primed for rapid p53 activation in response to diverse signaling pathways that can disrupt this equilibrium.

Response to Ionizing Radiation-Induced Genotoxic Stress

As the most thoroughly studied p53-activating pathway, the ATM-dependent response to ionizing radiation provides a paradigm for the role of MDM proteins in the cellular response to stress (reviewed in, for example, refs. 105, 127). Ionizing radiation rapidly leads to activation of ATM, which both phosphorylates p53 directly on serine 15, and promotes p53 phosphorylation at other residues in its N-terminal TAD1. Interaction between p53 and the de-ubiquitinating enzyme HAUSP is also increased.[128] Simplistically, these modifications favor the interaction between p53 and p300/CBP over its interaction with MDM2/MDMX, and consequent changes in the modification of C-terminal lysines from ubiquitination to acetylation. Phosphorylation of p53 is not, however, sufficient for its activation; ATM-dependent modifications to MDM2 and MDMX are also required. ATM activation results in the extremely rapid degradation of MDM2.[114] MDMX is also degraded with somewhat slower kinetics.[19,116] For both proteins, this increased degradation is a consequence of their reduced association with, and de-ubiquitination by, HAUSP, which in turn appears to be due to their phosphorylation by ATM.[117,127] For MDMX, CHK2-dependent phosphorylation of serine 367, which generates a 14-3-3 binding site, is also required for its destabilization,[113,129] Thus, ionizing radiation both activates p53, and promotes the destruction of its primary negative regulators. Finally, it is interesting to consider the role of MDM proteins during the recovery from stress; following the initial degradation of MDM2, by approximately 3 hours post-irradiation, p53-dependent MDM2 transcription results in MDM2 protein accumulating to levels much higher than in unstressed cells, whereas MDMX levels remain low due to MDM2-dependent degradation.[19] Thus MDM2 homodimers are presumably responsible for p53 degradation during this phase. If MDMX is in excess of MDM2 during this period; MDMX monomers could inhibit the rapid clearance of the excess p53 that is required to fully attenuate the p53 response.

Conclusion and Perspectives

MDM2 and MDMX act in concert to regulate the level of p53 activity in the cell. Each protein can impinge upon p53 function through more than one distinct mechanism; which of these mechanisms or combinations thereof, are prevalent in any particular cell under a defined set of conditions is highly dependent upon, amongst other things, the absolute and relative abundance of the two proteins. As well as making the interpretation of conventional transient over-expression experiments very difficult, this has important implications for understanding of both how cancer develops and how it may be treated. For example when considering the potential impact on tumorigenesis of changes in the expression of MDM2 or MDMX, caused either by inherited variations such as polymorphisms in promoter regions, or cancer associated changes (i.e., gene amplification, alternate splicing, or enhanced transcription) it is will become important to consider the effects on the relative abundance and activity of the two proteins in the cell. This is well illustrated by the *mdm2*^{Tr} mouse model, in which Mdm2 is expressed at 3-4 times normal levels such that MdmX is redundant for the survival of the animal.[35] MEFs from these animals fail to accumulate p53 in response to genotoxic (UV-irradiation) stress. Is this simply a consequence of greater levels of p53-inhibitory proteins per se, or is it more relevant that the balance between Mdm2 and MdmX is perturbed such that Mdm2 homodimers, rather than heterodimers, are the primary p53 E3 ubiquitin-ligases in the cell? The role of MdmX in cancer development in these mice is also the opposite to that of what might be expected, as it functions as a p53-dependent tumor suppressor gene. Is this because, in the presence of elevated cellular concentrations of MDM2, MDMX potentially has a net negative effect on cellular MDM2 activity within a tumor? Many currently used and cancer therapies under development act, at least in part, through the activation of the p53 pathway. The cellular response to at least one such therapy, ionizing radiation, involves the coordinated regulation of MDM2 and MDMX abundance and function. Enhanced MDM2-dependent degradation of MDMX is also an important component of the mechanism of action of pharmacological inhibitors of MDM2; cancer cells that apparently lack this response show resistance to these compounds.[119] Understanding these response pathways in normal and tumor cells remains an important goal in the development and optimization of cancer therapeutic strategies.

Acknowledgements

Work on MDM2 and MDMX in the author's laboratory is funded by the Association for International Cancer Research.

References

1. Fakharzedeh SS, Trusko SP, George DL. Tumorigenic potential associated with enhanced expression of a gene that is amplified in a mouse tumor cell line. EMBO J 1991; 10:1565-1569.
2. Momand J, Zambetti GP, Olson DC et al. The mdm-2 oncogene product forms a complex with the p53 protein and inhibits p53-mediated transactivation. Cell 1992; 69:1237-1245.
3. Shvarts A, Steegenga WT, Riteco N et al. MDMX: a novel p53-binding protein with some functional properties of MDM2. EMBO J 1996; 15:5349-5357.
4. Xirodimas DP, Chisholm J, Desterro JM et al. P14ARF promotes accumulation of SUMO-1 conjugated (H)Mdm2. FEBS Lett 2002; 528(1-3):207-211.
5. Cheng TH, Cohen SN. Human MDM2 isoforms translated differentially on constitutive versus p53-regulated transcripts have distinct functions in the p53/MDM2 and TSG101/MDM2 feedback control loops. Mol Cell Biol 2007; 27(1):111-119.
6. Sharp DA, Kratowicz SA, Sank MJ et al. Stabilization of the MDM2 oncoprotein by interaction with the structurally related MDMX protein. J Biol Chem 1999; 274(53):38189-38196.
7. Tanimura S, Ohtsuka S, Mitsui K et al. MDM2 interacts with MDMX through their RING finger domains. FEBS Lett 1999; 447(1):5-9.

8. Dang J, Kuo ML, Eischen CM et al. The RING domain of Mdm2 can inhibit cell proliferation. Cancer Res 2002; 62(4):1222-1230.
9. Poyurovsky MV, Priest C, Kentsis A et al. The Mdm2 RING domain C-terminus is required for supramolecular assembly and ubiquitin ligase activity. EMBO J 2007; 26(1):90-101.
10. Uldrijan S, Pannekoek WJ, Vousden KH. An essential function of the extreme C-terminus of MDM2 can be provided by MDMX. EMBO J 2007; 26(1):102-112.
11. Kostic M, Matt T, Martinez-Yamout MA et al. Solution structure of the Hdm2 C2H2C4 RING, a domain critical for ubiquitination of p53. J Mol Biol 2006; 363(2):433-450.
12. Kawai H, Lopez-Pajares V, Kim MM et al. RING domain-mediated interaction is a requirement for MDM2's E3 ligase activity. Cancer Res 2007; 67(13):6026-6030.
13. Singh RK, Iyappan S, Scheffner M. Hetero-oligomerization with MdmX rescues the ubiquitin/Nedd8 ligase activity of RING finger mutants of Mdm2. J Biol Chem 2007; 282(15):10901-10907.
14. Linke K, Mace PD, Smith CA et al. Structure of the MDM2/MDMX RING domain heterodimer reveals dimerization is required for their ubiquitylation in trans. Cell Death Differ 2008.
15. Honda R, Tanaka H, Yasuda H. Oncoprotein MDM2 is a ubiquitin ligase E3 for tumor suppressor p53. FEBS Lett 1997; 420(1):25-27.
16. Fang S, Jensen JP, Ludwig RL et al. Mdm2 is a RING finger-dependent ubiquitin protein ligase for itself and p53. J Biol Chem 2000; 275(12):8945-8951.
17. Honda R, Yasuda H. Activity of MDM2, a ubiquitin ligase, toward p53 or itself is dependent on the RING finger domain of the ligase. Oncogene 2000; 19(11):1473-1476.
18. de Graaf P, Little NA, Ramos YF et al. Hdmx protein stability is regulated by the ubiquitin ligase activity of Mdm2. J Biol Chem 2003; 278(40):38315-38324.
19. Kawai H, Wiederschain D, Kitao H et al. DNA damage-induced MDMX degradation is mediated by MDM2. J Biol Chem 2003; 278(46):45946-45953.
20. Saville MK, Sparks A, Xirodimas DP et al. Regulation of p53 by the ubiquitin-conjugating enzymes UbcH5B/C in vivo. J Biol Chem 2004; 279(40):42169-42181.
21. Vander Kooi CW, Ohi MD, Rosenberg JA et al. The Prp19 U-box crystal structure suggests a common dimeric architecture for a class of oligomeric E3 ubiquitin ligases. Biochemistry 2006; 45(1):121-130.
22. Badciong JC, Haas AL. MdmX is a RING finger ubiquitin ligase capable of synergistically enhancing Mdm2 ubiquitination. J Biol Chem 2002; 277(51):49668-49675.
23. Jackson MW, Berberich SJ. MdmX protects p53 from Mdm2-mediated degradation. Mol Cell Biol 2000; 20(3):1001-1007.
24. Stad R, Ramos YF, Little N et al. Hdmx stabilizes Mdm2 and p53. J Biol Chem 2000; 275(36):28039-28044.
25. Migliorini D, Danovi D, Colombo E et al. Hdmx recruitment into the nucleus by Hdm2 is essential for its ability to regulate p53 stability and transactivation. J Biol Chem 2002; 277(9):7318-7323.
26. Marine JC, Francoz S, Maetens M et al. Keeping p53 in check: essential and synergistic functions of Mdm2 and Mdm4. Cell Death Differ 2006; 13(6):927-934.
27. Marine JC, Jochemsen AG. Mdmx as an essential regulator of p53 activity. Biochem Biophys Res Commun 2005; 331(3):750-760.
28. Toledo F, Wahl GM. Regulating the p53 pathway: in vitro hypotheses, in vivo veritas. Nat Rev Cancer 2006; 6(12):909-923.
29. Montes de Oca Luna R, Wagner DS, Lozano G. Rescue of early embryonic lethality in mdm2-deficient mice by deletion of p53. Nature 1995; 378(6553):203-206.
30. Jones SN, Roe AE, Donehower LA et al. Rescue of embryonic lethality in Mdm2-deficient mice by absence of p53. Nature 1995; 378:206-208.
31. Parant J, Chavez-Reyes A, Little NA et al. Rescue of embryonic lethality in Mdm4-null mice by loss of Trp53 suggests a nonoverlapping pathway with MDM2 to regulate p53. Nat Genet 2001; 20:20.
32. Migliorini D, Denchi EL, Danovi D et al. Mdm4 (Mdmx) regulates p53-induced growth arrest and neuronal cell death during early embryonic mouse development. Mol Cell Biol 2002; 22(15):5527-5538.
33. Finch RA, Donoviel DB, Potter D et al. mdmx is a negative regulator of p53 activity in vivo. Cancer Res 2002; 62(11):3221-3225.
34. Itahana K, Mao H, Jin A et al. Targeted inactivation of Mdm2 RING finger E3 ubiquitin ligase activity in the mouse reveals mechanistic insights into p53 regulation. Cancer Cell 2007; 12(4):355-366.
35. Steinman HA, Hoover KM, Keeler ML et al. Rescue of Mdm4-deficient mice by Mdm2 reveals functional overlap of Mdm2 and Mdm4 in development. Oncogene 2005; 24(53):7935-7940.

36. Francoz S, Froment P, Bogaerts S et al. Mdm4 and Mdm2 cooperate to inhibit p53 activity in proliferating and quiescent cells in vivo. Proc Natl Acad Sci U S A 2006; 103(9):3232-3237.
37. Grier JD, Xiong S, Elizondo-Fraire AC et al. Tissue-specific differences of p53 inhibition by Mdm2 and Mdm4. Mol Cell Biol 2006; 26(1):192-198.
38. Xiong S, Van Pelt CS, Elizondo-Fraire AC et al. Loss of Mdm4 results in p53-dependent dilated cardiomyopathy. Circulation 2007; 115(23):2925-2930.
39. Boesten LS, Zadelaar SM, De Clercq S et al. Mdm2, but not Mdm4, protects terminally differentiated smooth muscle cells from p53-mediated caspase-3-independent cell death. Cell Death Differ 2006; 13(12):2089-2098.
40. Mendrysa SM, McElwee MK, Michalowski J et al. mdm2 Is critical for inhibition of p53 during lymphopoiesis and the response to ionizing irradiation. Mol Cell Biol 2003; 23(2):462-472.
41. Alt JR, Greiner TC, Cleveland JL et al. Mdm2 haplo-insufficiency profoundly inhibits Myc-induced lymphomagenesis. EMBO J 2003; 22(6):1442-1450.
42. Mendrysa SM, O'Leary KA, McElwee MK et al. Tumor suppression and normal aging in mice with constitutively high p53 activity. Genes Dev 2006; 20(1):16-21.
43. Blaydes JP, Wynford-Thomas D. The proliferation of normal human fibroblasts is dependent upon negative regulation of p53 function by mdm2. Oncogene 1998; 16(25):3317-3322.
44. Finlay CA. The mdm-2 oncogene can overcome wild-type p53 suppression of transformed cell growth. Mol Cell Biol 1993; 13:301-306.
45. Sigalas I, Calvert AH, Anderson JJ et al. Alternatively spliced mdm2 transcripts with loss of p53 binding domain sequences: transforming ability and frequent detection in human cancer. Nature Med 1996; 2:912-917.
46. Fridman JS, Hernando E, Hemann MT et al. Tumor promotion by Mdm2 splice variants unable to bind p53. Cancer Res 2003; 63(18):5703-5706.
47. Danovi D, Meulmeester E, Pasini D et al. Amplification of Mdmx (or Mdm4) directly contributes to tumor formation by inhibiting p53 tumor suppressor activity. Mol Cell Biol 2004; 24(13):5835-5843.
48. Laurie NA, Donovan SL, Shih CS et al. Inactivation of the p53 pathway in retinoblastoma. Nature 2006; 444(7115):61-66.
49. Ganguli G, Wasylyk B. p53-independent functions of MDM2. Mol Cancer Res 2003; 1(14):1027-1035.
50. Jones SN, Hancock AR, Vogel H et al. Overexpression of Mdm2 in mice reveals a p53-independent role for Mdm2 in tumorigenesis. Proc Natl Acad Sci U S A 1998; 95(26):15608-15612.
51. Onel K, Cordon-Cardo C. MDM2 and prognosis. Mol Cancer Res 2004; 2(1):1-8.
52. Momand J, Jung D, Wilczynski S et al. The MDM2 gene amplification database. Nucleic Acids Res 1998; 26(15):3453-3459.
53. Bond GL, Hu W, Bond EE et al. A single nucleotide polymorphism in the MDM2 promoter attenuates the p53 tumor suppressor pathway and accelerates tumor formation in humans. Cell 2004; 119(5):591-602.
54. Bond GL, Levine AJ. A single nucleotide polymorphism in the p53 pathway interacts with gender, environmental stresses and tumor genetics to influence cancer in humans. Oncogene 2007; 26(9):1317-1323.
55. Schmidt MK, Reincke S, Broeks A et al. Do MDM2 SNP309 and TP53 R72P interact in breast cancer susceptibility? A large pooled series from the breast cancer association consortium. Cancer Res 2007; 67(19):9584-9590.
56. Vassilev LT. MDM2 inhibitors for cancer therapy. Trends Mol Med 2007; 13(1):23-31.
57. Dey A, Verma CS, Lane DP. Updates on p53: modulation of p53 degradation as a therapeutic approach. Br J Cancer 2008; 98(1):4-8.
58. Shangary S, Qin D, McEachern D et al. Temporal activation of p53 by a specific MDM2 inhibitor is selectively toxic to tumors and leads to complete tumor growth inhibition. Proc Natl Acad Sci U S A 2008; 105(10):3933-3938.
59. Vogelstein B, Lane D, Levine AJ. Surfing the p53 network. Nature 2000; 408(6810):307-310.
60. Wang S, El-Deiry WS. p73 or p53 directly regulates human p53 transcription to maintain cell cycle checkpoints. Cancer Res 2006; 66(14):6982-6989.
61. Yin Y, Stephen CW, Luciani MG et al. p53 Stability and activity is regulated by Mdm2-mediated induction of alternative p53 translation products. Nat Cell Bio l 2002; 4(6):462-467.
62. Wawrzynow B, Zylicz A, Wallace M et al. MDM2 chaperones the p53 tumor suppressor. J Biol Chem 2007; 282(45):32603-32612.
63. Espinosa JM, Verdun RE, Emerson BM. p53 functions through stress- and promoter-specific recruitment of transcription initiation components before and after DNA damage. Mol Cell 2003; 12(4):1015-1027.

64. An W, Kim J, Roeder RG. Ordered cooperative functions of PRMT1, p300, and CARM1 in transcriptional activation by p53. Cell 2004; 117(6):735-748.
65. Oliner JD, Pietenpol JA, Thiagalingam S et al. Oncoprotein MDM2 conceals the activation domain of tumor suppressor p53. Nature (London) 1993; 362:857-860.
66. Kussie PH, Gorina S, Marechal V et al. Structure of the MDM2 oncoprotein bound to the p53 tumor suppressor transactivation domain [comment]. Science 1996; 274(5289):948-953.
67. Uesugi M, Verdine GL. The alpha-helical FXXPhiPhi motif in p53: TAF interaction and discrimination by MDM2. Proc Natl Acad Sci U S A 1999; 96(26):14801-14806.
68. Thut CJ, Chen J-L, Klemm R et al. p53 transcriptional activation mediated by coactivators TAF{-II}40 and TAF{-II}60. Science 1995; 267:100-104.
69. Brooks CL, Gu W. Ubiquitination, phosphorylation and acetylation: the molecular basis for p53 regulation. Curr Opin Cell Biol 2003; 15(2):164-171.
70. Minsky N, Oren M. The RING domain of Mdm2 mediates histone ubiquitylation and transcriptional repression. Mol Cell 2004; 16(4):631-639.
71. White DE, Talbott KE, Arva NC et al. Mouse double minute 2 associates with chromatin in the presence of p53 and is released to facilitate activation of transcription. Cancer Res 2006; 66(7):3463-3470.
71a. Tang Y, Zhao W, Chen Y et al. Acetylation is indispensable for p53 activation. Cell 2008; 133(4):612-626.
72. Bottger V, Bottger A, Garcia-Echeverria C et al. Comparative study of the p53-mdm2 and p53-MDMX interfaces. Oncogene 1999; 18(1):189-199.
73. Chen JD, Lin JY, Levine AJ. Regulation of transcription functions of the p53 tumor suppressor by the mdm-2 oncogene. Mol Med 1995; 1:141-152.
74. Rallapalli R, Strachan G, Cho B et al. A novel MDMX transcript expressed in a variety of transformed cell lines encodes a truncated protein with potent p53 repressive activity. J Biol Chem 1999; 274(12):8299-8308.
75. Dornan D, Shimizu H, Perkins ND et al. DNA-dependent acetylation of p53 by the transcription coactivator p300. J Biol Chem 2003; 278(15):13431-13441.
76. Sabbatini P, McCormick F. MDMX inhibits the p300/CBP-mediated acetylation of p53. DNA Cell Biol 2002; 21(7):519-525.
77. Kobet E, Zeng X, Zhu Y et al. MDM2 inhibits p300-mediated p53 acetylation and activation by forming a ternary complex with the two proteins. Proc Natl Acad Sci U S A 2000; 97(23):12547-12552.
78. Ito A, Kawaguchi Y, Lai CH et al. MDM2-HDAC1-mediated deacetylation of p53 is required for its degradation. EMBO J 2002; 21(22):6236-6245.
79. Thut CJ, Goodrich JA, Tjian R. Repression of p53-mediated transcription by MDM2: a dual mechanism. Genes Dev 1997; 11(15):1974-1986.
80. Mirnezami AH, Campbell SJ, Darley M et al. Hdm2 recruits a hypoxia sensitive corepressor to negatively regulate p53-dependent transcription. Curr Biol 2003; 13:1234-1239.
81. Linares LK, Hengstermann A, Ciechanover A et al. HdmX stimulates Hdm2-mediated ubiquitination and degradation of p53. Proc Natl Acad Sci U S A 2003; 100(21):12009-12014.
82. Yu GW, Rudiger S, Veprintsev D et al. The central region of HDM2 provides a second binding site for p53. Proc Natl Acad Sci U S A 2006; 103(5):1227-1232.
83. Ma J, Martin JD, Zhang H et al. A second p53 binding site in the central domain of Mdm2 is essential for p53 ubiquitination. Biochemistry 2006; 45(30):9238-9245.
84. Wallace M, Worrall E, Pettersson S et al. Dual-site regulation of MDM2 E3-ubiquitin ligase activity. Mol Cell 2006; 23(2):251-263.
85. Meulmeester E, Frenk R, Stad R et al. Critical role for a central part of Mdm2 in the ubiquitylation of p53. Mol Cell Biol 2003; 23(14):4929-4938.
86. Nakamura S, Roth JA, Mukhopadhyay T. Multiple lysine mutations in the C-terminal domain of p53 interfere with MDM2-dependent protein degradation and ubiquitination. Mol Cell Biol 2000; 20(24):9391-9398.
87. Rodriguez MS, Desterro JM, Lain S et al. Multiple C-terminal lysine residues target p53 for ubiquitin-proteasome- mediated degradation. Mol Cell Biol 2000; 20(22):8458-8467.
88. Xirodimas DP, Saville MK, Bourdon JC et al. Mdm2-mediated NEDD8 conjugation of p53 inhibits its transcriptional activity. Cell 2004; 118(1):83-97.
89. Chan WM, Mak MC, Fung TK et al. Ubiquitination of p53 at multiple sites in the DNA-binding domain. Mol Cancer Res 2006; 4(1):15-25.
90. Li M, Brooks CL, Wu-Baer F et al. Mono- versus polyubiquitination: differential control of p53 fate by Mdm2. Science 2003; 302(5652):1972-1975.

91. Grossman SR, Deato ME, Brignone C et al. Polyubiquitination of p53 by a ubiquitin ligase activity of p300. Science 2003; 300(5617):342-344.
92. Brooks CL, Gu W. p53 ubiquitination: Mdm2 and beyond. Mol Cell 2006; 21(3):307-315.
93. Brooks CL, Li M, Gu W. Mechanistic studies of MDM2-mediated ubiquitination in p53 regulation. J Biol Chem 2007; 282(31):22804-22815.
94. Carter S, Bischof O, Dejean A et al. C-terminal modifications regulate MDM2 dissociation and nuclear export of p53. Nat Cell Biol 2007; 9(4):428-435.
95. Haupt Y, Maya R, Kazaz A et al. Mdm2 promotes the rapid degradation of p53. Nature 1997; 387(6630):296-299.
96. Kubbutat MH, Jones SN, Vousden KH. Regulation of p53 stability by Mdm2. Nature 1997; 387(6630):299-303.
97. Xirodimas DP, Stephen CW, Lane DP. Cocompartmentalization of p53 and Mdm2 is a major determinant for Mdm2-mediated degradation of p53. Exp Cell Res 2001; 270(1):66-77.
98. Shirangi TR, Zaika A, Moll UM. Nuclear degradation of p53 occurs during down-regulation of the p53 response after DNA damage. FASEB J 2002; 16(3):420-422.
99. Brooks CL, Li M, Gu W. Monoubiquitination: the signal for p53 nuclear export? Cell Cycle 2004; 3(4):436-438.
100. Xiong S, Van Pelt CS, Elizondo-Fraire AC et al. Synergistic roles of Mdm2 and Mdm4 for p53 inhibition in central nervous system development. Proc Natl Acad Sci U S A 2006; 103(9):3226-3231.
101. Bottger A, Bottger V, Sparks A et al. Design of a synthetic Mdm2-binding mini protein that activates the p53 response in vivo. Curr Biol 1997; 7(11):860-869.
102. Vassilev LT, Vu BT, Graves B et al. In vivo activation of the p53 pathway by small-molecule antagonists of MDM2. Science 2004; 303(5659):844-848. .
103. Steinmeyer K, Maacke H, Deppert W. Cell cycle control by p53 in normal (3T3) and chemically transformed (Meth A) mouse cells. I. regulation of p53 expression. Oncogene 1990; 5:1691-1699.
104. Fu L, Benchimol S. Participation of the human p53 3'UTR in translational repression and activation following gamma-irradiation. EMBO J 1997; 16(13):4117-4125.
105. Wahl GM. Mouse bites dogma: how mouse models are changing our views of how P53 is regulated in vivo. Cell Death Differ 2006; 13(6):973-983.
106. Brignone C, Bradley KE, Kisselev AF et al. A post-ubiquitination role for MDM2 and hHR23A in the p53 degradation pathway. Oncogene 2004; 23(23):4121-4129.
107. Sdek P, Ying H, Chang DL et al. MDM2 promotes proteasome-dependent ubiquitin-independent degradation of retinoblastoma protein. Mol Cell 2005; 20(5):699-708.
108. Stad R, Little NA, Xirodimas DP et al. Mdmx stabilizes p53 and Mdm2 via two distinct mechanisms. EMBO Rep 2001; 2(11):1029-1034.
109. Gu J, Kawai H, Nie L et al. Mutual dependence of MDM2 and MDMX in their functional inactivation of p53. J Biol Chem 2002; 277(22):19251-19254.
110. Feng J, Tamaskovic R, Yang Z et al. Stabilization of Mdm2 via decreased ubiquitination is mediated by protein kinase B/Akt-dependent phosphorylation. J Biol Chem 2004; 279(34):35510-35517.
111. Gilkes DM, Pan Y, Coppola D et al. Regulation of MDMX expression by mitogenic signaling. Mol Cell Biol 2008.
112. Li C, Chen L, Chen J. DNA damage induces MDMX nuclear translocation by p53-dependent and -independent mechanisms. Mol Cell Biol 2002; 22(21):7562-7571.
113. LeBron C, Chen L, Gilkes DM et al. Regulation of MDMX nuclear import and degradation by Chk2 and 14-3-3. EMBO J 2006; 25(6):1196-1206.
114. Stommel JM, Wahl GM. Accelerated MDM2 auto-degradation induced by DNA-damage kinases is required for p53 activation. EMBO J 2004; 23(7):1547-1556.
115. Chen L, Gilkes DM, Pan Y et al. ATM and Chk2-dependent phosphorylation of MDMX contribute to p53 activation after DNA damage. EMBO J 2005; 24(19):3411-3422.
116. Pan Y, Chen J. MDM2 promotes ubiquitination and degradation of MDMX. Mol Cell Biol 2003; 23(15):5113-5121.
117. Meulmeester E, Maurice MM, Boutell C et al. Loss of HAUSP-mediated deubiquitination contributes to DNA damage-induced destabilization of Hdmx and Hdm2. Mol Cell 2005; 18(5):565-576.
118. Rallapalli R, Strachan G, Tuan RS et al. Identification of a domain within MDMX-S that is responsible for its high affinity interaction with p53 and high-level expression in mammalian cells. J Cell Biochem 2003; 89(3):563-575.
119. Wade M, Wong ET, Tang M et al. Hdmx modulates the outcome of p53 activation in human tumor cells. J Biol Chem 2006.
120. Agarwal ML, Ramana CV, Hamilton M et al. Regulation of p53 expression by the RAS-MAP kinase pathway. Oncogene 2001; 20(20):2527-2536.

121. Lin AW, Lowe SW. Oncogenic ras activates the ARF-p53 pathway to suppress epithelial cell transformation. Proc Natl Acad Sci U S A 2001; 98(9):5025-5030.
122. Zauberman A, Flusberg D, Barak Y et al. A functional p53-responsive intronic promoter is contained within the human mdm2 gene. Nucleic Acids Res 1995; 23:2584-2592.
123. Phelps M, Darley M, Primrose JN et al. p53-independent activation of the hdm2-P2 promoter through multiple transcription factor response elements results in elevated hdm2 expression in estrogen receptor alpha positive breast cancer cells. Cancer Res 2003; 63(10):2616-2623.
124. Ries S, Biederer C, Woods D et al. Opposing effects of Ras on p53: transcriptional activation of mdm2 and induction of p19ARF. Cell 2000; 103:321-330.
125. Phillips A, Blaydes JP. MNK1 and EIF4E are downstream effectors of MEKs in the regulation of the nuclear export of HDM2 mRNA. Oncogene 2007.
126. Ashcroft M, Ludwig RL, Woods DB et al. Phosphorylation of HDM2 by Akt. Oncogene 2002; 21(13):1955-1962.
127. Meulmeester E, Pereg Y, Shiloh Y et al. ATM-mediated phosphorylations inhibit Mdmx/Mdm2 stabilization by HAUSP in favor of p53 activation. Cell Cycle 2005; 4(9):1166-1170.
128. Li M, Chen D, Shiloh A et al. Deubiquitination of p53 by HAUSP is an important pathway for p53 stabilization. Nature 2002; 416(6881):648-653.
129. Okamoto K, Kashima K, Pereg Y et al. DNA damage-induced phosphorylation of MdmX at serine 367 activates p53 by targeting MdmX for Mdm2-dependent degradation. Mol Cell Biol 2005; 25(21):9608-9620.

CHAPTER 7

Regulation and Function of the Original p53-Inducible *p21* Gene

Jennifer A. Fraser*

Introduction

P21 is a well known regulator of cell cycle progression through its inhibitory actions on Cyclin dependent kinases, (Cdk)/cyclin complexes, and DNA replication via its binding to proliferating cell nuclear antigen (PCNA). p21 also has a role in many diverse cellular processes including modulation of apoptosis, regulation of Rho Kinase and modification of cytoskeletal structures, as well as cellular senescence and differentiation.[1-6] Due to its multiple and wide ranging effects on key cellular processes, intracellular p21 levels are tightly regulated. p53 is a major transcriptional regulator of p21, and p53-dependent transcription of p21 in response to DNA damage is well characterized. However, p53-independent transcriptional pathways also exist.[7] More recently, post-translational mechanisms that influence p21 steady state levels have been identified, including modulation of p21 binding interactions, phosphorylation status, subcellular localization, and trafficking to the proteasome. Post-translational regulation of p21, particularly at its COOH terminus has a significant impact on p21 stability and abundance and therefore is an important determinant of intracellular p21 concentration.

Studies using *Wnt-1* transgenic mice genetically engineered to carry varying gene dosages of p21 showed that although p21 levels had no effect on the age at which mammary tumors occurred, the tumor morphology, size, and proliferative capacity of the tumors were markedly affected by the dosage of p21 (Fig. 1).[8] Measurements of tumor volume showed that complete loss of p21 had very little effect on tumor expansion. However, tumors derived from heterozygous mice grew significantly faster than those found in wild-type and null mice. Furthermore, analysis of tumor sections showed that significantly greater numbers of cells in tumors derived from p21 heterozygous mice were undergoing mitosis and that they were actively replicating (as determined by BrdU incorporation) compared with tumors from p21 wild-type and p21 null mice.[8] This increase in cellular proliferation of heterozygous p21 cells was accompanied by a two-fold increase in cyclin D/Cdk activity towards retinoblastoma protein (Rb), while levels of cyclin D/Cdk activity in null cells were comparable to those in wild-type cells. Furthermore, the level of cyclin/Cdk activity was closely correlated with tumor growth rates. These findings therefore showed that reduced cellular p21 content may have a selective advantage during tumorigenesis and can influence tumor formation and progression. This was attributed to the possibility that in cells containing a single p21 allele, the intracellular p21 concentration would be such that p21 could promote cyclin/Cdk complex formation yet be unable to effectively inhibit Cdk activity thus favouring cellular proliferation. In contrast, cells lacking p21 are unable to promote assembly of cyclin/Cdk complexes and therefore cell proliferation and tumor growth would be

*Jennifer A. Fraser—Cell Signalling Unit, Edinburgh Cancer Research Centre, University of Edinburgh, Crewe Road South, Edinburgh, EH4 2XU, UK. Email: jennyfraser21@hotmail.com

p53, edited by Ayeda Ayed and Theodore Hupp. ©2010 Landes Bioscience and Springer Science+Business Media.

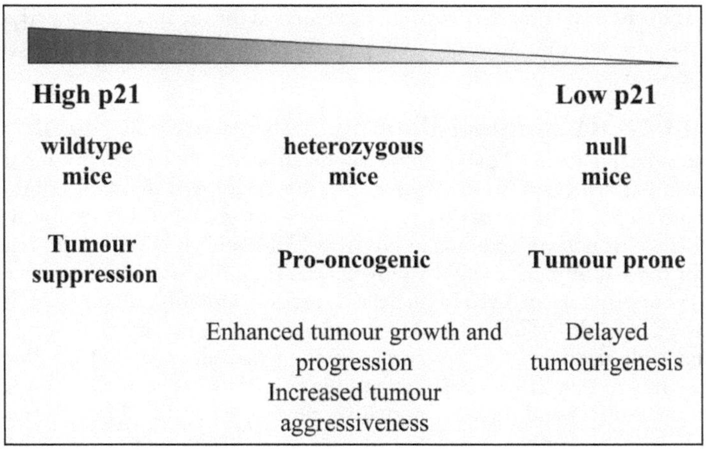

Figure 1. Summary of the effect *p21* gene dosage on murine tumor formation.

retarded. Indeed, more recent studies looking at tumor formation in mice lacking p21 over a 2 year period found that p21 null mice were in fact tumor prone, developing a wide range of spontaneous tumors.[9] In support of the findings by Jones et al,[8] tumor formation in these mice was delayed with an onset age of approximately 16 months.

Since changes in intracellular p21 concentrations have such a significant impact on cellular events such as growth and tumorigenesis, it is essential to understand the importance of both transcriptional and post-translational mechanisms involved in regulating the steady state p21 protein levels; small fluctuations in stability and intracellular turnover may drastically influence intracellular protein concentrations and therefore p21 cellular function. In this chapter, we focus on the post-translational mechanisms which influence intracellular p21 steady state levels and abundance.

p21: Unstructured and Disordered

p21 is a short-lived protein with an approximate half life in vivo of 30-60 minutes.[10,11] Turnover of p21 is thought to occur via active proteolytic degradation via the proteasome; the addition of proteasomal inhibitors in vivo, such as MG-132, lactacystin and LLnL[10,11] leads to the accumulation of p21 protein levels and a dramatic increase in the half-life and stability of the p21 protein. p21 is highly soluble and stable in solution. However, studies using NMR and Circular Dichroism (CD) analyses suggest it is an intrinsically disordered polypeptide with little evidence of secondary or tertiary structure.[12] It is thought that the lack of three-dimensional structure may target p21 for proteolytic degradation by 'default' by the availability of exposed stretches of hydrophobic residues or degrons[13] which may serve as signals for proteasomal degradation. p21 is, however, rarely found uncomplexed in vivo and is usually found as part of a quaternary complex containing cyclins, Cdks and PCNA.[14-16] Indeed, p21 interacts with a plethora of intracellular proteins.[17] It has also been shown to be highly plastic, assuming various conformations under different environmental conditions[18] including interaction with its binding partners. For example, p21 adopts a more defined and ordered structure upon binding to Cdk2.[12] Furthermore, the COOH terminal region of p21 adopts an extended β sheet conformation upon binding to PCNA and possibly the oncoprotein SET, while binding to Calmodulin causes the COOH region of p21 to form an alpha helical structure.[18] Such plasticity therefore enables p21 to interact with multiple diverse ligands and permits p21 to have many biological functions. The resulting increase in p21 order and secondary structure upon association with its binding partners may in theory reduce the accessibility of its hydrophobic sequences and protect p21 from degradation. Indeed, in vivo studies using mutant forms of p21 lacking key residues required for the interaction with PCNA and Cdk show that an association with its binding partners can dramatically modulate p21 stability and abundance.[10]

Impact of COOH Terminal Binding Proteins on p21 Stability

The C-terminal region of p21 is an important determinant of p21 stability and proteasomal degradation; protein-protein interactions within this region can dramatically influence the rate of p21 proteasomal turnover. One of the unique features of p21 is its ability to bind to PCNA, a nuclear protein which binds to DNA and functions in DNA synthesis and replication through the recruitment of DNA polymerases δ and ε, and in DNA repair, through interactions with proteins involved in nucleotide excision and mismatch repair.[19] Binding to p21 inhibits PCNA's ability to bind to DNA and assist DNA polymerases, inhibiting DNA synthesis and replication and therefore progression of the cell cycle. p21 binding to PCNA occurs via hydrogen bonding and electrostatic interactions between the COOH terminal 140-160 residues of p21 and the inter-domain connector loop of PCNA.[20] Interaction with PCNA has a significant impact upon p21 abundance and appears to actively promote p21 stability, as p21 mutants lacking the key residues required for PCNA binding (Met^{147}, Asp^{149} Phe^{150}, ref. 20) demonstrated decreased steady state levels and increased sensitivity to proteasomal degradation in vivo compared to wild-type p21.[10] Indeed, transient transfection of PCNA results in stabilization of p21 in vivo.[21] PCNA is thought to achieve such stabilization by masking the regions of p21 which target it for degradation and thus protect p21 from proteasomal degradation.

p21 can directly interact with C8 alpha subunit of the 20S proteasome through residues 156-161 of its COOH terminus.[21] The C8 subunit itself does not posses proteolytic activity, but is required for the correct assembly and stabilization of the proteasome.[22] Binding at the COOH terminal is sufficient to target p21 for degradation; loss of residues 156-161 prevents p21 from interacting with C8 and its subsequent degradation by purified 20S proteasomes in vitro. Furthermore, mutant forms of p21 lacking residues 156-161 display greater stability in vivo.[21,23] Interaction with the C8 subunit therefore recruits p21 to the proteasome directly without the need for prior ubiquitination. As the C8 interaction motif of p21 overlaps the region required for PCNA binding, it is possible that the observed increase in p21 stability in the presence of PCNA is due to masking of the C8 binding site, thus preventing the association with C8 and the proteolytic degradation of p21.

Cyclin D1 has also been shown to protect p21 from proteolytic degradation by disrupting C8 association with p21.[24] p21 contains two cyclin binding sites (Cy1, residues 17-24 and Cy2, residues 153-159) located at the NH_2 and COOH termini, respectively.[23] Cyclin D1 is able to interact with both Cy1 and Cy2 sites within p21 and, like PCNA, bind to the COOH terminus of p21. Furthermore, cyclin D1 competes with C8 for binding to p21 and can prevent purified C8 from interacting with p21 in a concentration dependent manner.[24] The presence of cyclin D1 was also able to protect p21 from degradation by purified 20S proteasomal subunits in vitro. The effects on p21 stability were specific for cyclin D1 as the highly homologous cyclin K was unable to protect p21 from degradation in the presence of purified 20S subunits. Transformation of 3T3 cells with oncogenic Ras causes the post-translational accumulation of p21.[25] The observed increase in p21 protein levels was paralleled by the accumulation of several proteins, including cyclin D1. The effects of Ras on p21 abundance in 3T3 cells could be mimicked by the addition of lactacystin and the transfection of cyclin D1 alone, suggesting that the observed stabilization of p21 may result from a cyclin D1 dependent decrease in proteasomal turnover of p21, highlighting the importance of cyclin D1 in influencing p21 stability under certain conditions in vivo.

Cyclin E can also stabilize p21;[26] co-expression of cyclin E in the presence of Cdk2 in Soas-2 cells increased the half life of p21 to over 3 hours. Through the use of p21 deletion mutants, it was shown that cyclin E/Cdk2 dependent stabilization was mediated through binding to the N-terminal Cy1 and Cdk binding sites on p21 as cyclin E/Cdk2 was unable to stabilize p21 in their absence. Surprisingly, stabilization of p21 in the presence of cyclin E was more pronounced when the C-terminal Cy2 site was deleted, indicating that cyclin E dependent stabilization of p21 is inhibited by the Cy2 site.[26] p21 was also far more stable in the presence of a kinase dead form of Cdk2, Cdk2 NFG, suggesting that Cdk2 activity may destabilize p21. Furthermore, the ability of Cdk2 to destabilize p21 required the presence of a potential Cdk2 phosphorylation site, Ser^{130}, and the Cy2 binding motif.[26] These studies therefore demonstrate that cyclins can stabilize p21 via different mechanisms. While the interaction with cyclin E may promote p21 stability, it can also destabilize p21 through its interaction with Cdk2. Recruitment of Cdk2 by cyclin E increases Cdk2 binding to p21 via the Cy2 site and Cdk2 dependent phosphorylation of p21 at Ser^{130}, and this in turn leads to the active destabilization of p21. p21 can, however, inactivate Cdk2 and as a result protect itself from degradation by preventing Cdk2 dependent phosphorylation at $Ser.^{130}$ High intracellular levels of p21 would therefore promote its own stability through inactivation of Cdk2 and the inhibition of active degradation. The relative stability and turnover of p21 will therefore be determined by the intracellular abundance of cyclin E, Cdk2 and p21.

In contrast, Mdm2 has been shown to negatively regulate p 21 stability by facilitating p21's interaction with the C8 subunit and subsequently promoting proteasomal degradation of p21.[27] Knockdown of Mdm2 protein expression using antisense oligonucleotides and siRNA showed an inverse correlation between p21 stability and Mdm2 levels which is independent of p53 status. The decrease in p21 stability in the presence of elevated Mdm2 could be partially overcome by inhibition of proteasomal activity with MG-132 suggesting that Mdm2 was modulating p21 turnover by the proteasome. Mdm2 is a well-known E3 ubiquitin ligase which facilitates p53 ubiquitination and degradation.[28] However a mutant form of Mdm2 lacking E3 ligase activity was still able to destabilize p21, suggesting that its effects on p21 degradation are not a result of p21 ubiquitination and enhanced trafficking to the proteasome. Instead, Mdm2 was shown to directly interact with p21 and enhance its interaction with C8 subunit, thus directly promoting p21's degradation.[27]

The many binding partners of p21 can have a significant impact on the structure, conformation and stability of p21. As many of these compete for interaction with the COOH terminal region of p21, the relative concentration and abundance of each binding partner will determine the fate of p21 by tipping the balance in favour of either stability or degradation. It should be noted, however, that p21 mutants lacking C8 binding sites still accumulate in vivo following the addition of lactacystin.[21] This suggests that p21 may always be destined for degradation by the proteasome and that the presence of binding partners such as PCNA, cyclin D and Mdm2 only serve to modulate the overall rate of p21 degradation.

Due to its impact on p21 stability and the multitude of binding partners which interact with p21 through its COOH terminus, this region has been viewed by many as a potential therapeutic drug target.[29-31] Several studies have shown that small peptide molecules corresponding to the COOH terminal region of p21 are sufficient to disrupt the function of many p21 binding partners both in vitro and in vivo. Peptide 10, corresponding to residues 141-160 of p21, was able to bind and specifically inhibit the activity of cyclin D/Cdk4 towards Rb at concentrations which were comparable to full length p21.[30] Cell penetrating forms of peptide 10 were also found to be biologically active in vivo, preventing hyper-phosphorylation of Rb and inducing cell cycle arrest in HaCaT cells.[30] Indeed, an 8 amino acid truncated version of this peptide containing the key residues required to inhibit cyclin D/Cdk4 activity (RRLIF) was able to mimic the activity of full length p21 and induce growth arrest.[30] The p21 RRLIF peptide is a highly potent inhibitor and is sufficient for the inhibition of the activity of cyclin A/Cdk2.[32] It has been successfully crystallized with the cyclinA/Cdk2 complex[31] (Fig. 2). Other studies show that COOH terminal peptide forms of p21 can also disrupt PCNA function in vitro[33] and can induce apoptotosis when fused to cell penetrating sequences such as HIV-Tat.[29]

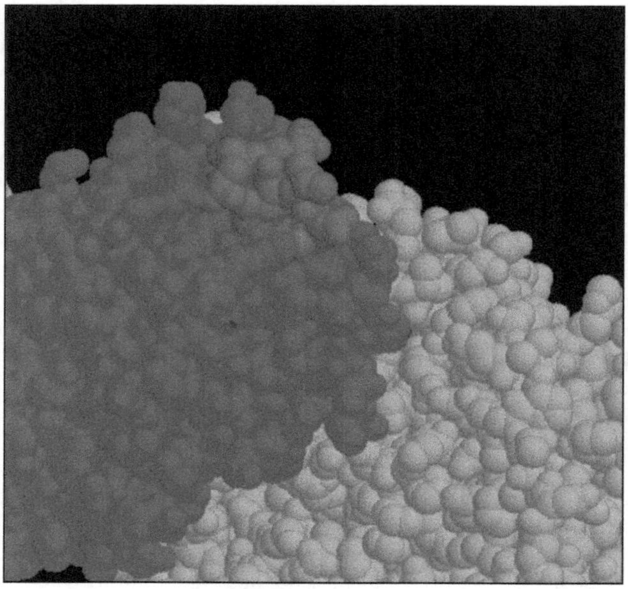

Figure 2. p21 peptide RRLIF complexed with Cyclin A/Cdk2 (Protein database code 1oKV, visualized using FirstGlance in Jmol and adapted from Kontopidis et al 2003). The RRLIF peptide is shown in brown bound to Cyclin A, shown in dark blue. Cdk2 is shown in pale blue. A color version of this figure is available online at www.landesbioscience.com/curie. Portions adapted from FirstGlance in Jmol (bioinformatics.org/firstglance).

These studies therefore show how small peptide ligands corresponding to the COOH-terminus of p21 can mimic the function of the full-length protein and act as potent inhibitors of large protein complexes such as cyclin/Cdk. Such peptide mimetics are therefore powerful tools to study and delineate protein function, and may also represent a potential angle for pharmacological intervention in the treatment of transformed cells in which the function of p21 may be re-instated as a way to halt tumor progression and development.

The Role of p21 NH_2 Terminus in Regulating Stability

Protein-protein interactions at the NH_2 terminal of p21 also regulate p21 stability. The interaction between cyclin/Cdk complexes and p21 requires binding to one, or both of the Cy sites, as well as binding to the Cdk binding site located in the centre of p21 (residues 49-71). p21 mutants lacking residues required for Cdk binding (Trp^{49}, Phe^{51} and Asp^{52}) exhibited decreased sensitivity to proteasomal degradation and resulted in enhanced p21 stability,[10] suggesting that the interaction with Cdk may actively promote p21 degradation.

More recent studies have identified novel binding proteins which interact with the NH_2 terminus of p21 and that also regulate p21's stability. A novel p21 binding protein, WISp39 (*W*AF1/*C*IP1 *s*tabilizing *p*rotein 39), interacts with residues 28-56 of p21, and was identified using yeast two hybrid analysis.[34] Modulation of intracellular WISp39 levels in COS cells using transient transfection significantly increased the abundance of p21 and increased its half life while, siRNA targeted knock down of WISp39 mRNA in U2OS and HeLa cells dramatically destabilized endogenous p21. Pulse chase analysis showed that the stability of metabolically labeled p21 was dramatically reduced in WISp39 siRNA treated U2OS cells, suggesting that WISp39 specifically targets newly synthesized p21. Since proteasome inhibitors lactacystin and LLnL could overcome the destabilizing effect of WISp39 mediated siRNA, it is thought that WISp39 may protect newly synthesized endogenous p21 from proteasomal turnover. WISp39 is a TPR containing protein which can bind heat shock protein 90 (Hsp90); gel filtration analysis of U2OS extracts show that WISp39 associates with Hsp90 and p21 as part of a 400 kDa trimeric complex. Interaction with Hsp90 is essential for WISp39 stabilization of p21 since inhibition of Hsp90 ATPase activity in NIH 3T3 cells using geldanamycin decreased endogenous p21 protein levels. Hsp90 is a well-known molecular chaperone which binds disordered polypeptides and promotes correct protein folding. Newly synthesized p21 would be a prime target for chaperone binding due to its inherent lack of secondary structure and the availability of hydrophobic degrons. Interaction with WISp39 and Hsp90 would therefore play an essential role in stabilizing newly synthesized p21 by shielding it from the proteasome until it interacts with its other cellular binding partners.

Exposure to ionizing radiation switches on the ATM:p53 pathway, leading to activation of p53, p53-dependent transcription of p21, and accumulation of p21 protein.[35] This is necessary in order to activate a checkpoint response so as to arrest the cell cycle and permit repair of any damaged DNA. WISp39 and Hsp90 appear to be essential for stabilizing p21 protein levels in response to genotoxic stress; inhibition of Hsp90 activity with geldanamycin or siRNA targeting of WISp39 prevents p21 accumulation following exposure to ionizing radiation.[34] Moreover, WISp39 siRNA treated HCT-116 cells fail to arrest in G2 following ionizing radiation and progress into mitosis, suggesting that the loss of WISp39 and Hsp90 activity compromises the normal p21 checkpoint response following DNA damage. WISp39 and Hsp90 are therefore instrumental in maintaining the DNA damage check point in response to genotoxic stress via their ability to regulate p21 stability.

Effect of Post-Translational Modifications on p21 Abundance

p21 stability is also influenced by post-translational modifications that modulate p21 turnover by perturbing p21 interactions with its binding partners. Phosphorylated forms of p21 have been isolated in vitro and in vivo,[36,37] and several kinases, including Akt, PKC, PKA, GSK (see below and ref. 38 for a recent review) and DAPK[39] are reported to target p21. PKA and PKC were shown to phosphorylate p21 at Thr^{145} and Ser^{146} respectively and these modifications were shown by ELISA to be sufficient to disrupt p21 binding to PCNA.[37] Thr^{145} and Ser^{146} are part of an amino acid cluster which adopts an alpha helical structure (3_{10} helix) when bound to PCNA which fits into PCNA's hydrophobic pocket.[20] These residues also form hydrogen binding interactions that stabilize the 310 helix structure and p21's interaction with PCNA. Phosphorylation of Thr^{145} and Ser^{146} may therefore destabilize the secondary structure of the p21 and perturb key hydrophobic interactions required for p21 binding to PCNA, accounting for the observed decrease in PCNA binding.

The atyptical PKC, PKCζ, phosphorylates p21 at Ser^{146} in vivo; activation of PKCζ signalling in HCT-116 cells by treatment with insulin or transient over-expression of upstream kinase PDK destabilized endogenous p21.[40] This was attributed to decreased binding to PCNA and the resultant increase in p21 proteasomal turnover as lactacystin was able to overcome PKCζ's effects on p21 stability. Activation of PKC thus permits cell growth in response to insulin signalling through targeted phosphorylation and degradation of p21. This would be sufficient to reduce the inhibition of PCNA and Cdk2 leading to the promotion of DNA synthesis and cell cycle progression.

The serine threonine kinase Akt has also been shown to phosphorylate p21 in vitro.[36] Activation of the Akt pathway enhanced levels of phospho-Thr^{145} p21 forms in vivo.[41] As observed with phosphorylation at Ser^{146}, Thr^{145} phosphorylation was correlated with decreased p21 interaction with endogenous PCNA[42] and transfection of phospho-mimetic forms of p21 showed that the capacity of the mutant Thr^{145}Asp p21 for binding to PCNA was reduced compared to that of Thr^{145}Ala and wild-type p21 forms.[36] However, in contrast to the destabilizing effect of Ser^{146} phosphorylation on p21, activation of Akt and the resultant p21 phosphorylation at Thr^{145} correlated with increased p21 stability and abundance in vivo.[41,42] This appears to be highly contradictory considering phosphorylated p21's compromised capacity to bind PCNA and the importance of PCNA in stabilizing p21.[10] However, activation of the Akt pathway leads to modulation of many intracellular proteins.[43] In addition to targeting p21, Akt can also positively regulate the abundance of cyclin D1. Akt achieves this both directly, through transcriptional upregulation of cyclin D1 mRNA,[24] and indirectly, via modulation of glycogen synthase kinase 3 (GSK-3) activity.[44] GSK-3 targets and phosphorylates cyclin D1 at Thr^{286}. Phosphorylation here enhances cyclin D1 interaction with CRM1, the nuclear pore protein, and promotes cyclin D1 translocation from the nucleus to the cytoplasm where it is subsequently targeted for ubiquitin mediated degradation.[44] GSK-3 activity therefore enhances cyclin D1 turnover and degradation. GSK-3b can also destabilize endogenous p21 in endothelial cells by phosphorylating p21 at Thr^{57}.[45] Mutation of the Thr^{57} consensus phosphorylation site or inhibiting GSK using LiCl prevents degradation of endogenous and exogenous p21.[45] As GSK-3 is phosphorylated and inactivated by Akt, activation of the Akt pathway can indirectly contribute to p21 stability via modulation of p21 and cyclin D1 abundance. Furthermore, p21 phosphorylation at Thr^{145} by Akt can promote cyclinD/Cdk4 complex formation in vitro,[45] while activation of Akt in vivo was correlated with increased cyclin D1 binding to p21 and enhanced assembly of active cyclinD1/Cdk4 complexes.[42] Therefore, despite the decrease in p21's protective interaction with PCNA following Thr^{145} phosphorylation, Akt's capacity to both enhance cyclin D1 abundance and promote its interaction with p21 (resulting in the blockade of C8 binding and proteasomal targeting) would account for the observed increase in p21 stability.

By targeting p21 and its binding partners, activation of the Akt pathway promotes cellular growth and proliferation through enhanced DNA synthesis due to the release of PCNA from p21 inhibition, as well as increased transit through the cell cycle due to increased cyclin D1/Cdk4 assembly and increased Cdk2 activity. As further proof, transfection of Thr^{145}Asp phospho-mimetic forms of p21 recapitulates Akt proliferative effects in endothelial and 3T3 cells.[36,41] Findings show that modification of p21 phosphorylation status can impact key cellular processes such as growth. The relative abundance and intensity of signalling will thus determine the outcome.

The Role of Ubiquitination in Proteasomal p21 Degradation

Post translational modification of cellular proteins by attachment of polyubiquitin chains targets proteins for degradation by the proteasome. While there is little doubt that p21 undergoes rapid proteolytic degradation by the proteasome, the exact mechanisms involved in targeting p21 for degradation remain hotly-debated, and much controversy exists as to whether p21 requires prior ubiquitination in order to mark p21 for degradation. This is partly due to the ability of p21 to interact with the C8 subunit of the 20S proteasome and undergo degradation by isolated 20S subunits in the absence of ubiquitin[21,46,47] suggesting that p21 may bypass the need for ubiquitin targeting by direct association with the proteasome.

Ubiquitinated forms of both exogenous and endogenous p21 have been identified in vitro and in vivo, and the abundance of such intermediates increases following inhibition of proteasomal function,[10,11,48] suggesting that ubiquitination of p21 is required prior to targeting it for degradation. In support of this, rapid degradation of in vitro translated p21 by purified 26S proteasomal subunits could be inhibited in the presence of methylated ubiquitin, a chemically modified form of ubiquitin which terminates polyubiquitin chains.[14] Furthermore, transfection of a mutant form of ubiquitin lacking lysine residues (Ub(K0)) caused endogenous p21 to accumulate in 293T cells.[14] As both methylated ubiquitin and Ub(K0) are unable to form polyubiquitin chains, it was concluded that poly-ubiquitination is required for p21 recognition and degradation by 26S proteasomes in vitro and in vivo. In contrast, over-expression of ubiquitin lysine mutants did not affect p21 turnover and basal expression, and the abundance of exogenous p21 in 293 cells and its sensitivity to proteasomal inhibition by MG-132 p21 was unaffected by over-expression of the ubiquitin mutant UbR7.[11] This was in contrast to other known ubiquitinated substrates, such as cyclin E, the abundance and half life of which was dramatically increased by the presence of UbR7 following proteasomal inhibition.[11] Such findings imply that prior multi-ubiquitination may not be required for efficient degradation of p21.

Studies using mutant forms of p21 lacking lysine residues and attachment sites for ubiquitin chains also suggest that p21 undergoes ubiquitin-independent means of degradation. The kinetics of turnover of p21 K6R, a p21 mutant containing lysine to arginine mutations, was analogous to that of exogenously expressed wild type p21. Both forms of p21 were equally sensitive to proteasomal inhibition in vivo by MG-132.[11] This was in contrast to the lysine-free cyclin E mutant whose abundance was no longer sensitive to proteasome inhibition. In vitro analysis of the turnover of p21 WT and p21(K0), a p21 mutant lacking lysine residues, in the presence of purified 26S proteasomal subunits showed that the rates of degradation and the half lives of both p21 forms were comparable,[14] suggesting that p21 does not require prior ubiquitination in order to be degraded by the 26S proteasome. In contrast, Bloom et al[14] found that the half lives of both wild type and p21(K0) increased in the presence of methylated ubiquitin and found traces of ubiquitinated forms of both wild type and p21 lysine mutant accumulate over time in vitro in the presence of purified 26S proteasomes. Transfection of lysine free ubiquitin (Ub(K0) also caused a dose dependent increase in the cellular levels of p21 K0 mutant, and ubiquitinated forms of p21 K0 were identified and isolated from NIH 3T3 and 293 cell extracts. From this, it was concluded that polyubiquitination and an intact ubiquitin system was in fact essential for the efficient degradation of both wild type and mutant p21 by the proteasome. Since p21 K0 behaved like wild type p21, the authors suggested that ubiquitin attachment may occur at residues other than lysines; ubiquitin attachment to the NH_2 terminal residue of p21 could mediate ubiquitin dependent targeting of p21 to the proteasome.[14]

Due to the low abundance and difficulty in detecting endogenous ubiquitinated p21, many researchers have studied p21 ubiquitination using exogenous forms of p21 and ubiquitin, often using plasmid DNA concentrations greatly in excess of physiological levels. The kinetics of exogenously expressed p21 protein turnover and stability do not resemble that of endogenous p21,[10,40] suggesting exogenous p21 may not undergo the same physiological regulation as endogenous p21. Mass spectrometric analyses of endogenous p21 and exogenously expressed HA-tagged p21 isolated from U2OS cells suggest they are also subject to differential post-translational modifications in vivo.[49] The majority of endogenous p21 was found to be N-terminal acetylated, an irreversible modification that would prevent further attachment of additional post-translational modifications such as ubiquitin. In contrast, exogenously expressed HA-p21 was mostly unacetylated at its NH_2 terminus, probably due to the presence of the HA tag altering p21 packaging and modification. The increased abundance of unacetylated HA-p21 may therefore permit spurious non-physiological attachment of ubiquitin at the unmodified NH_2 terminus of p21 and account for the NH_2 terminal ubiquitination observed by Bloom.[14] These findings highlight the effect protein tagging can have on post translational modification of proteins and underscore the potential pitfalls of using exogenous expression constructs when studying cellular processes as sensitive and complex as protein stability.

More recent studies have shown the importance of ubiquitin mediated degradation in p21 turnover through the targeted knockdown of expression of specific proteasomal subunits. Several forms of the proteasome complex exist within the cell, depending on the type of 'lid' protein complexed to the 20S catalytic core.[50] Lid proteins such as the 19S and 11S subunits regulate substrate recognition and gate protein access into the 20S catalytic core for degradation. The 26S proteasome consists of two regulatory 19S subunits complexed with a 20S subunit and as the regulatory 19S subunit confers specificity for ubiquitinated substrates the 26S proteasome is thought to selectively degrade proteins labeled with poly-ubiquitin chains. In contrast, the 11S subunits, composed of REGα, β or γ subunits, can activate 20S catalytic activity and permit substrate access to the 20S core in the absence of ubiquitin attachment.

By using siRNA to target and down regulate the expression of the 19S subunit of the proteasome, and thus uncouple ubiquitin dependent and independent proteasomal protein degradation, recent studies revealed that the rate of endogenous p21 degradation is unaffected by loss of 19S subunits.[46] This is in contrast to p27, whose degradation was dramatically reduced in the absence of 19S subunits. Down regulation of REGγ expression using siRNA did however dramatically stabilize p21 protein levels, causing a 4 fold increase in p21 half life[46,51] suggesting proteasomal degradation of p21 occurs via a REGγ dependent pathway and as such does not require prior ubiquitination. Interestingly, p21 binds to REGγ subunits[51] through residues 156-161 in the COOH terminus of p21, the same residues known to mediate interaction with the C8 subunit.[21,46]

The Impact of Ubiquitinating Enzyme Manipulation on p21 Stability In Vivo

The relative importance of the ubiquitin system in regulating the proteasomal targeting of p21 has also been studied using thermolabile cell lines, such as tsBN75 and ts20TG, which express temperature sensitive mutants of the E1 ubiquitin-activating enzyme. In such cell lines, the E1 ubiquitin activating enzyme is inactivated by shifting the cell growth conditions from ~35 to ~39 °C. Analysis of endogenous p21 turnover in ts20TG and tsBN75 cells grown at the restrictive temperature suggests that p21 stability is unaffected by the inactivation of the E1 enzyme; the half life of p21 at the restrictive temperature was comparable to its half life at the permissive temperature.[49,52] This was not due to incomplete inactivation of E1 since the half lives of several intracellular proteins known to be targeted for ubiquitination, such as cyclin D1, p53, and p27, were dramatically increased.[49] A recent study revealed that p21 degradation in ts85 cells growing at the restrictive temperature was only prevented by knock down of REGγ subunit using siRNA,[51] suggesting that proteasomal degradation of p21 under such conditions does not require a functional ubiquitin system, yet does require the presence of a REGγ proteasomal lid to gate p21 access to the catalytic core.

The potential role of cullin dependent ubiquitin ligases in the ubiquitination of p21 has also been studied using cell lines containing a temperature sensitive mutant of the Nedd-8 activating enzyme, APP-BP1.[14,49] Nedd-8 conjugation to cullins is required for their activity as ubiquitin ligases but at restrictive temperatures, Nedd-8 conjugation is inhibited, rendering cullins less active as ubiquitin ligases. Loss of APP-BP1 did not effect the rate of endogenous p21 turnover in ts41 cells and proteasomal degradation of p21 continued unperturbed.[49] These findings suggest that ubiquitin targeting of p21 may be dispensable for proteasomal degradation and that p21 is efficiently degraded in its absence. In contrast to the findings by Chen,[49] p21 accumulation in tsBN75 and ts41 cell lines following incubation at ~39 °C was observed by Bloom et al;[14] however, these studies examined turnover of exogenously expressed p21. Since exogenous p21 does not behave in the same manner as endogenous p21, excessive over-expression of exogenous p21 may quite easily swamp the normal endogenous degradation pathways of p21. Such conditions may lead to cellular stress due to protein accumulation and trigger degradation of excessive p21 via non-physiological or incidental ubiquitination. Furthermore, studies using such thermolabile cell lines have been compounded by the fact that endogenous p21 accumulates following growth at the restrictive temperature. This is due to induction of p53 dependent transcription of p21 that causes a secondary accumulation of p21 protein.[49]

There is obviously a great deal of contradictory evidence present in the literature as to the role of ubiquitin modification in p21 proteasomal turnover. It has been suggested that the discrepancies between studies may be due to factors such as insufficient expression of mutant forms of ubiquitin, preventing saturation and thus blockade of the endogenous ubiquitination system; or excessive expression of mutant forms of p21 swamping the physiological degradation pathways, forcing degradation by alternative means. Inhibition and manipulation of the endogenous ubiquitin system in vivo would also drastically perturb the abundance of many p21 binding partners, indirectly affecting p21 turnover and causing secondary accumulation

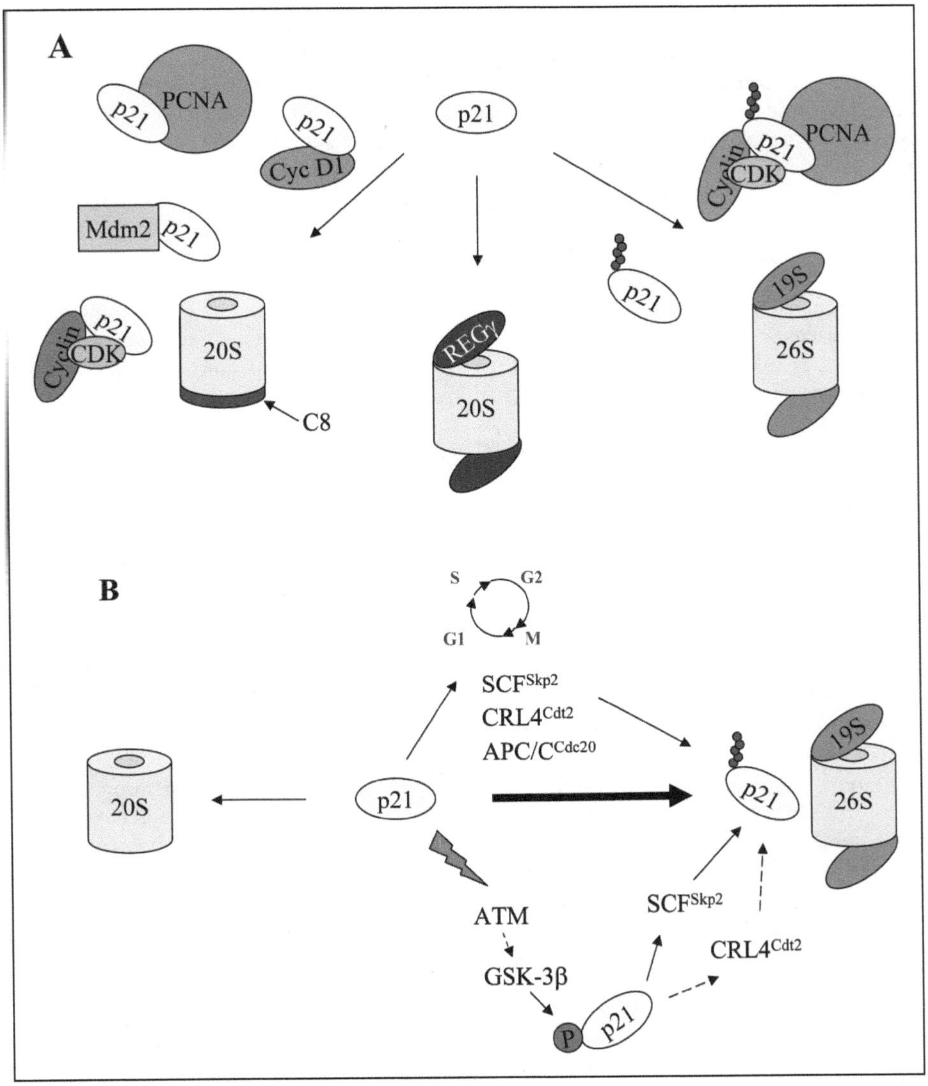

Figure 3. The many routes of p21 degradation. A) Literature suggests ubiquitin dependent and independent mechanisms regulate p21 turnover and that the presence and abundance of p21's binding proteins determine the path chosen and the flux of p21 through each path. B) The flux of p21 trafficking to the proteasome for degradation may alter under conditions of stress and at various points in the cell cycle.

of p21. Ubiquitination is however a reversible protein modification. As such, it is possible that the inability to identify ubiquitinated forms of p21 may not necessarily be due to the absence of such modification, but instead may be due to the action of deubiquitinating enzymes depleting ubiquitinated species to below detectable levels. There is therefore a considerable amount of evidence supporting both ubiquitin dependent and independent means of p21 protein degradation and it is most likely that both are involved in regulating the physiological turnover of p21 (Fig. 3A). The relative flux of p21 through each pathway, and therefore each pathway's importance, may fluctuate depending on various cellular factors, such as the stage of the cell cycle and conditions of stress (Fig. 3B).

Physiological Destabilization and Degradation of p21

p21 protein levels fluctuate throughout the cell cycle, peaking during G1 and G2. This facilitates the formation of cellular checkpoints necessary for correct entry into S phase and mitosis. In order to permit the efficient progression from G1 into S phase, and G2 into mitosis, cyclin/Cdk complexes must be liberated and co-ordinately activated. This is achieved by the active destabilization of the cyclin dependent kinase inhibitors, p21, p27 and p57. p27 is the principle substrate of the ubiquitin E3 ligase SCF^{Skp2}, the activity of which is essential for regulating p27's abundance and degradation.[53] SCF^{Skp2} levels fluctuate throughout the cell cycle and peak during S phase. The increase in SCF^{Skp2} activity coincides with degradation of p27.[53]

MEFs derived from $SCF^{Skp2-/-}$ mice enhance the steady state protein levels and half life of p21, compared to wild-type MEFS.[54] p21 protein levels are also lower in cells in which Skp2 and its co-factor Cul1 are down-regulated,[55] suggesting that the ubiquitin targeting of p21 by SCF^{Skp2} may influence p21 stability and turnover. Indeed, in vitro ubiquitination analyses have shown p21 to be a good substrate for SCF^{Skp2} and that ubiquitination of p21 could be reconstituted by the SCF^{Skp2} ubiquitin ligase complex in the presence of Cks1 and Cdk2/cyclin E.[54,56] Ubiquitination of p21 by SCF^{Skp2} differs slightly from that of p27; Cdk dependent phosphorylation of p21 is not essential for SCF^{Skp2} targeting of p21 in vitro, unlike p27, where phosphorylation at Thr^{187} is essential for it to be targeted by SCF^{Skp2}.[53,57] However, like p27, p21 turnover during the cell cycle may be controlled by the activity of the SCF^{Skp2} ubiquitin ligase.[54] Studies using MEFs synchronized in S phase have shown that p21 was rapidly degraded as cells transited through S phase, with a half life of ~1 hour. In contrast, p21 levels in $SCF^{Skp2-/-}$ MEFs are significantly higher than wild-type MEFs and p21 stability following release into S phase is far greater than in wild-type cells, with a half life of 2-4 hours. The reduced rate of p21 degradation in the absence of SCF^{Skp2} suggests that active degradation of p21 occurs during S-phase, in an ubiquitin dependent manner and under the control the SCF^{Skp2} E3 ligase. This is essential in order to permit the full activation of Cdk/cyclin complexes and transit through S phase, as well as the release of PCNA from p21 dependent inhibition, permitting DNA synthesis and replication. The authors do not mention whether $SCF^{Skp2-/-}$ MEFs show delayed transit through the cell cycle, although this would be expected since stabilization of p21 in the absence of $SCF^{Skp2-/-}$ would impede the timely activation of the relative cyclin/Cdk complexes and retard cell cycle progression.

More recent work suggests that additional ubiquitin E3 ligase complexes may also regulate p21 degradation during S-phase. In U20S cells, p21 protein levels are significantly stabilized following siRNA depletion of members of the $CRL4^{Ctd2}$ E3 ubiquitin ligase complex.[58] Furthermore, over expression of CRL4's substrate specificity factor, Ctd2, significantly lowered p21 levels. Analysis of synchronized U20S cells following treatment with nocodazole showed Ctd2 specifically decreased p21 levels during S phase. Ctd2 over-expression also resulted in counter stabilization of p21 following siRNA depletion of SCF^{Skp2}, suggesting that a degree of redundancy exists between these ligases in regulating p21 turnover during S-phase.[58]

Intracellular p21 protein levels peak again during G2 before undergoing rapid degradation prior to mitosis. Degradation of p21 was unaffected by the loss of SCF^{Skp2}, and instead, another E3 ligase, APC/C^{Cdc20}, was found to mediate the rapid degradation of p21 required for entry into mitosis.[59] Of the 21 human F box proteins tested, only SCF^{Skp2} and APC/C^{Cdc20} were shown to co-immunoprecipitate endogenous p21 from HEK293 cells.[59] APC/C^{Cdc20} is thought to interact with p21 through an RxxL recognition motif, known as the destruction or D-box. This motif is commonly found in substrates targeted by APC/C^{Cdc20}; a putative D-box is found in p21 at residues 86-94. While loss of this region had no effect on the ability of SCF^{Skp2} to rapidly degrade p21, the p21 D-box mutant was refractory to degradation in the presence of APC/C^{Cdc20}.[59] Furthermore, deletion of this region prevented effective ubiquitination of in vitro translated p21 in the presence of the APC/C^{Cdc20} complex, suggesting that APC/C^{Cdc20} specifically targets p21 for ubiquitination and subsequent proteasomal degradation via the D-box. These findings show that a second E3 ligase is important in degradation of p21 and is required to target and degrade p21 in early mitosis in an ubiquitin dependent manner. This rapid APC/C^{Cdc20} dependent degradation of p21 is essential in order to permit full activation of Cdk1/cyclin A2 and Cdk1/cyclin B1 complexes and for mitosis to proceed. Together these findings show that ubiquitin mediated targeting of p21 to the proteasome for degradation by the SCF^{skp2} and APC/C^{Cdc20} E3 ubiquitin ligase complexes is essential for rapid p21 turnover at various points of the cell cycle and underscore the importance of ubiquitin dependent pathways in regulating physiological turnover of p21 (Fig. 3B).

The presence and abundance of p21's binding partners can have a huge impact on its stability and turnover. Furthermore, many p21 binding partners are subject to regulation via ubiquitin mediated proteasomal degradation, including PCNA, cyclins and Cdk. Some researchers have argued that the loss of E3 ligases such as SCF^{Skp2} would perturb normal ubiquitination and proteasomal trafficking of many of p21's binding partners and may indirectly stabilize p21. Indeed, the abundance of cyclin E is enhanced in SCF$^{Skp2\ -/-}$ cells, and the restoration of cyclin E levels to wild-type SCF^{Skp2} levels using shRNA was found to restore the rapid degradation rate of p21.[49] However, few studies addressing the role of ubiquitin in regulating p21 degradation and stability use synchronized populations of cells, nor analyze p21 protein turnover at particular stages in the cell cycle. This may therefore account for the observed discrepancies and contradictory findings present in the literature. The stage of cell growth will determine the relative importance of ubiquitin dependent and independent pathways in p21 turnover. For example, p21 was only found to interact with APC/C^{Cdc20} in cell lysates derived from premitotic cells and not unsynchronized populations.[59] Furthermore, siRNA mediated knockdown of APC/C^{Cdc20} only impacted p21 stability during the early stages of mitosis and not during G1, G1/S and G2, while the impact of SCF^{Skp2} loss on p21 turnover was only apparent in cells synchronized in S-phase and not G2 or pro-metaphase.[59] These findings highlight the fact that p21 may be subject to several degradation pathways, and that the importance of these may only become apparent under certain physiological conditions and at particular points throughout the cell cycle.

Ubiquitin-Mediated p21 Degradation under Conditions of Stress

Many researchers believe that basal p21 turnover and degradation occurs independently of the ubiquitination system, and that the ubiquitination of p21 is switched on under conditions of cellular stress (Fig. 3B). Exposure to low doses of UV irradiation triggers rapid proteasomal degradation of endogenous p21 in many cell lines.[52,60] The rapid reduction of p21 concentration was prevented in the presence of caffeine and Cre recombinase deletion of ATR[52], or by siRNA knockdown of ATR protein levels[61], suggesting that the activation of the ATR pathway by UV irradiation triggers p21 degradation. ATR is thought to achieve this via activation of GSK-3β; transient increase in GSK-3β activity was observed in as little as 15 minutes after UV treatment of SK-MEL-1 cells, and this was prevented by treatment with caffeine or siRNA depletion of ATR.[61] Pre-treatment of SK-MEL-1 and MCF-7 cells with GSK-3β inhibitors, LiCl and TDZD-8, or by the depletion of GSK-3β protein levels using siRNA also prevented the rapid degradation of p21 following exposure to UV[61], suggesting that GSK-3β signals downstream of ATR destabilize p21. Destabilization of p21 in response to UV is thought to occur via GSK-3β phosphorylation of p21 at Ser^{114}; mutation of this site to alanine prevented UV induced degradation of p21. How phosphorylation at this residue triggers proteasomal degradation of p21 is unknown. However, it is possible that this phosphorylation creates a docking site which promotes p21's interaction with ubiquitin ligases; recent work by Abbas et al showed that the efficiency of p21 ubiquitination by the CRL4^{Ctd2} complex was greatly increased using a phospho-mimetic mutation of Ser^{114}.[58]

Several lines of evidence suggest that the rapid turnover of p21 following UV occurs in an ubiquitin dependent manner. His-ubiquitin forms of p21 could be isolated from U2OS cells following exposure to UV yet not under basal conditions.[52] Furthermore while WT and K6R mutant forms of p21 displayed similar kinetics of turnover under basal conditions, p21 K6R failed to undergo rapid degradation following UV damage and displayed much greater stability.[52] Indeed expression of dominant negative Lys[48] forms of ubiquitin, which are unable to form polyubiquitin chains, also prevented p21 degradation following UV treatment, suggesting that polyubiquitin attachment to p21 is required to induce rapid p21 turnover. ts20TG cells were also used to demonstrate the requirement of the ubiquitin system in p21 turnover as part of the cellular response to UV damage. While p21 turnover under basal conditions was unaffected by the inactivation of E1, the rapid turnover of p21 observed following UV damage was prevented when ts20TG cells were grown at the restrictive temperature under conditions in which E1 was inactivated.[52]

UV induced ubiquitination of p21 is thought to be dependent upon the SCF^{Skp2} E3 ligase. Although p21 does not interact with SCF^{Skp2} under basal conditions, p21/SCF^{Skp2} complexes could be isolated from U2OS cells following treatment with UV. Indeed, disruption of SCF^{Skp2} in MEFs prevented p21 UV induced degradation while p21 turnover in unstressed cells was unaffected and comparable to wild-type MEFs.[52] These findings suggest that p21 may be an in vivo substrate of SCF^{Skp2} under certain conditions, and that UV irradiation may switch the mode of p21 degradation from ubiquitin independent to ubiquitin dependent. As p21 degradation following UV treatment appears to be dependent upon the ATR pathway, it appears that ATR mediates this switch in p21 protein degradation. The $CRL4^{Ctd2}$ E3 ubiquitin ligase complex is also involved in ubiquitin dependent degradation of p21 in response to DNA damage, since siRNA depletion of Cul4, DDB4 and Ctd2 in HCT116 $p53^{-/-}$ cells prevented p21 degradation following UV irradiation.[58] PCNA has been shown to target $CRL4^{Ctd2}$ substrates for destruction and that the loss of PCNA binding prevented ubiquitination of p21 and degradation following UV damage.[58] Increased ubiquitination of p21 lacking the PCNA PIP binding site following UV treatment was not observed and this form of p21 was resistant to degradation even after high doses of UV.[58] These findings suggest that PCNA is an active co-factor in p21 ubiquitination by the $CRL4^{Ctd2}$ complex and that it promotes efficient ubiquitination and degradation of p21 in response to DNA damage.[58] Rapid degradation of p21 following low doses of UV irradiation is thought to be essential in order to permit efficient DNA repair. Following irradiation, PCNA associates with chromatin in a triton non-extractable fraction of the cell which regulates DNA repair.[52] This association was inhibited by the presence of high levels of p21. The p21 K6R form, which was not subject to rapid ubiquitination and degradation following UV damage, remained bound to PCNA, inhibiting its interaction with and the repair of damaged DNA. Persistent p21 levels are therefore detrimental to the efficient repair of UV induced DNA damage. ATR mediated targeting and turnover of p21 is thus essential for the cellular response to UV induced DNA damage.

Rapid degradation of p21 may also be an important stage in the induction of apoptosis in response to DNA damage. Elevated levels of p21 trigger cell cycle arrest and inhibit apoptosis. However in response to lethal doses of DNA damaging agents such as adriamycin, p21 levels are decreased despite activation of p53.[62,63] This is thought to relieve p21's block on the cell cycle and its inhibition of apoptotic proteins, such as pro-caspase 3, and permit the activation of the apoptotic machinery. Recent work suggests that the loss of p21 prior to apoptosis may occur due to ubiquitin mediated degradation of p21 through a novel E3 ligase, MKRN1, since depletion of MKRN1 prevents the destabilization of p21 following adriamycin treatment of U2OS cells and significantly attenuates the apoptotic response.[63] Formation of ubiquitinated p21 species was markedly increased in the presence of MKRN1, while proteasomal inhibition prevented MKRN1 mediated p21 destabilization, indicating that MKRN1 promotes p21 turnover through ubiquitin dependent proteasomal degradation. Stable over-expression of MKRN1 potently enhanced the sensitivity of U20S cells to adriamycin induced apoptotic cell death, and this correlated with destabilization of p21—in the absence of any change in p53 stability—and activation of the apoptotic cascade.[63] Efficient ubiquitin mediated degradation and turnover of p21 thus appear to be essential for the induction of the early phases of apoptotic cell death in response to DNA damage.

Ubiquitin Independent Protein Turnover

Although ubiquitin mediated protein turnover is the major mechanism regulating the stability and abundance of many intracellular proteins, ubiquitin independent turnover is not unknown. Ornithine decarboxylase and the transcription factor c-Fos are both known to be physiologically degraded in the absence of ubiquitin.[64,65] In prokaryotic systems such as bacteria, ubiquitination is not required for targeting proteins for proteolytic degradation even though proteasomal like proteases exist. This suggests that eukaryotic proteasomes must have evolved to use ubiquitin as a marker for degradation, possibly as a mechanism for specificity and quality control in order to prevent indiscriminate protein turnover.[13] The requirement of ubiquitination for the degradation of the majority of eukaryotic intracellular proteins does not, however, rule out the possibility that the proteasome still retains the capacity for protein recognition. Well known ubiquitin regulated proteins such as p53 and p73 may also undergo ubiquitin independent proteasomal degradation.[66,67] It is therefore possible that many cellular proteins may not fit the dogma that proteasomal degradation via ubiquitination is the sole mechanism for regulating protein turnover (reviewed by ref. 68). Indeed, several cellular proteins, including the cell cycle regulators p16 and p14, lack lysine residues and thus attachment sites for post translational modification by ubiquitin. As such, these proteins must undergo ubiquitin independent protein degradation. Recent studies have shown that this is mediated via the REGγ/20S proteasomal complexes.[46] Different subpopulations of proteins such as p21 may also display differing sensitivities to the various pathways of degradation present within the cell. For example, it has been suggested that nascent p21 protein levels may be regulated via 'default degradation' through type 1 degrons and that ubiquitination may be required to specifically remove p21 proteins which are stably associated as part of larger complexes. In support of this, REGγ dependent degradation of p21 was delayed in the presence of purified cyclin E/Cdk2 subunits suggesting that REGγ does not target p21 for degradation when it is part of a larger complex and instead targets unbound/free p21.[46,51]

Ubiquitination is a very complex and sophisticated post-translational modification and occurs in many form in vivo. Ubiquitin itself contains lysine residues which can be ubiquitinated, a modification that modulates its activity. Lys^{48} attached polyubiquitin chains represent the signal for proteasomal targeting and degradation. However, not all polyubiquitin chains target proteins to the proteasome and other forms of conjugations such as Lys^{63} can influence cellular processes such as protein kinase activation, receptor internalization, and DNA repair.[69] Ubiquitinated forms of p21 have been identified following inhibition of proteasomal function;[10,11,48] however, the type of p21 ubiquitin modification observed in vivo has not been classified and only changes in p21 stability and rate of turnover have been investigated. This is possibly due to the low abundance and variability of the observed ubiquitinated p21 forms. It is possible that the observed ubiquitin modifications may alter p21s function independently of the rate of p21 proteolytic degradation. p21 K6R or p21 K0 are known to be folded properly and still act as Cdk inhibitors.[11] However, additional p21 functions such as such as transcriptional activity or protein-protein interactions may be perturbed by the inhibition of proper ubiquitin mediated modification of function.

In summary, if both ubiquitin dependent and independent means of protein degradation co-exist within the cell and are both involved in regulating the turnover of proteins such as p21 (Fig. 3). p21 turnover can also be modified by posttranslational modifications such as phosphorylation. Various forms of the proteasome exist, including 20S, 26S and the REGγ/20S proteasomal complexes that have different means of targeting proteins for degradation. The relative abundance of various isoforms of the proteasome may vary between cell types, within intracellular compartments, and under conditions of cellular stress.[50] For example, expression of the REGγ subunits is restricted to the nucleus and thus may only impact turnover of proteins located within the nuclear compartment. REGγ-mediated degradation of p21 was also found to be cell type specific.[51] The relative importance of each form of the proteasome and the flux of p21 protein through each pathway may therefore fluctuate depending upon intracellular conditions, subcellular localization of the substrate protein and extracellular stimuli. Further studies into this area are necessary for a fuller understand the complex and intricate mechanisms that control intracellular p21 abundance and function.

References

1. Besson A, Assoian RK, Roberts JM. Regulation of the cytoskeleton: an oncogenic function for CDK inhibitors? Nat Rev Cancer 2004; 4:948-55.
2. Besson A, Dowdy SF, Roberts JM. CDK inhibitors: cell cycle regulators and beyond. Dev Cell 2008; 14:159-69.
3. Denicourt C, Dowdy SF. Cip/Kip proteins: more than just CDKs inhibitors. Genes Dev 2004; 18:851-5.
4. Okuyama R, LeFort K, Dotto GP. A dynamic model of keratinocyte stem cell renewal and differentiation: role of the p21WAF1/Cip1 and Notch1 signaling pathways. J Investig Dermatol Symp Proc 2004; 9:248-52.
5. Roninson IB. Oncogenic functions of tumour suppressor p21(Waf1/Cip1/Sdi1): association with cell senescence and tumour-promoting activities of stromal fibroblasts. Cancer Lett 2002; 179:1-14.
6. Sherr CJ, Roberts JM. CDK inhibitors: positive and negative regulators of G1-phase progression. Genes Dev 1999; 13:1501-12.
7. Gartel AL, Tyner AL. Transcriptional regulation of the p21((WAF1/CIP1)) gene. Exp Cell Res 1999; 246:280-9.
8. Jones JM, Cui XS, Medina D, Donehower LA. Heterozygosity of p21WAF1/CIP1 enhances tumor cell proliferation and cyclin D1-associated kinase activity in a murine mammary cancer model. Cell Growth Differ 1999; 10:213-22.
9. Martin-Caballero J, Flores JM, Garcia-Palencia P, Serrano M. Tumor susceptibility of p21(Waf1/Cip1)-deficient mice. Cancer Res 2001; 61:6234-8.
10. Cayrol C, Ducommun B. Interaction with cyclin-dependent kinases and PCNA modulates proteasome-dependent degradation of p21. Oncogene 1998; 17:2437-44.
11. Sheaff RJ, Singer JD, Swanger J et al. Proteasomal turnover of p21Cip1 does not require p21Cip1 ubiquitination. Mol Cell 2000; 5:403-10.
12. Kriwacki RW, Hengst L, Tennant L et al. Structural studies of p21Waf1/Cip1/Sdi1 in the free and Cdk2-bound state: conformational disorder mediates binding diversity. Proc Natl Acad Sci U S A 1996; 93:11504-9.
13. Verma R, Deshaies RJ. A proteasome howdunit: the case of the missing signal. Cell 2000; 101:341-4.
14. Bloom J, Amador V, Bartolini F et al. Proteasome-mediated degradation of p21 via N-terminal ubiquitinylation. Cell 2003; 115:71-82.
15. Li Y, Jenkins CW, Nichols MA, Xiong Y. Cell cycle expression and p53 regulation of the cyclin-dependent kinase inhibitor p21. Oncogene 1994; 9:2261-8.
16. Zhang H, Xiong Y, Beach D. Proliferating cell nuclear antigen and p21 are components of multiple cell cycle kinase complexes. Mol Biol Cell 1993; 4:897-906.
17. Dotto GP. p21(WAF1/Cip1): more than a break to the cell cycle? Biochim Biophys Acta 2000; 1471:M43-56.
18. Esteve V, Canela N, Rodriguez-Vilarrupla A et al. The structural plasticity of the C terminus of p21Cip1 is a determinant for target protein recognition. Chembiochem 2003; 4:863-9.
19. Maga G, Hubscher U. Proliferating cell nuclear antigen (PCNA): a dancer with many partners. J Cell Sci 2003; 116:3051-60.
20. Gulbis JM, Kelman Z, Hurwitz J et al. Structure of the C-terminal region of p21(WAF1/CIP1) complexed with human PCNA. Cell 1996; 87:297-306.
21. Touitou RJ, Richardson S, Bose M et al. A degradation signal located in the C-terminus of p21WAF1/CIP1 is a binding site for the C8 alpha-subunit of the 20S proteasome. EMBO J 2001; 20:2367-75.
22. Bose S, Stratford FL, Broadfoot KI et al. Phosphorylation of 20S proteasome alpha subunit C8 (alpha7) stabilizes the 26S proteasome and plays a role in the regulation of proteasome complexes by gamma-interferon. Biochem J 2004; 378:177-84.
23. Chen J, Saha P, Kornbluth S et al. Cyclin-binding motifs are essential for the function of p21CIP1. Mol Cell Biol 1996; 16:4673-82.
24. Coleman ML, Marshall CJ, Olson MF. Ras promotes p21(Waf1/Cip1) protein stability via a cyclin D1-imposed block in proteasome-mediated degradation. EMBO J 2003; 22:2036-46.
25. Coleman ML, Densham RM, Croft DR, Olson MF. Stability of p21Waf1/Cip1 CDK inhibitor protein is responsive to RhoA-mediated regulation of the actin cytoskeleton. Oncogene 2006; 25:2708-16.

26. Zhu H, Nie L, Maki CG. Cdk2-dependent inhibition of p21 stability via a C-terminal cyclin-binding motif. J Biol Chem 2005; 280:29282-8.
27. Zhang Z, Wang H, Li M et al. MDM2 is a negative regulator of p21WAF1/CIP1, independent of p53. J Biol Chem 2004; 279:16000-6.
28. Brooks CL, Gu W. p53 ubiquitination: Mdm2 and beyond. Mol Cell 2006; 21:307-15.
29. Bake RD, Howl J, Nicholl ID. A sychnological cell penetrating peptide mimic of p21(WAF1/CIP1) is pro-apoptogenic. Peptides 2007; 28:731-40.
30. Ball KL, Lain S, Fahraeus R et al. Cell-cycle arrest and inhibition of Cdk4 activity by small peptides based on the carboxy-terminal domain of p21WAF1. Curr Biol 1997; 7:71-80.
31. Kontopidis G, Andrews MJ, McInnes C et al. Insights into cyclin groove recognition: complex crystal structures and inhibitor design through ligand exchange. Structure 2003; 11:1537-46.
32. Zheleva DI, McInnes C, Gavine AL et al. Highly potent p21(WAF1)-derived peptide inhibitors of CDK-mediated pRb phosphorylation: delineation and structural insight into their interactions with cyclin A. J Pept Res 2002; 60:257-70.
33. Warbrick E, Lane DP, Glover DM, Cox LS. A small peptide inhibitor of DNA replication defines the site of interaction between the cyclin-dependent kinase inhibitor p21WAF1 and proliferating cell nuclear antigen. Curr Biol 1995; 5:275-82.
34. Jascur T, Brickner H, Salles-Passador I et al. Regulation of p21(WAF1/CIP1) stability by WISp39, a Hsp90 binding TPR protein. Mol Cell 2005; 17:237-49.
35. Kurz EU, Lees-Miller SP. DNA damage-induced activation of ATM and ATM-dependent signaling pathways. DNA Repair (Amst) 2004; 3:889-900.
36. Rossig L, Jadidi AS, Urbich C et al. Akt-dependent phosphorylation of p21(Cip1) regulates PCNA binding and proliferation of endothelial cells. Mol Cell Biol 2001; 21:5644-57.
37. Scott MT, Morrice N, Ball KL. Reversible phosphorylation at the C-terminal regulatory domain of p21(Waf1/Cip1) modulates proliferating cell nuclear antigen binding. J Biol Chem 2000; 275:11529-37.
38. Child ES, Mann DJ. The intricacies of p21 phosphorylation: protein/protein interactions, subcellular localization and stability. Cell Cycle 2006; 5:1313-9.
39. Fraser JA, Hupp TR. Chemical genetics approach to identify peptide ligands that selectively stimulate DAPK-1 kinase activity. Biochemistry 2007; 46:2655-73.
40. Scott MT, Ingram A, Ball KL. PDK1-dependent activation of atypical PKC leads to degradation of the p21 tumour modifier protein. EMBO J 2002; 21:6771-80.
41. Zhou BP, Liao Y, Xia W et al. HER-2/neu induces p53 ubiquitination via Akt-mediated MDM2 phosphorylation. Nat Cell Biol 2001; 3:973-82.
42. Li Y, Dowbenko D, Lasky LA. AKT/PKB phosphorylation of p21Cip/WAF1 enhances protein stability of p21Cip/WAF1 and promotes cell survival. J Biol Chem 2002; 277:11352-61.
43. Liang J, Slingerland JM. Multiple roles of the PI3K/PKB (Akt) pathway in cell cycle progression. Cell Cycle 2003; 2:339-45.
44. Diehl JA, Cheng M, Roussel MF, Sherr CJ. Glycogen synthase kinase-3beta regulates cyclin D1 proteolysis and subcellular localization. Genes Dev 1998; 12:3499-511.
45. Rossig L, Badorff C, Holzmann Y et al. Glycogen synthase kinase-3 couples AKT-dependent signaling to the regulation of p21Cip1 degradation. J Biol Chem 2002; 277:9684-9.
46. Chen X, Barton LF, Chi Y et al. Ubiquitin-independent degradation of cell-cycle inhibitors by the REGgamma proteasome. Mol Cell 2007; 26:843-52.
47. Liu CW, Corboy MJ, DeMartino GN, Thomas PJ. Endoproteolytic activity of the proteasome. Science 2003; 299:408-11.
48. Maki CG, Howley PM. Ubiquitination of p53 and p21 is differentially affected by ionizing and UV radiation. Mol Cell Biol 1997; 17:355-63.
49. Chen XY, Chi A, Bloecher R et al. N-acetylation and ubiquitin-independent proteasomal degradation of p21(Cip1). Mol Cell 2004; 16:839-47.
50. Rivett AJ, Bose S, Brooks P, Broadfoot KI. Regulation of proteasome complexes by gamma-interferon and phosphorylation. Biochimie 2001; 83:363-6.
51. Li X, Amazit L, Long W et al. Ubiquitin- and ATP-independent proteolytic turnover of p21 by the REGgamma-proteasome pathway. Mol Cell 2007; 26:831-42.
52. Bendjennat M, Boulaire J, Jascur T et al. UV irradiation triggers ubiquitin-dependent degradation of p21(WAF1) to promote DNA repair. Cell 2003; 114:599-610.
53. Carrano AC, Eytan E, Hershko A, Pagano M. SKP2 is required for ubiquitin-mediated degradation of the CDK inhibitor p27. Nat Cell Biol 1999; 1:193-9.
54. Bornstein G, Bloom J, Sitry-Shevah D et al. Role of the SCFSkp2 ubiquitin ligase in the degradation of p21Cip1 in S phase. J Biol Chem 2003; 278:25752-7.

55. Yu ZK, Gervais JL, Zhang H. Human CUL-1 associates with the SKP1/SKP2 complex and regulates p21(CIP1/WAF1) and cyclin D proteins. Proc Natl Acad Sci U S A 1998; 95:11324-9.
56. Wang W, Nacusi L, Sheaff RJ, Liu X. Ubiquitination of p21Cip1/WAF1 by SCFSkp2: substrate requirement and ubiquitination site selection. Biochemistry 2005; 44:14553-64.
57. Vlach J, Hennecke S, Amati B. Phosphorylation-dependent degradation of the cyclin-dependent kinase inhibitor p27. EMBO J 1997; 16:5334-44.
58. Abbas T, Sivaprasad U, Terai K et al. PCNA dependent regulation of p21 ubiquitulation and degradation via the $CRL4^{Cdt2}$ ubiquitin ligase complex. Genes Dev 2008; 22:2496-2506.
59. Amador V, Ge S, Santamaria PG et al. APC/C(Cdc20) controls the ubiquitin-mediated degradation of p21 in prometaphase. Mol Cell 2007; 27:462-73.
60. Lee H, Zeng SX, Lu H. UV Induces p21 rapid turnover independently of ubiquitin and Skp2. J Biol Chem 2006; 281:26876-83.
61. Lee JY, Yu SJ, Park YG et al. Glycogen synthase kinase 3beta phosphorylates p21WAF1/CIP1 for proteasomal degradation after UV irradiation. Mol Cell Biol 2007; 27:3187-98.
62. Martinez LA, Yang J, Vazquez ES et al. p21 modulates threshold of apoptosis induced by DNA-damage and growth facto withdrawal in preostrate cancer cells. Carcinogenesis 2002; 23:1289-1296.
63. Lee EW, Lee MS, Camus S et al. Differential regulation of p53 and p21 by MKRN1 E3 ligase controls cell cycle arrest and apoptosis. EMBO J 2009; 28: 2100-2113.
64. Bossis G, Ferrara P, Acquaviva C et al. c-Fos proto-oncoprotein is degraded by the proteasome independently of its own ubiquitinylation in vivo. Mol Cell Biol 2003; 23:7425-36.
65. Kahana C, Asher G, Shaul Y. Mechanisms of protein degradation: an odyssey with ODC. Cell Cycle 2005; 4:1461-4.
66. Asher G, Shaul Y. p53 proteasomal degradation: poly-ubiquitination is not the whole story. Cell Cycle 2005; 4:1015-8.
67. Asher G, Tsvetkov P, Kahana C, Shaul Y. A mechanism of ubiquitin-independent proteasomal degradation of the tumor suppressors p53 and p73. Genes Dev 2005; 19:316-21.
68. Asher G, Reuven N, Shaul Y. 20S proteasomes and protein degradation "by default". Bioessays 2006; 28:844-9.
69. Sun L, Chen ZJ. The novel functions of ubiquitination in signaling. Curr Opin Cell Biol 2004; 16:119-26.

Chapter 8

p53 Localization

Carl G. Maki*

Introduction

Inactivation of the p53 tumor suppressor pathway is essential for the development of most or all human cancers. Over 50% of cancers harbor missense mutations in p53 that destroy its normal function.[1] In cancers that retain wild-type p53, the protein is often inactivated through other means, including being abnormally sequestered in the cytoplasm, over-expression of MDM2 (the key negative regulator of p53), and deletion of p14/Arf (which normally inhibits MDM2 function). P53 undergoes nuclear-cytoplasmic shuttling and, in most unstressed cells, is expressed at low levels localized in both the nucleus and the cytoplasm. In response to DNA damage and other stresses, p53 is subject to various post-translational modifications that result in its stabilization, accumulation in the nucleus, and activation as a transcription factor. While most p53 accumulates in the nucleus following stress, recent studies indicate a significant fraction remains in the cytoplasm, and that both nuclear and cytoplasmic p53 participate in its tumor suppressor program.[2] Notably, certain post-translational modifications may direct p53 to specific sub-cellular locales (Table 1), and this appear to be important in unleashing p53's full growth suppressive capabilities. For example, at least some nuclear p53 that accumulates following certain stresses is directed to sub-nuclear domains (PML bodies) where it is subjected to further activating modifications. Similarly, cytoplasmic p53 is directed to the mitochondria following stress where it interacts with pro- and anti-apoptotic members of the Bcl2 family, resulting in the release of factors from the mitochondria that drive apoptosis. This chapter will review studies of p53 localization control including its nuclear-cytoplasmic shuttling, movement to PML bodies and to the mitochondria.

p53 Nuclear Import

Middeler et al (1997) provided the first evidence that p53 undergoes active import and export from the nucleus.[3] In their study, human p53 was expressed in *E. coli*, purified, and labeled with fluorescein iodoacetamide (IAF). This IAF-p53 was then used for micro-injection into either the cytoplasm or nucleus of NIH-3T3 cells. A portion of IAF-p53 injected into the cytoplasm translocated to the nucleus within minutes of injection. Similarly, a portion of IAF-p53 injected into the nucleus was translocated in to the cytoplasm within minutes of injection. Further analysis indicated that transport to and from the nucleus were energy-dependent processes. The conclusions of these findings were that p53 is subject to both nuclear import and nuclear export, and that both processes are rapid and energy-dependent.

Studies of endogenous wild-type p53 localization in growth-stimulated murine fibroblasts found that p53 is cytoplasmic in G1 phase, accumulates in the nucleus during S-phase, and cycles back to the cytoplasm after DNA synthesis in S-phase is initiated.[4] This suggested nuclear import and/or nuclear export of p53 are regulated during the cell-cycle. Sequence analysis

*Carl G. Maki—Department of Radiation and Cellular Oncology, University of Chicago, 5841 S. Maryland Ave., MC1105, Chicago, Illinois 60637, USA. Email: cmakilab@mac.com

p53, edited by Ayeda Ayed and Theodore Hupp. ©2010 Landes Bioscience and Springer Science+Business Media.

Table I. Effects of post-translational modifications on p53 sub-cellular trafficking

	Modification	Proposed/Supported Effect on p53 Localization	References
Nuclear p53	Mono-ubiquitination of C-terminal and DBD lysines	Nuclear export via C-terminal NES	13,31-34,42
	Acetylation of multiple C-terminal lysines	* Promote nuclear export ** Inhibit nuclear export by preventing C-terminal ubiquitination	*59 **60
	Phosphorylation of N-terminal sites	Inhibit nuclear export that occurs via N-terminal NES	27
	Sumoylation in N and C-terminus	Direct or recruit human p53 to yeast PML-NB like nuclear foci	Di Ventura et al PLoS ONE Jan 30 (2008)
		Direct or recruit drosophila p53 to PML-NB like nuclear foci	25
Cytoplasmic p53	Ubiquitination	Promote p53 movement to mitochondria	2

identified three putative nuclear localization signals (NLSs) at the p53 C-terminus.[5] Sequences containing NLSI were shown to promote nuclear import when fused to a normally cytoplasmic heterologous protein (pyruvate kinase),[6] and mutation/localization studies confirmed that NLSI (between residues 318-322) is the primary NLS for p53 nuclear localization.[5] NLSII (residues 378-382) and NLSIII (residues 386-390) are weaker than NLSI and contribute less to p53 nuclear import. Wild-type p53 can inhibit cell cycle progression when over-expressed, and certain cancer-derived p53s can cooperate with oncogenic ras to promote transformation. In these first studies, wild-type p53s deficient in nuclear import were unable to inhibit cell cycle progression, and cancer-derived p53s deficient in nuclear import were unable to promote transformation.[7,8] Thus, nuclear import is essential for these p53 functions.

A series of studies by Liang and Clarke examined sequences around NLSI that affect p53 localization. They found that NLSI is bipartite in structure and includes not only residues 318-322 but also a lysine (K) at position 305 and arginine (R) at position 306.[9-11] Conversion of K305 (or R306) to any other amino acid caused p53 to be excluded from the nucleus. Interestingly, deletion of residues 326-355 that include the oligomerization domain restored nuclear localization to K305 or R306-mutated p53.[9] These results suggested that K305/R306 are required for p53 nuclear localization, while residues 326-355 can sequester p53 in the cytoplasm. The cytoplasmic localization of K305 or R306-mutated p53 could be explained by either a lack of nuclear import or excessive nuclear export. In fact, a CRM1-dependent nuclear export signal (NES) has since been recognized within the 326-355 region between residues 340-351.[12] To examine these possibilities, the investigators used leptomycin B (LMB), an inhibitor of CRM1-dependent nuclear export. In their studies, K305-mutated p53 remained cytoplasmic in LMB-treated cells while cyclin-B1 (their positive control) accumulated in the nucleus.[9] Thus, the cytoplasmic localization of K305-mutated p53 appeared not to result from excessive nuclear export. Next, the investigators tested the ability of various GST p53 C-terminal fragments to bind the nuclear import factor importin-α. They found that mutation of K305 inhibited the ability of p53 residues 300-393 to bind importin-α, but that importin-α binding was restored upon deletion of residues 326-355. Based on these findings it was concluded that K305-mutated p53 was cytoplasmic due to defective nuclear import, and that residues 326-355 inhibited nuclear import by somehow blocking the association between NLSI and importin-α. The investigators suggested that residues 326-355 might interact with NLSI sequences to inhibit importin-α binding. Post-translational modifications of K305 could affect p53 localization by altering the association between NLSI and residues 326-355, should this association exist. These conclusions are somewhat at odds with more recent reports. For example, Gu et al[13] and O'keefe et al[14] both reported that K305-mutated p53 relocalized to the nucleus when critical leucines in the NES between 340-351 were mutated. This led them to conclude that the cytoplasmic localization of K305-mutated p53 resulted from excessive nuclear export via the NES. Whether the NES- mutations affects any putative interaction between NLSI and residues 326-355 is unknown. At any rate, the evidence supports cross-talk and potential interaction between the C-terminal NES and the bipartite NLSI of p53. This cross-talk/interaction appears to control exposure of the NLSI and NES and, subsequently, p53 localization.

Multiple factors have been identified that facilitate p53 nuclear import. For example, Fojo and colleagues[15] reported in vitro and in vivo binding between p53 and tubulin, and the colocalization of p53 with microtubules. Micrrotubule binding required the first 25 amino acids of p53, which include the MDM2-binding domain. Importantly, microtubule-stabilizing or destabilizing drugs such as taxol and vincristine diminished p53 nuclear accumulation following DNA damaging stress. Moreover, inhibition of the microtubule-associated motor protein dynein also inhibited p53 nuclear accumulation following stress. These findings suggested that functional microtubules and the dynein motor protein facilitate the transport of cytoplasmic p53 to the nucleus and its accumulation in the nucleus in stressed cells. The binding between p53 and dynein appears to be indirect. For example, studies by Ljungman and colleagues demonstrated the presence of p53-Hsp90-Immunophillin-dynein complexes in cells.[16] Through analysis of these complexes in the presence of different competing peptides, the investigators concluded that immunophillins link the p53-Hsp90 complex to the dynein motor protein for transport into the nucleus. Later studies by Trostel et al[17] reported that oligomerization of p53 in the cytoplasm was required for its interaction with dynein, and that association with dynein occurred independently of microtubules. Taken together, these data suggest that cytoplasmic p53 oligomers, in association with Hsp90 and immunophillins, bind to the dynein motor protein. This is followed by association of dynein with microtubules and transport of the complex to the nuclear pore. At the pore, p53 is released from the complex and transferred to importin-α for nuclear import.[17]

Sub-Nuclear Trafficking of p53

Nuclear accumulation after stress allows p53 to carry out its role as a transcription factor. Nuclear p53 inhibits growth by activating the expression of genes such as *P21*, *PUMA*, and *Bax* that signal growth arrest or apoptosis.[18] However, nuclear accumulation is not always sufficient to fully activate nuclear p53 activity. Instead, full p53 activation is thought, at least in some cases, to require its trafficking to specific sub-nuclear domains where it is subjected to further activating modifications. This is evidenced by studies examining the link between p53 and PML, another tumor suppressor protein. PML localizes in distinct nuclear foci termed PML nuclear bodies (PML-NBs). PML-NBs are multi-protein complexes, and PML is required for their formation and for the recruitment of other proteins to these bodies.[19] A relationship between PML and p53 was suggested by the finding that p53 could interact directly with PML and was recruited by PML into PML-NBs.[20] Subsequent studies have shown that PML co-recruits both p53 and CBP/p300 to PML-NBs. CBP/p300 then promotes the acetylation of p53, which increases p53 DNA binding affinity and thus leads to an activation of p53-responsive genes.[21,22] Perhaps the most compelling evidence linking PML and p53 growth-suppressive pathways comes from studies of oncogene-induced senescence. In these studies, p53 activity was measured in $PML^{+/+}$ and $PML^{-/-}$ cells infected with retroviruses expressing an activated Ras oncogene (Ras V12). Ras V12 expression in $PML^{+/+}$ cells resulted in the recruitment of p53 into PML-NBs, the acetylation and activation of p53, and the induction of premature senescence. In contrast, p53 was neither activated nor recruited into PML-NBs in $PML^{-/-}$ cells, and the cells were resistant to Ras-induced senescence.[21,22] These studies provided evidence that recruitment to sub-nuclear PML-NBs is required for the efficient activation of p53 in response to aberrant oncogene signaling and potentially other stresses.

What triggers p53 recruitment to PML-NBs? Our data suggest that p53 recruitment to PML-NBs is not simply a result of increased p53 levels in stressed cells. For example, we compared the extent of p53 recruitment to PML-NBs in U2OS cells treated with either IR or the MDM2 antagonist Nutlin-3a. Nutlin-3a caused a pronounced increase in p53 levels, and IR also caused increased p53 levels but to a lesser extent than Nutlin-3a. P53 localized to PML-NBs in about 25% of IR-treated cells 48 hrs after treatment, but in less than 5% of Nutlin-3a treated cells after 48 hrs after treatment. Thus, p53 recruitment to PML-NBs occurs in response to stress signals and is not simply a result of increased p53 levels. Many proteins recruited to PML-NBs (including PML and p53) are post-translationally modified by SUMO (small ubiquitin-like modifier).[23] There are at least 3 different SUMO isoforms (SUMO-1, 2, and 3). P53 is modified by SUMO-1 at lysine-386 (K386), and this sumoylation is reported to activate the transcriptional response of p53.[24] Human p53 with K386 mutated could be recruited to PML-NBs, suggesting sumoylation is not required for p53 recruitment to PML-NBs.[20] However, more recent reports suggest that sumoylation can affect p53 movement to PML-NBs. For example, Di Ventura and Serrano et al (PLoS 2008) monitored human p53 and MDM2 interactions when both proteins were expressed in budding yeast. They found that coexpression of p53 and MDM2 in yeast led to p53 sumoylation at lysine-386 and colocalization of p53 and MDM2 in PML-like nuclear foci. Mutation of lysine-386 inhibited p53 sumoylation and its localization in these foci. Interestingly, fusion of SUMO to the p53 C-terminus (to mimic p53 sumoylation) restored localization to nuclear foci, but only when MDM2 was coexpressed. The results suggested that both MDM2-binding and sumoylation at lysine-386 are required for p53 recruitment to yeast PML-NBs. More recently, Mauri et al[25] reported that drosophila p53 (DMp53) is sumoylated at two lysine residues. Mutation of these sites inhibited both recruitment of DMp53 to PML-NBs and its transcriptional activity. In contrast, a DMp53-SUMO fusion protein generated to mimic sumoylation localized in PML-NBs similar to the wild-type protein. Thus, as with human p53 expressed in yeast, sumoylation appears required for DMp53 recruitment to PML-NBs. SUMO-1 modified p53 accumulated in human cells exposed to UV radiation,[24] and SUMO-2/3 modified p53 accumulated in human cells stressed by H_2O_2 treatment.[26] Thus, one possibility is that sumoylation of p53 in a stressed cell stimulates or enhances its sub-nuclear trafficking to PML-NBs where it can then be modified for full activation.

p53 Nuclear Export

P53 is also subject to active nuclear export, and factors that can affect this nuclear export have been identified. P53 contains two nuclear export signals (NESs), one located in the N-terminal MDM2-binding domain of p53, and the second located in the C-terminus within the tetramerization domain.[12,27] Both of these NESs can function as autonomous export signals when fused to a heterologous protein. The notion that p53 nuclear export is required for its degradation was suggested by the observation that wild-type p53 accumulated in the nucleus and was stabilized in cells exposed to leptomycin B (LMB), a specific inhibitor of nuclear export.[12,28,29] Further studies reported that MDM2 contains an NES within its sequence, and that mutating this NES or blocking nuclear export by LMB treatment inhibited the ability of MDM2 to degrade p53.[29,30] These findings suggested a model in which MDM2 transports p53 from the nucleus to the cytoplasm for p53 to be degraded, and that this transport depends on a functional NES within MDM2. In conflict with this model was the finding that p53 contains its own NES located in the tetramerization domain, and that this NES is sufficient to promote p53 nuclear export even in cells that lack MDM2 expression.[12] Structural analysis indicated that this NES would be sequestered and inaccessible to the export machinery when p53 is a tetramer, but would be exposed and accessible when p53 is in either a dimeric or monomeric form. Taken together, these latter findings supported a second model in which p53 is exported from the nucleus via its own NES and independently of MDM2, and that this export may be regulated by changes in p53 tetramerization. Continuing studies attempted to clarify the role of MDM2 in p53 nuclear export. In these studies, it was found that MDM2 can promote p53 nuclear export in transiently transfected cells, and that this export requires the p53 NES but not the NES of MDM2.[31,32] Interestingly, mutations in the MDM2 RING-finger domain that inhibited p53 ubiquitination also inhibited p53 nuclear export. These results supported a third model for p53 nuclear export. In this model it was suggested that MDM2-mediated ubiquitination of p53 promotes p53 nuclear export, most likely by causing dissolution of p53 tetramers and exposure of the C-terminal NES to the export factor CRM1. Most subsequent studies have supported this third model. For example, mutations in C-terminal ubiquitination sites of p53 inhibited the nuclear export of p53 that is mediated by MDM2,[13,33] and a p53-ubiquitin fusion protein made to mimic ubiquitinated p53 displayed a cytoplasmic-only localization in p53 and MDM2 double-null cells.[34] In contrast, other studies reported that p53 that is ubiquitinated and exported from the nucleus when expressed with MDM2, as well as the p53-ubiquitin fusion protein, were still capable of forming tetramers.[13,35] Thus, the idea that ubiquitination causes p53 nuclear export by causing the dissociation of p53 tetramers does not appear to be correct.

MDM2-mediated ubiquitination leads to either degradation of p53 by the proteasome, or export of p53 from the nucleus. What determines whether ubiquitination leads to nuclear export or degradation? Polyubiquitination is required for proteins to be recognized and degraded by the proteasome.[36] Li et al[34] monitored ubiquitination, nuclear export, and degradation of p53 when expressed with increasing amounts of MDM2. Low levels of MDM2 promoted mono-ubiquitination and nuclear export of p53, but did not promote p53 degradation. In contrast, high amounts of MDM2 promoted p53 polyubiquitination and its degradation by the proteasome. Finally, fusion of ubiquitin to the p53 C-terminus (to mimic mono-ubiquitination) was sufficient to promote cytoplasmic localization (presumed to be nuclear export) of p53 in p53/MDM2 double-null cells. These findings led to the view that p53 fate (degradation vs nuclear export) depends on the relative level of MDM2. Low levels of MDM2 promote mono-ubiquitination and nuclear export of p53, whereas high levels of MDM2 promote p53 polyubiquitination and degradation. What is the basis of these differences? A potential clue comes from studies by Grossman et al.[37-39] In their studies, it was demonstrated that p300 can form a complex with p53 and MDM2, and that p300 has an intrinsic or associated ubiquitin ligase activity that can cooperate with MDM2 to promote the poly-ubiquitination and degradation of p53. One possibility is that high levels of MDM2 can efficiently recruit p300 to p53, leading to p53 polyubiquitination and degradation, while low levels of MDM2 cannot efficiently recruit p300 and therefore p53 is only monoubiquitinated and exported.

More recent studies have suggested a link between the conformation of p53 and its nuclear export. Wild-type p53 is a conformationally labile protein. Changes in p53 conformation can be monitored by reactivity with conformation-specific antibodies that recognize p53 in either a wild-type (pAb1620) or mutant (pAb240) conformation.[40,41] Nie et al[42] observed that certain cancer-derived p53s with a largely mutant (primary antibody 1620-/pAb240+) conformation localized to the cytoplasm to a greater extent and displayed increased susceptibility to ubiquitination than p53s with a more wild-type (primary antibody 1620+/pAb240-) conformation. The cytoplasmic localization of mutant p53s required the C-terminal NES and an intact ubiquitination pathway. Mutant p53 ubiquitination occurred at lysines in both the DNA-binding domain (DBD) and the C-terminus. Interestingly, Lys to Arg mutations that inhibited ubiquitination restored some nuclear localization to mutant p53 but had

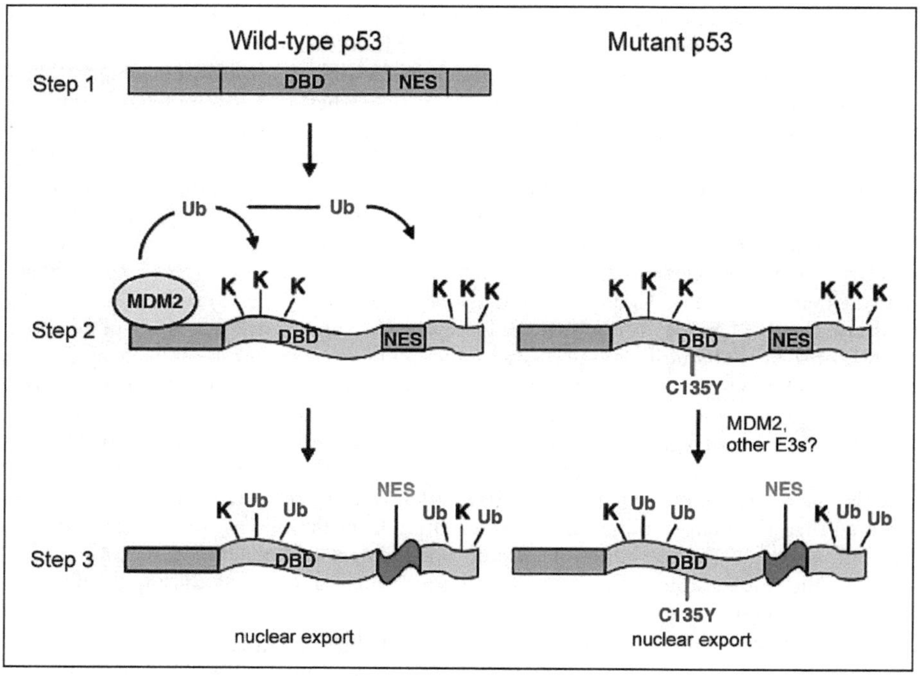

Figure 1. A stepwise model for p53 nuclear export. Right) Cancer-derived mutations (such as C135Y) alter p53 conformation, evidenced by exposure of the pAb240 epitope (Step 2). This altered conformation renders mutant p53 susceptible to ubiquitination at DBD and C-terminal lysines by MDM2 and/or other ubiquitin-protein ligases (E3s). Ubiquitination at DBD and C-terminal lysines promotes nuclear export of mutant p53s via the C-terminal NES (Step 3). Right) Wild-type p53 that is not ubiquitinated has a wild-type conformation (Step 1). MDM2 binding alters p53 conformation, evidenced by exposure of the pAb240 epitope (Step 2). Like mutant p53, this altered conformation probably renders p53 susceptible to ubiquitination at DBD and C-terminal lysines. Ubiquitination at DBD and C-terminal lysines promotes nuclear export of wild-type p53 via the C-terminal NES (Step 3). Reprinted with permission from: Nie et al. J Biol Chem 2007; 282:14616-14625;[42] ©2007 The American Society for Biochemistry and Molecular Biology.

no apparent effect on p53 conformation. This suggested that mutant p53 ubiquitination in the DBD and C-terminus promotes its nuclear export via the C-terminal NES. Through the use of cleavable p53s it was demonstrated that wild-type p53, like mutant p53, is ubiquitinated by MDM2 in both the DBD and C terminus and that ubiquitination in both regions contributes to its nuclear export. Finally, MDM2 binding was shown to induce a conformational change in wild-type p53, evidenced by increased exposure of the pAb240 epitope. However, this conformational change was insufficient to promote p53 nuclear export in the absence of MDM2 ubiquitination activity. Together, these results support a stepwise model for mutant and wild-type p53 nuclear export (Fig. 1). In this model, the conformational change induced by either the cancer-derived mutation or MDM2 binding precedes p53 ubiquitination. The addition of ubiquitin to DBD and C-terminal lysines then promotes nuclear export by exposing or in someway activating the C-terminal NES.

What is the purpose of p53 nuclear export? Because nuclear p53 functions as a transcription factor, it seems reasonable that promoting p53 nuclear export could represent a fail-safe mechanism that ensures that MDM2 can inhibit nuclear p53 function even if conditions for p53 degradation are not favorable. Original studies suggested that p53 is degraded by cytoplasmic proteasomes and that nuclear export is required for this degradation.[28-30] However, this model has fallen into disfavor since (1) it has been shown that p53 degradation can also occur in the nucleus,[43,44] and (2) blocking nuclear export by LMB treatment did not prevent the degradation of p53 in cells during recovery from stress treatment.[45] A second possibility is that nuclear export precedes p53 movement to the mitochondria, where it can interact with Bcl-2 family members to promote apoptosis. This possibility was suggested by the observation that the arginine-72 (R72) p53 polymorphic variant localized more to mitochondria and promoted more apoptosis than the proline-72 (P72) p53 variant, and was also more susceptible to MDM2-mediated ubiquitination and nuclear export.[46] However, this possibility was challenged by a recent study that found p53 ubiquitination by MDM2 in the cytoplasm promotes p53 movement to the mitochondria and subsequent apoptosis, completely independent of nuclear export.[2] Wild-type p53 is inactivated in certain cancers through an abnormal sequestration in the cytoplasm. Thus, notwithstanding its potential movement to mitochondria, a third possibility is that exported p53 is bound in the cytoplasm and held in an inactive state by one or more of these "sequestration" factors. Factors reported to sequester p53 in the cytoplasm include but are not limited to Parc, Mot-2, and glucocorticoid receptor (GR).[47-50] Notably, wild-type p53 sequestered in the cytoplasm of breast, liver, and neuroblastoma cancer cell lines shuttled back into the nucleus upon treatment with either MDM2 antisense oligonucleotides, or the nuclear export inhibitor leptomycin-B.[12,51,52] While there are other possible explanations, these findings are at least consistent with the possibility that the cytoplasmic localization of p53 in these cancers results from excessive nuclear export that is mediated by MDM2. It will be interesting in the future to determine whether mono-ubiquitinated p53 has higher affinity for Parc, Mot-1, GR, or other cytoplasmic binding partners than nonubiquitinated p53. Finally, p53 can reportedly interact with ribosomal protein L5 and 5.8S ribosomal RNA,[53,54] suggesting that p53 might play a role in protein translation by virtue of its association with active ribosomal complexes. Whether nuclear export enhances p53 interaction with the protein translation machinery is unknown.

p53 Trafficking to the Mitochondria

Wild-type p53 is stabilized and its levels increased following stress. The majority of stabilized p53 accumulates in the nucleus; however, a fraction also accumulates at the mitochondria.[2] Nuclear p53 induces apoptosis by promoting transcription of pro-apoptotic proteins, such as bax, PUMA, and Noxa.[18] Mitochondrial p53 promotes apoptosis by engaging in inhibitory and activating complexes with the anti- and pro-apoptotic members of the Bcl-2 family of proteins.[55] The net effect of these interactions is to increase mitochondrial membrane permeability, resulting in the release of pro-apoptotic triggers from the mitochondria such as cytochrome c. What is it that directs cytoplasmic p53 to the mitochondria in stressed cells? As mentioned earlier, previous work on conditional temperature-sensitive Arg/Pro72 p53 mutants proposed that polyubiquiquitinated p53 is shuttled from the nucleus to the mitochondria in an MDM2/CRM1-dependent manner.[46] However, more recent data suggest this is not the case. In particular, Marchenko et al[2] reported that the stress-induced accumulation of p53 at the mitochondria was inhibited neither by the nuclear export blocker LMB nor over-expression of the competitive CRM1 inhibitor protein Rex. These results suggested that the cytoplasm contains a separate p53 pool that is the major source for p53 translocation to the mitochondria. Importantly, the extent to which cytoplasmic p53 was mono-ubiquitinated by MDM2 correlated with its mitochondrial localization, and MDM2 did not appear to shuttle p53 to the mitochondria since MDM2-p53 complexes were not observed in mitochondrial preparations. This suggested a role for mono-ubiquitination in p53 translocation to mitochondria. Once at the mitochondria, the data suggested that stress induced p53 associated with mitochondrial HAUSP and was rapidly deubiquitinated, generating the apoptotically active, nonubiquitinated p53. Together, the data support the following model for p53 trafficking to the mitochondria:

under normal conditions, cytoplasmic MDM2 promotes polyubiquitination and degradation of cytoplasmic p53. Following stress, post-translational modifications to p53 and/or MDM2 inhibit polyubiquitination but do not apparently inhibit MDM2 binding. P53 mono-ubiquitination promotes its trafficking to mitochondria. Notably, mono-ubiquitination in the p53 C-terminus also causes dissociation of p53-MDM2 complexes,[56] providing an explanation for why MDM2 does not migrate to the mitochondria with p53. Mitochondrial p53 is deubiquitinated by HAUSP to generate apoptotically active p53.

Conclusion

Various cancers have been described in which p53 is inactivated due to its abnormal sequestration in the cytoplasm. Strategies that direct p53 to the nucleus, mitochondria, or both, in these cancers are expected to provide a strong therapeutic benefit. DNA damage and other stresses cause a pronounced nuclear accumulation of p53 and a less-pronounced but highly significant accumulation of p53 in the mitochondria. Stress-induced nuclear accumulation of p53 could result from increased nuclear import, diminished nuclear export, or both. Conceivably, dissociation of p53 from cytoplasmic sequestration complexes following stress could contribute to either the nuclear accumulation of p53 or its accumulation in mitochondria. The effects of stress on either p53 nuclear import or its association with cytoplasmic anchor proteins have not been widely studied. However, there is ample evidence that p53 nuclear export is inhibited in stressed cells. For example, DNA damaging stress induces p53 phosphorylation at N-terminal sites, including sites within or near the MDM2-binding domain (i.e., S15, S20). These phosphorylations are thought to inhibit or diminish MDM2 binding to p53's N-terminus. In studies from the Oren and Brune laboratories, wild-type p53 was phosphorylated at serine-15 (S15) and accumulated in the nucleus of cells treated with nitric oxide (NO).[57,58] To examine the effect of NO treatment on p53 nuclear export, the investigators used a heterokaryon assay and immunostaining with an antibody directed against phosphorylated S15. P53 phosphorylated at S15 did not undergo nuclear export following NO treatment, while nonphosphorylated p53 did. Thus, stress-induced phosphorylation can apparently inhibit p53 nuclear export, perhaps by diminishing MDM2-binding and nuclear export via the C-terminal NES. It is important that the N-terminal NES of p53 spans the MDM2-binding domain,[27] and that the same N-terminal phosphorylations thought to inhibit MDM2 binding also inhibit the activity of this N-terminal NES. Thus, DNA damage-induced phosphorylation may promote optimal nuclear accumulation of p53 by inhibiting MDM2 binding and nuclear export via the C-terminal NES, as well as nuclear export via the NES in the N-terminus.[59,60]

References

1. Hollstein M, Sidransky D, Vogelstein B et al. p53 mutations in human cancers. Science 1991; 253:49-53.
2. Marchenko ND, Wolff S, Erster S et al. Monoubiquitylation promotes mitochondrial p53 translocation. EMBO J 2007; 26:923-934.
3. Middeler G, Zerf K, Jenovai S et al. The tumor suppressor p53 is subject to both nuclear import and export, and both are fast, energy-dependent and lectin-inhibited. Oncogene 1997; 14:1407-1417.
4. Shaulsky G, Ben-Ze'ev A, Rotter V. Subcellular distribution of the p53 protein during the cell cycle of Balb/c 3T3 cells. Oncogene 1990; 5:1707-1711.
5. Shaulsky G, Goldfinger N, Ben-Ze'ev A et al. Nuclear accumulation of p53 protein is mediated by several nuclear localization signals and plays a role in tumorigenesis. Mol Cell Biol 1990; 10:6565-6577.
6. Dang CV, Lee WM. Nuclear and nucleolar targeting sequences of c-erb-A, c-myb, N-myc, p53, HSP70, and HIV tat proteins. J Biol Chem 1989; 264:18019-18023.
7. Shaulsky G, Goldfinger N, Tosky MS et al. Nuclear localization is essential for the activity of p53 protein. Oncogene 1991; 6:2055-2065.
8. Shaulsky G, Goldfinger N, Peled A et al. Involvement of wild-type p53 protein in the cell cycle requires nuclear localization. Cell Growth Differ 1991; 2:661-667.

9. Liang SH, Clarke MF. A bipartite nuclear localization signal is required for p53 nuclear import regulated by a carboxyl-terminal domain. J Biol Chem 1999; 274:32699-32703.
10. Liang SH, Clarke MF. The nuclear import of p53 is determined by the presence of a basic domain and its relative position to the nuclear localization signal. Oncogene 1999; 18:2163-2166.
11. Liang SH, Hong D, Clarke MF. Cooperation of a single lysine mutation and a C-terminal domain in the cytoplasmic sequestration of the p53 protein. J Biol Chem 1998; 273:19817-19821.
12. Stommel JM, Marchenko ND, Jimenez GS et al. A leucine-rich nuclear export signal in the p53 tetramerization domain: regulation of subcellular localization and p53 activity by NES masking. EMBO J 1999; 18:1660-1672.
13. Gu J, Nie L, Wiederschain D et al. Identification of p53 sequence elements that are required for MDM2-mediated nuclear export. Mol Cell Biol 2001; 21:8533-8546.
14. O'Keefe K, Li H, Zhang Y. Nucleocytoplasmic shuttling of p53 is essential for MDM2-mediated cytoplasmic degradation but not ubiquitination. Mol Cell Biol 2003; 23:6396-6405.
15. Giannakakou P, Sackett DL, Ward Y et al. p53 is associated with cellular microtubules and is transported to the nucleus by dynein. Nat Cell Biol 2000; 2:709-717.
16. Galigniana MD, Harrell JM, O'Hagen HM et al. Hsp90-binding immunophilins link p53 to dynein during p53 transport to the nucleus. J Biol Chem 2004; 279:22483-22489.
17. Trostel SY, Sackett DL, Fojo T. Oligomerization of p53 precedes its association with dynein and nuclear accumulation. Cell Cycle 2006; 5:2253-2259.
18. Vousden KH, Lu X. Live or let die: the cell's response to p53. Nat Rev Cancer 2002; 2:594-604.
19. Jensen K, Shiels C, Freemont PS. PML protein isoforms and the RBCC/TRIM motif. Oncogene 2001; 20:7223-7233.
20. Fogal V, Gostissa M, Sandy P et al. Regulation of p53 activity in nuclear bodies by a specific PML isoform. EMBO J 2000; 19:6185-6195.
21. Pearson M, Carbone R, Sebastiani C et al. PML regulates p53 acetylation and premature senescence induced by oncogenic Ras. Nature 2000; 406:207-210.
22. Ferbeyre G, de Stanchina E, Querido E et al. PML is induced by oncogenic ras and promotes premature senescence. Genes Dev 2000; 14:2015-2027.
23. Seeler JS, Dejean A. SUMO: of branched proteins and nuclear bodies. Oncogene 2001; 20:7243-7249.
24. Rodriguez MS, Desterro JM, Lain S et al. SUMO-1 modification activates the transcriptional response of p53. EMBO J 1999; 18:6455-6461.
25. Mauri F, McNamee LM, Lunardi A et al. Modification of Drosophila p53 by SUMO Modulates Its Transactivation and Pro-apoptotic Functions. J Biol Chem 2008; 283:20848-20856.
26. Li T, Santockyte R, Shen RF et al. Expression of SUMO-2/3 induced senescence through p53- and pRB-mediated pathways. J Biol Chem 2006; 281:36221-36227.
27. Zhang Y, Xiong Y. A p53 amino-terminal nuclear export signal inhibited by DNA damage-induced phosphorylation. Science 2001; 292:1910-1915.
28. Freedman DA, Levine AJ. Nuclear export is required for degradation of endogenous p53 by MDM2 and human papillomavirus E6. Mol Cell Biol 1998; 18:7288-7293.
29. Roth J, Dobbelstein M, Freedman DA et al. Nucleo-cytoplasmic shuttling of the hdm2 oncoprotein regulates the levels of the p53 protein via a pathway used by the human immunodeficiency virus rev protein. EMBO J 1998; 17:554-564.
30. Tao W, Levine AJ. Nucleocytoplasmic shuttling of oncoprotein Hdm2 is required for Hdm2-mediated degradation of p53. Proc Natl Acad Sci USA 1999; 96:3077-3080.
31. Geyer RK, Yu ZK, Maki CG. The MDM2 RING-finger domain is required to promote p53 nuclear export. Nat Cell Biol 2000; 2:569-573.
32. Boyd SD, Tsai KY, Jacks T. An intact HDM2 RING-finger domain is required for nuclear exclusion of p53. Nat Cell Biol 2000; 2:563-568.
33. Lohrum MA, Woods DB, Ludwig RL et al. C-terminal ubiquitination of p53 contributes to nuclear export. Mol Cell Biol 2001; 21:8521-8532.
34. Li M, Brooks CL, Wu-Baer F et al. Mono- versus polyubiquitination: differential control of p53 fate by Mdm2. Science 2003; 302:1972-1975.
35. Brooks CL, Li M, Gu W. Mechanistic studies of MDM2-mediated ubiquitination in p53 regulation. J Biol Chem 2007; 282:22804-22815.
36. Chau V, Tobias JW, Bachmair A et al. A multiubiquitin chain is confined to specific lysine in a targeted short-lived protein. Science 1989; 243:1576-1583.
37. Grossman SR, Deato ME, Brignone C et al. Polyubiquitination of p53 by a ubiquitin ligase activity of p300. Science 2003; 300:342-344.
38. Grossman SR, Perez M, Kung AL et al. p300/MDM2 complexes participate in MDM2-mediated p53 degradation. Mol Cell 1998; 2:405-415.

39. Grossman SR. p300/CBP/p53 interaction and regulation of the p53 response. Eur J Biochem 2001; 268:2773-2778.
40. Gannon JV, Greaves R, Iggo R et al. Activating mutations in p53 produce a common conformational effect. A monoclonal antibody specific for the mutant form. EMBO J 1990; 9:1595-1602.
41. Milner J, Cook A, Sheldon M. A new anti-p53 monoclonal antibody, previously reported to be directed against the large T antigen of simian virus 40. Oncogene 1987; 1:453-455.
42. Nie L, Sasaki M, Maki CG. Regulation of p53 nuclear export through sequential changes in conformation and ubiquitination. J Biol Chem 2007; 282:14616-14625.
43. Yu ZK, Geyer RK, Maki CG. MDM2-dependent ubiquitination of nuclear and cytoplasmic p53. Oncogene 2000; 19:5892-5897.
44. Xirodimas DP, Stephen CW, Lane DP. Cocompartmentalization of p53 and Mdm2 is a major determinant for Mdm2-mediated degradation of p53. Exp Cell Res 2001; 270:66-77.
45. Shirangi TR, Zaika A, Moll UM. Nuclear degradation of p53 occurs during down-regulation of the p53 response after DNA damage. FASEB J 2002; 16:420-422.
46. Dumont P, Leu JI, Della Pietra AC 3rd et al. The codon 72 polymorphic variants of p53 have markedly different apoptotic potential. Nat Genet 2003; 33:357-365.
47. Nikolaev AY, Li M, Puskas N et al. Parc: a cytoplasmic anchor for p53. Cell 2003; 112:29-40.
48. Sengupta S, Vonesch JL, Waltzinger C et al. Negative cross-talk between p53 and the glucocorticoid receptor and its role in neuroblastoma cells. EMBO J 2000; 19:6051-6064.
49. Sengupta S, Wasylyk B. Ligand-dependent interaction of the glucocorticoid receptor with p53 enhances their degradation by Hdm2. Genes Dev 2001; 15:2367-2380.
50. Wadhwa R, Takano S, Robert M et al. Inactivation of tumor suppressor p53 by mot-2, a hsp70 family member. J Biol Chem 1998; 273:29586-29591.
51. Lu W, Pochampally R, Chen L et al. Nuclear exclusion of p53 in a subset of tumors requires MDM2 function. Oncogene 2000; 19:232-240.
52. Rodriguez-Lopez AM, Xenaki D, Eden TO et al. MDM2 mediated nuclear exclusion of p53 attenuates etoposide-induced apoptosis in neuroblastoma cells. Mol Pharmacol 2001; 59:135-143.
53. Fontoura BM, Atienza CA, Sorokina EA et al. Cytoplasmic p53 polypeptide is associated with ribosomes. Mol Cell Biol 1997; 17:3146-3154.
54. Guerra B, Issinger OG. p53 and the ribosomal protein L5 participate in high molecular mass complex formation with protein kinase CK2 in murine teratocarcinoma cell line F9 after serum stimulation and cisplatin treatment. FEBS Lett 1998; 434:115-120.
55. Marchenko ND, Moll UM. The role of ubiquitination in the direct mitochondrial death program of p53. Cell Cycle 2007; 6:1718-1723.
56. Carter S, Bischof O, Dejean A et al. C-terminal modifications regulate MDM2 dissociation and nuclear export of p53. Nat Cell Biol 2007; 9:428-435.
57. Wang X, Zalcenstein A, Oren M. Nitric oxide promotes p53 nuclear retention and sensitizes neuroblastoma cells to apoptosis by ionizing radiation. Cell Death Differ 2003; 10:468-476.
58. Schneiderhan N, Budde A, Zhang Y et al. Nitric oxide induces phosphorylation of p53 and impairs nuclear export. Oncogene 2003; 22:2857-2868.
59. Kawaguchi Y, Ito A, Appella E et al. Charge modification at multiple C-terminal lysine residues regulates p53 oligomerization and its nucleus-cytoplasm trafficking. J Biol Chem 2006; 281: 1394-1400.
60. Li M, Luo J, Brooks CL et al. Acetylation of p53 inhibits its ubiquitination by Mdm2. J Biol Chem 2002; 277:50607-50611.

CHAPTER 9

Modes of p53 Interactions with DNA in the Chromatin Context

Vladana Vukojevic,* Tatiana Yakovleva and Georgy Bakalkin

Introduction

The tumor suppressor p53, a sequence-specific transcription factor playing a central role in cancer surveillance, exerts its main functions by activating or repressing transcription of its target genes.[1-4] Protein products of several genes activated by p53, including *p21*, *Bax*, *GADD45*, *Mdm2*, and *Cyclin G*, are required to regulate cell cycle progression and apoptosis. p53 activates or inhibits transcription by binding to specific DNA target sequences. The p53 binding site consists of two half-sites 5′-PuPuPuC(A/T)(T/A)GPyPyPy-3′, linked by a 0-13 nucleotide spacer. Each half-site decamer, comprising two copies of the pentamer sequence PuPuPuC(A/T) arranged head-to-head or head-to-tail,[1-3] binds a dimer of p53 to form the productive p53-tetramer-DNA complex. The degenerate nature of DNA sequences allows structural diversity that may be critical for plasticity of p53-mediated responses to extracellular signals and genotoxic stress.

The modular p53 structure can be divided into three main domains (Fig. 1): the N-terminal transactivation domain (amino acids 1-99) comprising two contiguous transcriptional activation subdomains (amino acids 1-42 and 43-63) and an adjacent proline-rich region (amino acids 62-91), the core domain (amino acids 100-300) that binds to specific DNA target sites, and the C-terminal domain (CTD; amino acids 301-393) including the tetramerization domain (amino acids 325-356) and a regulatory region (amino acids 363-393). The tetramerization domain allows oligomerization of this protein, and p53 binds to DNA with highest affinity when it is tetrameric. A unique property of p53 as a transcription factor is that it contains a second DNA binding site that is associated with the regulatory region in the CTD, which binds with high affinity to single-stranded DNA ends, insertion/deletion DNA mismatches, Holliday junctions and damaged DNA, and catalyzes DNA renaturation and strand exchange.[1,4]

In "normal" (i.e., nonstressed) cells, p53 is present at low levels and is latent.[1-7] Genotoxic stress or oncogenes activate p53-mediated gene transcription,[1,3,5-7] and stabilize its concentration by disrupting p53-MDM2 interactions (MDM2 targets p53 to degradation by the ubiquitin-proteasome system).[2,3] Both activation and stabilization of p53 appear to be essential for responses to cellular stress. However, they can be distinguished experimentally. For example, accumulation of p53 appears to be insufficient for the activation of p53-dependent apoptosis and growth arrest;[1-5,6] ultraviolet irradiation induces p53 accumulation, whereas p53-mediated transcription of genes such as *p21* is inhibited.[5,7]

Regulation of p53 transcription function is thought to occur through post-translational modifications of the p53 protein. As a result of DNA damage response and oncogene activation, p53 becomes heavily phosphorylated and acetylated at sites in the N-terminal domain

Corresponding Author: Vladana Vukojevic—Department of Clinical Neuroscience, Karolinska Institute, CMM L8:01, 17176 Stockholm, Sweden. Email: vladana.vukojevic@ki.se

p53, edited by Ayeda Ayed and Theodore Hupp. ©2010 Landes Bioscience and Springer Science+Business Media.

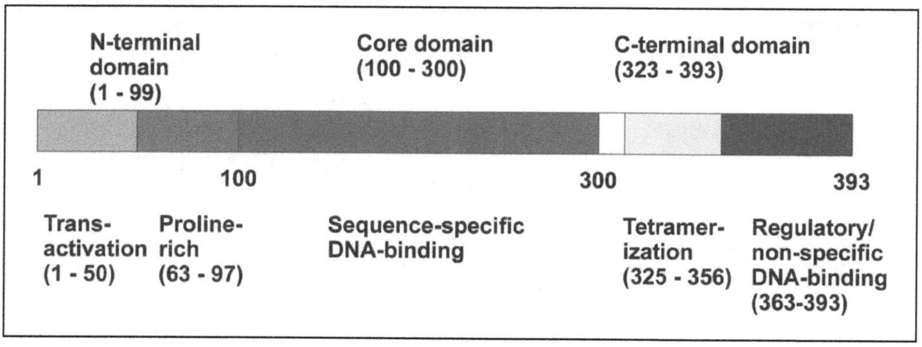

Figure 1. Domain organization of the tumor suppressor protein p53. Reprinted from: Yakovleva T, Pramanik A, Terenius L et al. p53 latency—out of the blind alley. Trends Biochem Sci 27:612-618, ©2002 with permission from Elsevier.[15]

and CTD.[1-5] Phosphorylation of the N-terminus leads to p53 stabilization and recruitment of p300/CBP/PCAF that acetylate the distant CTD. In vitro assays show that levels of these post-translational modifications correlate with the initiation of p53-mediated transcription.[1-5] However, several reports cast doubts as to whether these mechanisms operate in vivo. Mutations of all sites subjected to post-translational modifications which are expected to play important roles in p53 activation, generally fail to produce the anticipated changes in p53 activation.[2,3,8]

Allosteric/Conformational Mechanism

p53 transcription function appears to be critically regulated through p53-DNA-binding properties. The fact that p53 is a latent transcription factor in cells, and is activated in response to cellular stress, is supported by in vitro observations that p53 does not bind to its target sites in short DNA fragments unless the protein is modified by phosphorylation, acetylation or deletion of its CTD.[5,9] Factors that activate p53 for sequence specific DNA binding are targeted to the critical CTD regulatory segment p53(361-382). Thus, binding of the monoclonal antibody PAb421 to this segment activates sequence-specific DNA binding in vitro.[5] Modifications of this segment through phosphorylation or acetylation, or its deletion, activate p53 binding to specific targets in DNA fragments (Fig. 2A). Thus, modification of the CTD appears to dictate whether p53 is latent or active for binding to DNA.[5] It was proposed that unmodified CTD allosterically locks the p53 molecule in a state latent for binding to DNA by intramolecular interactions with the core domain (Fig. 2A). Signaling events in cells seem to alter the CTD regulatory region and, thereby, prevent interaction of the core domain with DNA target sites.

The CTD and CTD fragments including p53(361-382) have been found to be capable to activate binding of the latent p53 to DNA in vitro in a band shift assay.[5,9,10] The interpretation was that these fragments displace the CTD of the full-length p53 from its binding site within the core domain, and that this alters p53 conformation such that it activates p53 binding to DNA (Fig. 2, A1). Activation by CTD derived peptides has been reported to be specific since neither mutant p53(361-382) peptides nor nonhomologous fragments of the N-terminal p53 domain activate p53.[5,10,11] The CTD also displayed target specificity since the interaction of other transcription factors including SRF and GAL4-VP16 with DNA target was not altered by the CTD in a band shift assay.[9] It was also suggested that the regulatory region of CTD in latent p53 makes contacts with adjacent surfaces of the core and N-terminal domains including the proline rich region deletion of the proline-rich region leads to activation of p53 sequence specific DNA binding.[12] This region has been also described as a binding pocket for CTD derived peptides activating p53.[13]

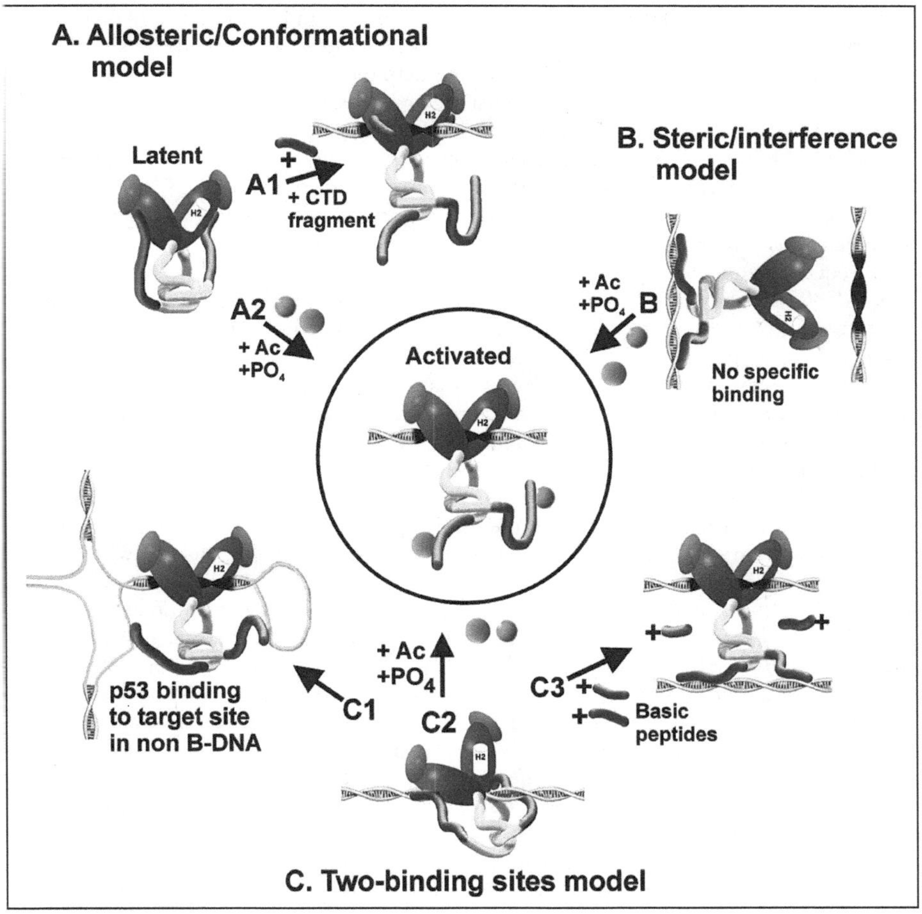

Figure 2. Models for the activation of sequence-specific p53 binding to DNA. For clarity, p53 dimers instead of tetramers are shown with the N-terminal, core and tetramerization domains in green, red and yellow, respectively, and the CTD basic regulatory region p53(363-393) in violet. H2 α-helix of one monomer in the p53 dimer is shown as a white cylinder on the surface, which contacts the DNA helix. p53 target sites in DNA are black. A) Allosteric/conformational model. Binding of the CTD regulatory region to the core domain prevents p53 interaction with DNA. Displacement of CTD from intramolecular complex with the core domain by short CTD fragments (arrow A1) or modification of the CTD regulatory sequence by acetylation (Ac) or phosphorylation (PO_4) (arrow A2) induces conformational changes in p53, allowing interaction of the core domain with specific DNA sites. B) Steric/interference model postulates that CTD binding to nonspecific DNA prevents interactions of the core domain with DNA target sites. Acetylation or phosphorylation of CTD prevents its interaction with DNA and, therefore, permis the p53 core binding to DNA (arrow B). C) The two-binding sites model. p53 binds via both its core domain and CTD to nontarget DNA. The binding of the core domain to target DNA sequences is prevented by CTD interactions with DNA due to steric hindrance. Nonmodified CTD binds to single-stranded or non B form DNA-segments in cruciforms (arrow C1). Modifications of the CTD basic sequences by acetylation or phosphorylation prevent the CTD interaction with DNA (arrow C2). Basic peptides aggregate DNA, allowing interactions of CTD with an adjacent DNA molecule (arrow C3). Modification of CTD (arrow C2) or binding of nonmodified CTD to either single-stranded DNA regions in non B-form of DNA (arrow C1) or to an adjacent DNA molecule (arrow C3) allows the proper positioning of the core domains and their binding to target DNA sites. Reprinted from: Yakovleva T, Pramanik A, Terenius L et al. p53 latency—out of the blind alley. Trends Biochem Sci 27:612-618, ©2002 with permission from Elsevier.[15] A color version of this figure is available at www.landesbioscience.com./curie.

Data generated by several laboratories were convincing, and the beautifully constructed conformational hypothesis[1-5,9-13] dominated the field for over 7 years in spite of lack of direct evidence. However, several recent observations undermine this hypothesis, and require interpretations that are out of its framework. First, Yakovleva et al reexamined the sequence- and target-specificity of interaction of the activating CTD fragment with the full-length p53 using a band shift assay and Fluorescence Correlation Spectroscopy (FCS).[14-16] p53 activation by CTD derived peptides has been described as sequence- and target-specific.[5,9-11] However, the basic character of these peptides has been neglected. Several nonhomologous, long basic peptides such as poly-lysine have been demonstrated to to activate binding of latent p53 to target DNA sequences in a band shift assay.[14] These peptides were even more potent activators than p53(361-382). The net positive charge and overall peptide length appeared to be inherent characteristics of the activating peptides. Furthermore, activation was not specific for p53; binding of transcription factor YY1 to DNA was activated 10-50-fold by p53(361-382), CTD, and full-length p53 but not by the p53 mutant lacking the CTD basic segment.[14] Thus, the basic segment in latent p53 is available for intermolecular interactions with YY1 but paradoxically, does not activate p53 itself, as it would do if the allosteric hypothesis was correct.

Basic peptides may activate p53 by interacting either with the p53 protein or with target DNA. To distinguish between them, FCS experiments were performed, demonstrating that activating peptides do not interact with the full-length p53 and the core domain, but rather induce aggregation of the target DNA.[14] The ability to activate p53 correlated with peptide potency to aggregate DNA, suggesting that p53 binds to target sequences upon interactions with tightly packed DNA in the aggregates. The facts that p53 is activated under conditions when target DNA is aggregated, and when the activating peptides do not interact with the p53 protein, demonstrated that p53 activation by basic peptides including p53(361-382) is not relevant for situations in vivo, but represents an in vitro phenomenon. This conclusion questions the earlier observed activation of p53-dependent transcription and apoptosis by p53(361-382)[10,11,13] and suggests that it is a consequence of nonspecific toxic effects of basic peptides on cells under conditions when ectopic wild-type or mutant p53 proteins are overexpressed. Cytotoxic effects of other exogenous polycations have been reported.[17,18] On the other hand, polycationic molecules such as histones and polyamines, present in the cell nuclei at high concentrations, produce no effects on p53-mediated cell death. Whatever the mechanism of activation, the effects of p53(361-382) and poly-lysine (as a control peptide) on wild-type and mutant p53 in tumor cells require further verification.

The second set of experiments was performed by NMR spectroscopy. Ayed et al[19] found that conformations of the full-length p53 protein and the CTD truncated p53, which, according to the alosteric hypothesis are the latent and active forms of p53, respectively, are identical. This observation is in compelling contradiction with the conformational hypothesis, which postulates that these conformations are different.[5] However, binding of the protein domain to a small molecule or to another protein domain does not always induce changes in conformation of rigid protein domains.[20] Therefore, identity in conformation of latent and activated p53 does not itself disprove the allosteric mechanism but poses certain limitations; the molecular architecture could be similar in both p53 states but by interacting with the core domain, CTD may create steric hindrances for the core domain-DNA interactions. However, this appears not to be the case.

The proposed interactions between the CTD and the core domain[5] have also been tested by NMR.[19] The conclusion was in line with the first NMR result: CTD does not interact with the p53 core. Independently, Friedler et al[21] and Klein et al[22] have tested whether the CTD derived peptides, which activate latent p53 for binding to target DNA, may interact with the p53 core domain when they are in phosphorylated and unphosphorylated forms and under various conditions (temperature, pH, ionic strength). Several NMR techniques and surface plasmon resonance have been applied to detect the CTD peptide interactions with p53.[12,13,22] Interactions were not observed between the activating CTD peptides and free or DNA bound p53 core, nor between the CTD peptide and the proline-rich region of the N-terminal extended p53 core domain, which was earlier proposed to be the binding site for the CTD peptide. None of these experiments produced any indication of molecular interactions between the activating CTD peptides and free or DNA-bound p53 core, or between the CTD peptide and the proline-rich region of the N-terminal extended p53 core domain, which earlier has been proposed to be the binding site for the CTD peptide.[12,13,22] The similar conformations of the active and latent p53, the lack of evidence for interactions between CTD and the core domain, and the sequence- and target-nonspecific p53 activation by the CTD peptides do not support the allosteric/conformational mechanism of p53 latency.

Steric/Interference Hypothesis

Overlapping of the basic regulatory region in CTD with the second DNA-binding site in the p53 molecule is the key feature of steric hypotheses,[23,24] postulating that p53 interactions with genomic DNA via the CTD prevents binding of the core to target sequences in gene promoters (Fig. 2B). Observations that long nonspecific DNA molecules inhibit sequence-specific binding of the full-length p53 to short target DNAs, but show no effects on p53 with deletion of a CTD basic sequence, supports this hypothesis. However, these experiments, performed with partially activated p53 produced in a baculovirus system,[25] have not been confirmed with truly latent bacterially-produced p53 protein.[14,19] Thus, Yakovleva et al[14] and Ayed et al[19] demonstrated that the latent full-length p53 does not bind to target DNA sites in the absence of nonspecific DNA competitors, in contradiction with the steric hypothesis of p53 latency.

The Novel Two-Binding Sites Hypothesis

The ability of p53, which is latent for binding to DNA target sites, to form complexes via its core domain with internal segments of long single-stranded and double-stranded DNA, DNA damaged by ionizing radiation, insertion/deletion DNA mismatches and supercoiled DNA1,[4,14,26-30] was overlooked for many years. Remarkably, the affinity of latent full-length p53 and isolated p53 core domain for negatively supercoiled plasmid DNA, which is lacking p53 target sites, is considerably greater than affinity of these proteins for target DNA sites in linear DNA fragments.[30] Nonspecific DNA binding requires both the CTD and the core domain.[14,27,30] Thus, p53 is a conditionally latent transcription factor—latent for binding to specific target sites in short DNA fragments and in long B-form DNAs, but interacting via its core domain and CTD with nontarget long linear and supercoiled DNA molecules.[14,27,30] This conclusion is in conflict with both the allosteric and steric hypotheses, because none of the hypotheses allow binding of the core domain of latent p53 to DNA.

To solve the paradox of "conditional latency" of p53, we proposed the two-binding sites model,[14,15] which postulates that binding of CTD of the full-length p53 to DNA prevents the interaction of the core domain with specific but not with nonspecific DNA sequences (Fig. 2C). Each p53 monomer may bind to DNA via its core domain and the CTD. Eight DNA-interacting sites of a p53 tetramer molecule, four in the CTD and four in the core domain, may make contacts with DNA when a tetramer assembles on a DNA segment. CTDs bound to DNA could affect the orientation and positioning of the core domains that are necessary to form a complex with target DNA sequences. CTDs are linked to core domain H2 α-helices which are critical for recognition of target sites.[31] When bound to DNA, CTD may pull these helices towards the CTD-DNA contact sites. Resulting changes in the position and orientation of the H2 α-helices, including a critical Arg^{280}, may prevent the binding to target sites but not to nonspecific DNA sequences. Contact of Arg^{280} with the guanidine of the invariant C-G pair of the target decamer sequence is critical for the formation of the p53-DNA complex.[32] However, Arg^{280} is not engaged in p53 binding to nontarget DNA as shown by crystallographic studies.[31] The CTD modified by phosphorylation or acetylation is not able to interact with DNA, and this allows rapprochement of core domains and their binding to the adjacent pentamers in p53 target DNA half-sites, or the interaction of H2 α-helices with these sequences (Fig. 2, C2).

The CTD, bound to DNA via its basic segment, could make contact with the adjacent surfaces of the core or N-terminal domains. It has been suggested that the CTD interacts with the proline rich region[12] and that this interaction could potentially determine positions of CTD-DNA contacts. Deletion of the proline-rich region leads to activation of p53 sequence specific DNA binding[12] possibly because the CTD of this p53 mutant protein is situated on DNA without creating steric clashes with the core domains or without inducing deformations of H2 α-helices in p53 tetramer-DNA complexes.

p53 Is Activated in DNA Aggregates

p53(361-382) and other basic peptides activate binding of p53 to specific DNA sequences without direct interaction with p53 but through the aggregation of target DNA.[14] In aggregates, DNA molecules are tightly packed, and the CTD and the core domain of the full-length p53 molecule can interact with juxtaposed DNA molecules, resulting in p53 activation (Fig. 2, C3). Electrostatic repulsion between negatively charged DNA molecules in the absence of basic peptides prevents the interaction of p53 domains with two DNA molecules, and therefore binding of both domains to one DNA molecule would be more favorable.

Surprisingly, FCS studies demonstrated that p53 may dissociate[14] and denature[14a] DNA aggregates formed in the presence of basic peptides or polyamines.[14] This novel p53 activity requires cooperation between the core and the CTD because none of them, taken separately, catalyzes this reaction. Moreover, this activity appears to be energy dependent since GTP, but not GDP, GMP and the nonhydrolyzable analog GMPPNP, strongly potentiates DNA disaggregation by p53.[14] It has been reported earlier that p53 has intrinsic ATPase and GTPase activities,[33] which may be involved in specifically the catalysis of DNA renaturation and strand exchange,[26,27] and generally in the regulation of the p53 binding to DNA.[33] DNA molecules may undergo intermolecular aggregation into ordered, highly condensed and thermodynamically stable states in solutions where DNA-DNA interactions prevail (for instance in the presence of polycations), or upon confinement of DNA into a limited space such as an intracellular highly crowded environment.[34] Liquid-crystalline DNA organization was observed in dinoflagellate chromosomes and sperm cells.[35] Many proteins involved in DNA recombination can promote DNA aggregation, and the formation of hybrid DNA intermediates may be facilitated in DNA aggregates.[36] We can speculate that inhibition of DNA recombination by p53[4] is based on its ability to inhibit DNA aggregation, in which strand exchange, a necessary event in DNA hybrid formation, is attenuated. Another possibility is that DNA aggregates, if they are formed in the cell nucleus, may activate p53 via the mechanism described in Figure 2, C3.

p53 Interactions with Chromatin

Several recent studies have demonstrated that in the cell nucleus, p53 is not a latent DNA binding factor.[37-40] Studies of p53 interactions with long DNA molecules showed that p53 binds to target sites in the *p21* promoter, and that deletion of the CTD does not affect p53 dissociation from the promoter and the DNase I protection of these sites.[37] Espinosa and Emerson[38] and Göhler at al.[39] observed that p53 can bind to two sites in the chromatin-assembled *p21* gene promoter and that modifications of the CTD are not critical for binding and promoter activation. The conclusion was that the CTD does not inhibit, but instead is required for p53 interactions with the *p21* promoter. Based on quantitative chromatin immunoprecipitation studies, Kaeser and Iggo[40] concluded that genotoxic stress, which activates p53 dependent transcription, induces only small changes in p53-assisted chromatin precipitation. Thus, p53 is already mainly present in DNA bound form in nonstressed cells. This indicates that an allosteric mechanism is not relevant for the regulation of p53-DNA interactions.

According to the steric model "interactions of the C terminus of p53 with genomic DNA in vivo would prevent p53 binding to specific promoters and that the cellular mechanisms to block C-terminal DNA binding would be required".[24] Observations that the CTD is necessary for p53 to activate transcription, and that CTD modifications are not required for p53 binding to the chromatin assembled *p21* promoter,[38,41] provide additional evidence against this model.

A distinction between the steric and the two-binding sites models, regardless of whether a p53 binding target is located in short DNA fragments or on chromatin, is that in the two-binding sites model the core domain is always engaged in interactions with DNA, whether or not p53 activates transcription (Fig. 2C). In contrast, in the steric model the p53 core is prevented from binding to DNA unless the CTD is modified or deleted (Fig. 2B).[23,24]

Distinct features of p53-binding sites in the *p21* promoter is that they are prone to form cruciform structures; in this form, they present targets for p53 binding.[39,42,43] Conformational shift from linear to cruciform structures can promote p53 binding to DNA. Modifications or deletion of the regulatory sequence in the CTD is required for p53 binding to target sites in B-form DNA. In contrast, unmodified CTD strengthens p53 interactions with cruciform targets.[39,42,43] In cruciforms, the CTD may bind to denatured single-stranded DNA regions (Fig. 2, C1) as it binds to four-way Holliday junctions. Such CTD binding may allow proper orientation of the core domains to facilitate their binding to the target sites, in contrast to sterically different binding to linear or supercoiled DNA (Fig. 2C) which result in steric clashes between core domains in the p53 tetramer-DNA complexes.

Formation of cruciform structures is often thermodynamically unfavorable in linear double-stranded DNA molecules. In the cell nucleus, several helicases and ATP-dependent chromatin remodeling complexes may extrude cruciform DNAs from chromatin.[44-46] Formation of noncanonical DNA structures is increased in vivo under conditions of active DNA replication, recombination and gene transcription.[39] It is, however, still unclear how basal and stimulated transcription from cruciform p53 targets is regulated.

p53 activation leads to distinct functional responses, including apoptosis, cell cycle arrest, cell differentiation or senescence. p53 also inhibits angiogenesis and activates expression of DNA damage repair genes.[1-5] The choice of response is determined by the subset of genes that is activated or inhibited by p53. The mechanism of coordinated activation and repression of these subsets is not clear. Allosterism of DNA binding sites is critical for alternative gene activation or repression by several transcription factors.[47] Thus, the POU domain factor Pit-1-mediated activation and repression of growth hormone expression depend on distinction in a two-base pair spacing in accommodation of the bipartite POU domains on its binding site.[47] The allosteric effect on Pit-1 results in the recruitment of a corepressor complex, including nuclear receptor corepressor N-CoR, which is required for the repression of the growth hormone gene in lactotropes but not in somatotropes. By analogy, the allosteric effects of the DNA binding elements on p53 configuration may serve as critical determinants for interactions with components of the coactivator/corepressor machinery as well as with chromatin remodeling and modifying complexes (Fig. 3).[38,39,45] These interactions could be essential for cooperative switching on or of the physiologically relevant groups of genes.

Covalent modifications such as phosphorylation and acetylation have been suggested to regulate p53 conformation.[1-5] Changes in protein conformation may attenuate p53 affinity for different promoter targets. Complementary, distinct p53 conformations may be assumed upon binding to structurally diverse DNA target sequences (Fig. 3).[39,42,43] Allosterism of p53 target DNA sites was described in terms of sequence degeneracy, variability in the number of half-sites composing these elements, difference in length of spacers connecting half-sites, and palindromic nature of these sites, allowing the formation of cruciform DNA structures.[1-5,39,48-50]

The general model depicted in Figure 3, is supported by several observations.[48-50] Thus, whether p53 represses or activates transcription has been shown to depend on the orientation of pentamer target sequences PuPuPuC(A/T) in half-site decamers, constituting a p53 binding site. p53 represses transcription when it binds to the pentamer sequences arranged "head-to-tail" in the MDR1 gene promoter, whereas p53 interactions with the pentamers in the "head-to-head" orientation in the same promoter context results in activation of transcription (Fig. 3).[48] Simple recruitment of p53 to the promoter appears not to be sufficient for making the decision whether the bound protein could activate or repress transcription. It is important to mention that p53 target sites in promoters of all genes activated by p53 are composed of pentamers in "head-to-head" orientation. It has been speculated that p53 in a conformation bound to the "head-to-tail"

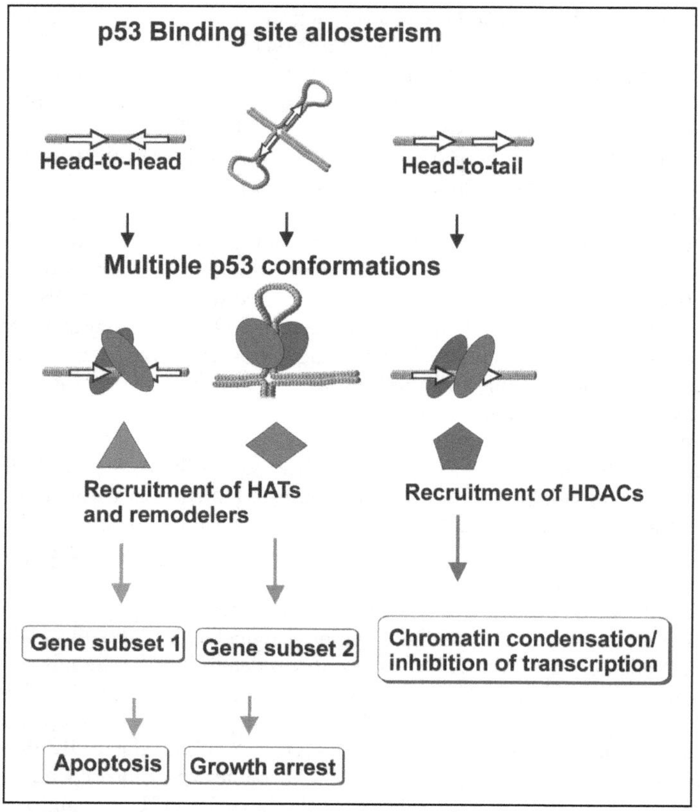

Figure 3. Model for allosterism of p53 DNA-binding sites and complementary conformations of the p53 protein. p53 conformations may critically determine interactions with components of coactivator/corepressor machinery, and chromatin remodeling and modifying complexes. These interactions are essential for cooperative switching on or off of the physiologically relevant groups of genes. Reprinted from: Yakovleva T, Pramanik A, Terenius L et al. p53 latency—out of the blind alley. Trends Biochem Sci 27:612-618, ©2002 with permission from Elsevier.[15]

structure preferentially recruits histone deacetylase complexes (HDACs) to chromatin,[48] whereas p53 in a conformation specific for the "head-to-head" DNA complex, may interact predominantly with histone acetyltransferases (HATs). Indeed, inhibition of transcription by p53 was described to be mediated by histone deacetylase mSin3a,[51,52] whereas p53 involves histone acetyltransferases such as p300/CBP, PCAF, TRRAP and ADA3 when it activates transcription.[53-55]

Another example suggesting that the activating or inhibiting conformations of p53 depend on the nature of the specific DNA binding element is provided by studying the influence of spacers connecting two p53 target half sites.[49,50] p53 consensus sites contain a spacer of 0-13 nucleotides,[1-5,49,50,56] whereas functionally defined p53 target sequences in p53-induced genes predominantly contain less than 2 nucleotide spacers. An increase in spacer length from 1 to 4 nucleotides eliminates the potential to initiate p53-dependent transcription. However, a 10 nucleotide spacer restores transactivation.[56] It has been proposed that p53 activates transcription when the dimers are on the same face of the DNA helix, whereas an increase of the spacer length changes the orientation of the p53 dimers relative to each other.[49,56] Interestingly, instead of being neutral for transactivation, p53-binding sites with 3-nucleotide spacers in the promoters of the anti-apoptotic surviving gene and *p202*, an interferon-inducible negative regulator of cell growth, have been identified as targets for p53-mediated repression.[49,50] Taken together, these data support the notion that p53 assumes different conformations when binding to the DNA target sites with diverse topological properties. Distinct p53 conformations may favor recruitment of specific coactivator constellations and corepressors, which would cooperatively modulate transcription of subsets of target genes critical for different physiological responses (Fig. 3). The general model, depicted in Figure 3, raises several questions: Do p53 target sites form cruciform structures in the cell nucleus? Does the activation of transcription by p53 depend on the transition of p53 target sites from B-form to cruciform structures? Does p53 assume different conformations in complex with diverse DNA targets? And do these p53 conformers differentially recruit chromatin-modifying and chromatin-remodeling complexes to each target promoter in a subset of genes, leading to distinct physiological change?

Conclusions regarding p53 conformational isomorphism are consistent with the long existing hypothesis on protein structure that is currently known as the "new view".[57] In contrast to the traditional standpoint on proteins as structurally restricted and functionally distinctive, the "new view" postulates the existence of proteins as an ensemble of conformational isomers.[57-63] In a smooth energy landscape, the free energy profile of protein folding exhibits a deep "energy valley" (Fig. 4A). Thus, there are high-energy structures (typically more than one, but only one is shown for simplicity), and one low-energy structure at the bottom of the "energy valley". A protein molecule that is in the "energy valley" is funneled to the lowest energy state, and the lower the energy the more closely the protein structure resembles the lowest energy structure. In the "new view" hypothesis, there is not a single, but rather several low-energy structures (Fig. 4B). The energy profile shows many "valleys", which do not have a single minimum, but rather several local extremes, thus the "energy landscape" is rugged. Conformational isomers belong to one "energy valley" comprising several "pits" (Fig. 4B). They are characterized by similar, but distinctive, values of relative Gibbs free energies (Fig. 4B), and exist in an equilibrium state. (This is a simplified interpretation of the "new view" hypothesis. See refs. 58-63 for a more detailed description).

Conformational isomorphism of proteins is invoked as regulatory mechanism in several recent studies.[64-67] Analysis of Horse Radish Peroxidase (HRP) at a single molecule level by FCS demonstrated that enzymatic activity varies broadly, and that this variation reflects alterations in the protein conformation arising due to thermodynamic fluctuations.[64] HRP molecules exist apparently as several conformations that are thermodynamically slightly different from one another, and each conformation exhibits different enzymatic activity. Single-molecule activity is less distributed at higher substrate concentrations, indicating that past intercourse with the substrate determines the conformation of a single enzyme molecule for a certain time period. Another example of conformational isomorphism is given by studies of specificity of monoclonal antibodies.[65] Monoclonal antibody raised against 2,4-dinitrophenyl can adopt two different

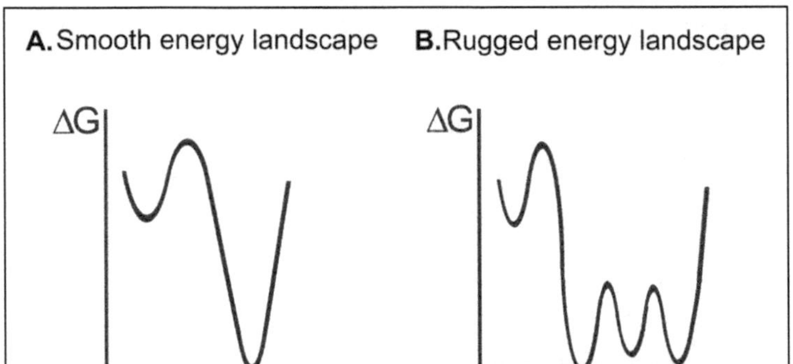

Figure 4. Free energy profile of protein folding. A) Smooth energy landscape. B) Rugged energy landscape.

conformations, existing in an equilibrium state. These conformations are independent of the presence of antigen, and antibodies in each conformation bind to a different antigen. This phenomenon may be biologically relevant, assuming that conformational isomers, which bind different haptens, increase the antibody repertoire. So far, about forty proteins with multi-functional properties based on conformational isomorphism have been identified.[66,67]

Conformational isomorphism of p53 relevant for interactions with specific DNA targets is depicted in Figure 5A. p53 may assume several conformations existing in equilibrium.

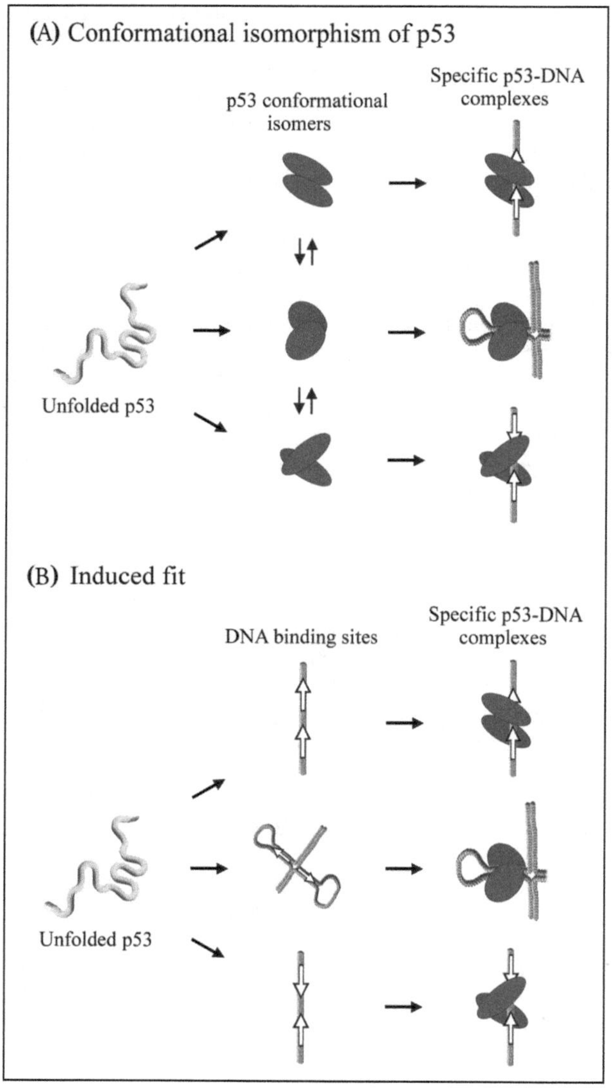

Figure 5. Models for formation of specific p53-DNA-binding complexes achieved through (A) p53 conformational isomorphism or (B) induced-fit by allosteric DNA-sites. Conformational isomorphism—p53 may assume several conformations existing in equilibrium. Conformational isomers display different affinity for different DNA target sites. In the induced-fit mechanism, binding to structurally diverse DNA-target sequences induces conformational changes in the unfolded/unstructured p53 molecules, resulting in formation of p53 conformers differing in configuration of DNA binding pockets.

Conformational isomers display different affinity for different DNA target sites, so that several specific p53-DNA complexes may be formed. For example, if a single DNA-target site is available, it would react with the p53 conformer having the highest affinity for this site, disrupting the equilibrium between the protein conformers. To counteract the decrease in concentration introduced by binding, and in order to maintain the equilibrium between the p53 conformational isomers, concentrations of other p53-conformers would change to favor the formation of the conformer with highest affinity for the specific DNA-binding site. Another mechanism for the recognition of specific DNA sequences by p53 may be based on the induced-fit mechanism (Fig. 5B). The p53 protein belongs to a class of partially structured and unfolded proteins.[68-71] Binding to structurally diverse DNA-target sequences may induce conformational changes in the unfolded/unstructured p53 molecules, resulting in the formation of p53 conformers differing in configuration in their DNA binding pockets. P53 conformational isomorphism may be critical for the regulation of different cellular functions such as apoptosis and growth arrest, or they may be beneficial as p53 conformers may correlate cellular activities, for example. However, if we think in terms of drug design, the conformational isomorphism of p53 is obviously an obstacle; a treatment that corrects one function of the protein impaired by mutations may not be sufficient to treat the disease.

Rescue of p53 Mutants by "Conformational" Drugs: A Proof of Principle Is Missing?

p53 is mutated or functionally inactivated in many types of human cancer, and pharmacological restoration of wild-type p53 conformation and function by small molecules targeting mutant p53 proteins was suggested as a strategy for the design of anti-cancer drugs.[10,71-73,77] To this end, CTD fragment p53 (361-382) linked with a cell penetrating peptide,[10,11,13] Pfizer compound CP-31398,[72] PRIMA-1[73-76] and the alkaloid ellipticine,[77] have been reported to function as effective anti-cancer agents capable of stabilizing and reactivating p53 mutants. However, reevaluation of these studies demonstrated that p53(361-382)[18] and CP-31398[78,79] do not interact with p53, but rather bind to DNA. Ellipticine intercalates and binds covalently to DNA,[80] whereas data on its interaction with p53 are missing. Cellular effects of CP-31398 appear to be mediated via stabilization and accumulation of transcriptionally active p53 in the cell nuclei, occurring due to inhibition of ubiquitination and induction of p53 relocalization in the cell.[79] PRIMA-1 is generally toxic—it induces cell death in all 55 tested cell lines with remarkably similar LC_{50} lethal doses ($LC_{50} = (8.7 \pm 1.5) \times 10^{-5}$ mol dm^{-3}) regardless of the p53 genotype (wild-type versus mutant) and levels of p53 expression (see the NCI base for Development of Therapeutic Programs at http://dtp.nci.nih.gov). These observations pose serious doubt as to the interpretation of cellular effects of CP-31398 and PRIMA-1 as achieved through binding to mutant p53 proteins and restoring wild-type function.

Another example of a small molecule interference with p53-dependent pathways is pifithrin-α, described initially as a compound that binds to wild-type p53 and inhibits its transcription functions.[81] However, pifithrin-α also suppresses heat shock and glucocorticoid receptor signaling in a p53-independent manner.[82] As such, it reduces activation of the heat shock transcription factor and increases cell sensitivity to heat, as well as reduces activation of glucocorticoid receptor and rescues mouse thymocytes from apoptosis after dexamethasone treatment. Molecular target for pifithrin-α common for all three transcription pathways, may comprise heat shock protein complexes that participate in holding inactive the heat shock factor 1, glucocorticoid receptor and p53 in the cytoplasm.[82] Structural similarity of pifithrin-α with quercetin, a flavanoid inhibitor of the shock protein complex, suggests this complex as an obvious target for pifithrin-α. Paradoxically, depending on the cellular context, pifithrin-α can either inhibit p53 or activate p53-mediated signaling and apoptosis.[82] It can also inhibit enzymatic activity of Firefly Luciferase,[83] suggesting that its cellular effects are mediated through multiple targets. Absence of biochemical evidence for binding of CP-31398, PRIMA-1, ellipticine and pifithrin-α to wild-type or mutant p53 protein, interactions of some of them with DNA and the toxicity of others, the p53-unspecific character of the effects of these compounds on cell death and proliferation do not support the notion that these compounds are "conformational" drugs that bind to mutant p53 and restore its wild-type functions.

It is easier to break than to mend. Clearly, it is easier to design a compound that impairs functions of a protein rather than to find a chemical substance that restores a wild-type activity lost due to mutations. A variety of small molecules have been found that bind to active sites or other protein domains and inhibit protein functions. We are, however, not aware of any synthetic compound that restores protein activities impaired by mutations other than those claimed for p53. For example, no small molecules have been identified that increase the affinity of mutant haemoglobin to oxygen or activity of mutant phenylalanine hydroxylase in genetic haemoglobin-related diseases and phenylketonuria, respectively. In the absence of such examples, the task to restore wild-type functions of mutant proteins appear to be difficult already for proteins with a single function such as haemoglobin and phenylalanine hydroxylase. Pharmacological tasks to restore wild-type activities of multifunctional proteins such as p53 whose functions are differently impaired by different mutations are challenging. Whether this is feasible, at least for a single p53 mutation and a single function, awaits future studies.

Conclusion

The view on p53 latency and activation is currently undergoing substantial changes. The emerging picture is that p53 is involved in multiple interactions with genomic DNA via both its core domain and CTD. To explain data accumulated for the past two years, we proposed the two-binding sites model of the p53 tetramer-DNA complexes in which CTD binding to DNA prevents the interactions of the core domain with specific B form DNA sequences but not with nonspecific DNA sequences (Fig. 2C). Interactions of p53 with nonspecific DNA sequences in the genome may be critical for searching for target sites in gene promoters. p53 associated with DNA may translocate towards target sequences by sliding along DNA molecules, cycles of dissociation/reassociation, or inter-segment transfer. Evidence for such mechanisms has been provided for other transcription factors,[84,85] and p53 sliding and dissociation/reassociation events have recently been observed by time-lapse atomic force microscopy.[86] The sequence nonspecific mode of p53-DNA interactions may also be essential for the second p53 function, the recognition and repair of damaged DNA.[4] This function may be performed via p53 interactions with components of the DNA repair/recombination machinery, p53 endonuclease and DNA disaggregating activities, and via p53 binding to single-stranded DNA and single-stranded DNA ends, internal DNA segments and Holliday junctions.[1,4,26,27] p53 bound to specific target sequences or damaged DNA may recruit components of the transcription machinery, chromatin remodeling and modifying complexes or DNA repair machinery. Initiation of transcription by p53 without preliminary protein modifications may be limited to interactions with cruciform targets.[44] p53, activated for binding to target sequences in B-form DNA via phosphorylation or acetylation of the CTD, may exploit different search strategies: jumping and hopping instead of sliding. Whether p53 bound to specific DNA target sequences activates or represses transcription may be dictated by the nature of these sequences, "head-to-head" or "head-to-tail" orientation of the quarter half-sites, the number of half-sites and their potential to form cruciform structures. Thus, p53 is a polyfunctional protein and its ability to regulate multiple processes is presumably based on its conformational isomorphism, allowing the protein to assume distinct conformations and form complexes with diverse DNA target sites and nonspecific DNA sequences.

Acknowledgements

This work was supported by grants from the Swedish Cancer Society and Swedish Medical Research Council to G. Bakalkin and the Wenner-Gren Foundations to V. Vukojevic.

References

1. Ko LJ, Prives C. p53: Puzzle and paradigm. Genes Dev 1996; 10:1054-1072.
2. Meek DW. Mechanisms of switching on p53: A role for covalent modification? Oncogene 1999; 18:7666-7675.
3. Oren M. Regulation of the p53 tumor suppressor protein. J Biol Chem 1999; 274:36031-36034.
4. Albrechtsen N, Dornreiter I, Grosse F et al. Maintenance of genomic integrity by p53: Complementary roles for activated and nonactivated p53. Oncogene 1999; 18:7706–7717.
5. Hupp T, Sparks A, Lane D. Small peptides activate the latent sequence-specific DNA binding function of p53. Cell 1995; 83:237-245.
6. Woo RA, Jack MT, Xu Y et al. DNA damage-induced apoptosis requires the DNA-dependent protein kinase, and is mediated by the latent population of p53. EMBO J 2002; 21:3000-3008.
7. Zhu Q, Wani MA, El-Mahdy M et al. Modulation of transcriptional activity of p53 by ultraviolet radiation: Linkage between p53 pathway and DNA repair through damage recognition. Mol Carcinog 2000; 28:215-224.
8. Ashcroft M, Kubbutat MHG, Vousden KH. Regulation of p53 function and stability by phosphorylation. Mol Cell Biol 1999; 19:1751-1758.
9. Jayaraman L, Prives C. Activation of p53 sequence-specific DNA binding by short single strands of DNA requires the C-terminus. Cell 1995; 81:1021-1029.
10. Selivanova G, Iotsova V, Okan I et al. Restoration of the growth suppression function of mutant p53 by a synthetic peptide derived from p53 C-terminal domain. Nat Med 1997; 3:632-638.
11. Selivanova G, Ryabchenko L, Jansson E et al. Reactivation of mutant p53 through interaction of a C-terminal peptide with the core domain. Mol Cell Biol 1999; 19:3395-3402.
12. Müller-Tiemann BF, Halazonetis TD, Elting JJ. Identification of an additional negative regulatory region for p53 sequence-specific DNA binding. Proc Natl Acad Sci USA 1998; 95:6079-6084.
13. Kim AL, Raffo AJ, Brandt-Rauf PW et al. Conformational and molecular basis for induction of apoptosis by a p53 C-terminal peptide in human cancer cells. J Biol Chem 1999; 274:34924-34931.
14. Yakovleva T, Pramanik A, Kawasaki T et al. p53 Latency: C-terminal domain prevents binding of p53 core to target but not to nonspecific DNA Sequences. J Biol Chem 2001; 276:15650-15658.

14a. Vukojevic V, Ykaovleva T, Terenius L et al. Denaturation of dsDNA by p53: Fluorescence correlation spectroscopy study. Biochem Biophys Res Commun 2004; 316:1150-1155.

15. Yakovleva T, Pramanik A, Terenius L et al. p53 latency—out of the blind alley, Trends Biochem Sci 2002; 27:612-618.
16. Vukojevic V, Pramanik A, Yakovleva T et al. Study of molecular events in cells by Fluorescence Correlation Spectroscopy. Submitted 2003.
17. Ekrami HM, Shen WC. Carbamylation decreases the cytotoxicity but not the drug-carrier properties of polylysines. J Drug Target 1995; 2:469-475.
18. Tan-No K, Cebers G, Yakovleva T et al. Cytotoxic effects of dynorphins through nonopioid intracellular mechanisms. Exp Cell Res 2001; 269:54-63.
19. Ayed A, Mulder FAA, Yi G-S et al. Latent and active p53 are identical in conformation. Nat Struct Biol 2001; 8:756-760.
20. Branden C, Tooze J. Introduction to protein structure. 2nd ed. Gurland Publishing Inc. 1999.
21. Friedler A, Hansson LO, Veprintsev DB et al. A peptide that binds and stabilizes p53 core domain: Chaperone strategy for rescue of oncogenic mutants. Proc Natl Acad Sci USA 2002; 99:937-942.
22. Klein C, Planker E, Diercks T et al. NMR spectroscopy reveals the solution dimerization interface of p53 core domains bound to their consensus DNA. J Biol Chem 2001; 276:49020-49027.
23. Bayle JH, Elenbaas B, Levine AJ. The carboxyl-terminal domain of the p53 protein regulates sequence-specific DNA binding through its nonspecific nucleic acid-binding activity. Proc Natl Acad Sci USA 1995; 92:5729-5733.
24. Anderson ME, Woelker B, Reed M et al. Reciprocal interference between the sequence-specific core and nonspecific C-terminal DNA binding domains of p53: Implications for regulation. Mol Cell Biol 1997; 17:6255-6264.
25. Hupp TR, Lane DP. Allosteric activation of latent p53 tetramers. Current Biol 1994; 4:865-875.
26. Bakalkin G, Yakovleva T, Selivanova G et al. p53 binds single-stranded DNA ends and catalyzes DNA renaturation and strand transfer. Proc Natl Acad Sci USA 1994; 91:413-417.
27. Bakalkin G, Selivanova G, Yakovleva T et al. p53 binds single-stranded DNA ends through the C-terminal domain and internal DNA segments via the middle domain. Nucl Acids Res 1995; 23:362-369.
28. Reed M, Woelker B, Wang P et al. The C-terminal domain of p53 recognizes DNA damaged by ionizing radiation. Proc Natl Acad Sci USA 1995; 92:9455-9459.

29. Szak ST, Pietenpol JA. High affinity insertion/deletion lesion binding by p53 - Evidence for a role of the p53 central domain. J Biol Chem 1999; 274:3904-3909.
30. Palecek E, Brázdová M, Brázda V et al. Binding of p53 and its core domain to supercoiled DNA. Eur J Biochem 2001; 268:573-581.
31. Cho Y, Gorina S, Jeffrey PC et al. Crystal structure of a p53 tumor suppressor-DNA complex: Understanding tumorigenic mutations. Science 1994; 265:346-355.
32. Nagaich AK, Zhurkin VB, Sakamoto H et al. Architectural accommodation in the complex of four p53 DNA binding domain peptides with the p21/waf1/cip1 DNA response element. J Biol Chem 1997; 272:14830-14841.
33. Okorokov AL, Milner J. An ATP/ADP-dependent molecular switch regulates the stability of p53-DNA complexes. Mol Cell Biol 1999; 19:7501-7510.
34. Bloomfield VA. DNA condensation by multivalent cations. Biopolymers 1997; 44:269-282.
35. Livolant F, Maestre MF. Circular dichroism microscopy of compact forms of DNA and chromatin in vivo and in vitro: Cholesteric liquid-crystalline phases of DNA and single dinoflagellate nuclei. Biochemistry 1988; 27:3056-3068.
36. Tseng M, Palaniyar N, Zhang W et al. DNA binding and aggregation properties of the vaccinia virus I3L gene product. J Biol Chem 1999; 274:21637-21644.
37. Cain C, Miller S, Ahn J et al. The N terminus of p53 regulates its dissociation from DNA. J Biol Chem 2000; 275:39944-39953.
38. Espinosa JM, Emerson BM. Transcriptional regulation by p53 through intrinsic DNA/chromatin binding and site-directed cofactor recruitment. Mol Cell 2001; 8:57-69.
39. Göhler T, Reimann M, Cherny D et al. Specific interaction of p53 with target binding sites is determined by DNA conformation and is regulated by the C-terminal domain. J Biol Chem 2002; 277:41192-41203.
40. Kaeser MD, Iggo RD. Chromatin immunoprecipitation analysis fails to support the latency model for regulation of p53 DNA binding activity in vivo. Proc Natl Acad Sci USA 2002; 99:95-100.
41. Barlev NA, Liu L, Chehab NH et al. Acetylation of p53 activates transcription through recruitment of coactivators/histone acetyltransferases. Mol Cell 2001; 8:1243-1254.
42. Kim E, Albrechtsen N, Deppert W. DNA-conformation is an important determinant of sequence-specific DNA binding by tumor suppressor p53. Oncogene 1997; 15:857-869.
43. Kim E, Rohaly G, Heinrichs S et.al. Influence of promoter DNA topology on sequence-specific DNA binding and transactivation by tumor suppressor p53. Oncogene 1999; 18:7310-7318.
44. Havas K, Flaus A, Phelan M et al. Generation of superhelical torsion by ATP-dependent chromatin remodeling activities. Cell 2000; 103:1133-1142.
45. Emerson BM. Specificity of gene regulation. Cell 2002; 109:267-270.
46. Narlikar GJ, Fan H-Y, Kingston RE. Cooperation between complexes that regulate chromatin structure and transcription. Cell 2002; 108:475-487.
47. Scully KM, Jacobson EM, Jepsen K et al. Allosteric effects of Pit-1 DNA sites on long-term repression in cell type specification. Science 2000; 290:1127-1131.
48. Johnson RA, Ince TA, Scotto KW. Transcriptional repression by p53 through direct binding to a novel DNA element. J Biol Chem 2001; 276:27716-27720.
49. D'Souza S, Xin H, Walter S et al. The gene encoding p202, an interferon-inducible negative regulator of the p53 tumor suppressor, is a target of p53-mediated transcriptional repression. J Biol Chem 2001; 276:298-305.
50. Hoffman WH, Biade S, Zilfou JT et al. Transcriptional repression of the anti-apoptotic survivin gene by wild-type p53. J Biol Chem 2002; 277:3247-3257.
51. Zilfou JT, Hoffman WH, Sank M et al. The corepressor mSin3a interacts with the proline-rich domain of p53 and protects p53 from proteasome-mediated degradation. Mol Cell Biol 2001; 21:3974-3985.
52. Murphy M, Ahn J, Walker KK et al. Transcriptional repression by wild-type p53 utilizes histone deacetylases, mediated by interaction with mSin3a. Genes Dev 1999; 13:2490-2501.
53. Ard PG, Chatterjee C, Kunjibettu S et al. Transcriptional regulation of the mdm2 oncogene by p53 requires TRRAP acetyltransferase complexes. Mol Cell Biol 2002; 22:5650-5661.
54. Wang T, Kobayashi T, Takimoto R et al. hADA3 is required for p53 activity. EMBO J 2001; 20:6404-6413.
55. Langley E, Pearson M, Faretta M et al. Human SIR2 deacetylates p53 and antagonizes PML/p53-induced cellular senescence. EMBO J 2002; 21:2383-2396.
56. Tokino T, Thiagalingam S, el-Deiry WS et al. p53 tagged sites from human genomic DNA. Hum Mol Genet 1994; 3:1537-1542.
57. James LC, Tawfik DS. Conformational diversity and protein evolution – a 60-year-old hypothesis revisited. TRENDS Biochem Sci 2003; 28:361-368.

58. Frauenfelder H, Sligar SG, Wolynes PG. The energy landscapes and motions of proteins. Science 1991; 254:1598-1603.
59. Bryngelson J, Onuchic J, Socci N et al. Funnels, pathways, and the energy landscape of protein folding: A synthesis. Proteins: Struct Funct Genet 1995; 21:167-195.
60. Frauenfelder H, McMahon BH, Austin RH et al. The role of structure, energy landscape, dynamics and allostery in the enzymatic function of myoglobin. PNAS 2001; 98:2370-2374.
61. Chan HS, Dill KA. Protein folding in the landscape perspective: Chevron plots and nonArrhenius kinetics. Proteins 1998; 30:2-33.
62. Tsai C-J, Kumar S, Ma B et al. Folding funnels, binding funnels, and protein function. Prot Sci 1999; 8:1181-1190.
63. Leeson DT, Gai F, Rodriguez HM et al. Protein folding and unfolding on a complex energy landscape. PNAS 2000; 97:2527-2532.
64. Edman L, Rigler R. Memory landscapes of single-enzyme molecules. PNAS 2000; 97:8266-8271.
65. James LC, Roversi P, Tawfik DS. Antibody multispecifity mediated by conformational diversity. Science 2003; 299:1362-1367.
66. Jeffrey CJ. Multifunctional proteins: Examples of gene sharing. Ann Med 2003; 35:28-35.
67. Copley SD. Enzymes with extra talents: Moonlightning functions and catalytic promiscuity. Curr Opin Chem Biol 2003; 7:265-272.
68. Dawson R, Müller L, Dehner A et al. The N-terminal domain of p53 is natively unfolded. J Mol Biol 2003; 332:1131-1141.
69. Bell S, Klein C, Müller L et al. p53 contains large unstructured regions in its native state. J Mol Biol 2002; 322:917-927.
70. Wright PE, Dyson HJ. Intrinsically unstructured proteins: Reassesing the protein structurefunction paradigm. J Mol Biol 1999; 293:321-331.
71. Lane TD, Hupp TR. Drug discovery and p53. Drug discov Today 2003; 8:347-355.
72. Foster BA, Coffey HA, Morin MJ et al. Pharmacological rescue of mutant p53 conformation and function. Science 1999; 286:2507-2510.
73. Bykov VJN, Issaeva N, Shilov A et al. Restoration of the tumor suppressor function to mutant p53 by a low-molecular-weight compound. Nat Med 2002; 8:282-288.
74. Bykov VJN, Issaeva N, Shilov A et al. Restoration of the tumor suppressor function to mutant p53 by a low-molecular-weight compound. Nat Med 2002; 8:282-288.
75. Bykov VJN, Issaeva N, Selivanova G et al. Mutant p53-dependent growth suppression distinguishes PRIMA-1 from known anticancer drugs: A statistical analysis of information in the National Cancer Institute database. Carcinogenesis 2002; 23:2011-2018.
76. Bykov VJN, Selivanova G, Wiman KG. Small molecules that reactivate mutant p53. Eur J Cancer 2003; 39:1828-1834.
77. Peng Y, Li C, Chen L et al. Rescue of mutant p53 transcription function by Ellipticine. Oncogene 2003; 22:4478-4487.
78. Rippin TM, Bykov VJN, Freund SMV et al. Characterization of the p53-rescue drug CP-31398 in vitro and in living cells. Oncogene 2002; 21:2119-2129.
79. Wang W, Takimoto R, Rastinejad F et al. Stabilization of p53 by CP-31398 inhibits ubiquitination without altering phosphorylation at serine 15 or 20 or MDM2 binding. Mol Cell Biol 2003; 23:2171-2181.
80. Stiborova M, Stiborova-Rupertova M, Borek-Dohalska L et al. Rat microsomes activating the anticancer drug Ellipticine to species covalently binding to deoxyguanosine in DNA are a suitable model mimicking Ellipticine bioactivation in humans. Chem Res Toxicol 2003; 16:38-47.
81. Komarov PG, Komarova EA, Kondratov RV et al. A chemical inhibitor of p53 that protects mice from the side effects of cancer therapy. Science 1999; 285:1733-1737.
82. Komarova EA, Neznanov N, Komarov PG et al. p53 inhibitor pifithrin-α can suppress heat shock and glucocorticoid signaling pathways. J Biol Chem 2003; 278:15465-15468.
83. Rocha S, Campbell KJ, C Roche KC et al. The p53-inhibitor Pifithrin-α inhibits Firefly Luciferase activity in vivo and in vitro. BMC Molecular Biology 2003; 4:9.
84. Lieberman BA, Nordeen SK. DNA intersegment transfer, how steroid receptors search for a target site. J Biol Chem 1997; 272:1061-1068.
85. Viadiu H, Aggarwal AK. Structure of BamHI bound to nonspecific DNA: A model for DNA sliding. Mol Cell 2000; 5:889-895.
86. Jiao Y, Keegstra W, Brisson A. Dynamic interactions of p53 with DNA in solution by time-lapse atomic force microscopy. J Mol Biol 2001; 314:233-243.

Chapter 10

p53's Dilemma in Transcription: Analysis by Microarrays

Karuppiah Kannan,* Gideon Rechavi and David Givol

Abstract

Protection against cancer by p53 is due mainly to its activity as a transcription factor. The function of p53 in transactivation of target genes is analyzed here with emphasis on the dilemma between cell growth arrest and apoptosis pathways. The question as to which of these p53 functions is required for tumor suppression in vivo is revisited in light of new studies that renew the focus on senescence and growth arrest mechanisms in the protection against cancer. Global gene expression analysis by microarrays, employing either transcription or conformation switches for p53 activation were utilized to distinguish primary (direct) p53 targets from secondary (indirect) ones, and to probe pathways of inhibition of p53-induced apoptosis. The profile of gene expression indicates that p53 is a central node in the cellular network of growth control modulation and its activation results in altered expression of more than a thousand genes. Some of these are co-activators of p53 and may be involved in the decision making of the choice between p53 functions. The major conclusion is that the response to p53 activation is heterogeneous and is mainly dependent on the cellular context, which is evident from the pattern of p53-induced genes in different cell types and in various organs in response to irradiation. The analysis of gene expression profiles following activation or suppression of apoptosis by either chemotherapy or cytokines, respectively, may facilitate the identification of ways to bypass the loss of p53 activity and to design new modalities for cancer treatment.

Introduction

After the discovery of *p53* in 1979, it became essential to clone the *p53* cDNA in order to understand its cellular function. The first murine p53 cDNA was cloned in 1983 at the Weizmann Institute. In 1981 M. Oren came back from A. Levine's lab in Stony Brook and joined the lab of D. Givol with the purpose of cloning *p53*. Initially a small fragment (~300 bp) from the 3' non-coding region was obtained, followed by a full-length clone that contained the entire cDNA sequence and the translated amino acid sequence.[1] The mouse genome also contains a p53 pseudogene that showed a similar size to that of the cDNA and was used to characterize the isolated p53 gene by heteroduplex analysis. Later on, the entire p53 gene containing all the exons spanning 13 kbp was characterized and sequenced to determine the exon structure and intron borders.[2] This gene was cloned into an expression vector and used to study the transforming activity of *p53* in combination with *Ras*. Apparently this gene contained a point mutation at position 135 (Ala→Val) that rendered it transforming for NIH3T3 cells at 37°C.[3] However, it was later found that at 32°C this mutant p53 acquired the wild-type (wt) conformation and exhibited inhibition of oncogene (e.g., Myc and Ras)

*Corresponding Author: Karuppiah Kannan—Cancer Pharmacology, Millennium Pharmaceuticals Inc., 40 Landsdowne Street, Cambridge, MA 02139, USA. Email: karuppiah.kannan@mpi.com

p53, edited by Ayeda Ayed and Theodore Hupp. ©2010 Landes Bioscience and Springer Science+Business Media.

-mediated focus formation.[4-5] This construct or variants of it made up of a chimera composed of cDNA and genomic fragments was used since then as the temperature-sensitive (ts) p53, denoted p53val135, that can shift from transforming (mutant) to suppressing (wild-type) p53 by a temperature shift from 37°C to 32°C.[5]

The human full-length *p53* was cloned in 1984 by D. Givol from a cDNA library of SV-40 transformed fibroblasts[6] and in hindsight could have been expected to be wild-type p53 without mutations[7] since the p53 in these cells was inactivated by the binding of SV-40 T antigen. Indeed this human clone did not transform NIH3T3 cells but this line of research was unfortunately abandoned because of lack of interest in non-oncogenic p53 at that time. An important advance was made when this human cDNA was used to map the chromosomal localization of *p53* to chromosome 17p13[6,8] because this information linked genetics with cancer mutations. Two years later, Vogelstein's lab determined the loss of heterozygosity (LOH) sites in colon cancer, and showed that colon carcinoma have high incidence (75%) of LOH at 17p13.[9] Using human *p53* cDNA to determine whether the *p53* chromosomal location coincides with the LOH site on chromosome 17, they found that *p53* was deleted from one allele and that the second allele had a mutation.[10] This is precisely what the Knudson's two-hit hypothesis predicted for tumor suppressor genes[11-12] and was sufficient to rename and define *p53* as a tumor suppressor. Similarly, the lab of J. Minna used this p53 clone to analyze tumors from the lung cancer NSCLC and found homozygous deletions, abnormal sized mRNA for p53 or loss of expression, along with a variety of mutations, reinforcing the tumor suppressor nature of p53.[13] Concomitant in vitro experiments showed that wild-type p53 could inhibit the transformation of NIH3T3 cells by oncogenes.[4,14] These studies merged genetic and clinical data with functional information to conclude that p53 is a tumor suppressor.

p53 as a Transcription Factor

The molecular function of p53 was defined to be that of a transcription factor[15] by showing that it binds specifically to its target DNA fragments that had a symmetrical structure of two 10 bp palindromes, separated by a spacer of 0-14 bp as follows:

PuPuPu(CA/TA/TG)PyPyPy (0-14N) PuPuPu(CA/TA/TG)PyPyPy.

The three dimensional structure of p53 with its target DNA molecule demonstrated that the core domain of p53 is involved in DNA binding[16] and that the amino acid residues of p53 that make contact with the DNA exhibited the highest mutation rate in cancer; p53 that is mutated at these positions fail to bind the target DNA. These results defined the DNA sequence element with which p53 interacts in order to activate transcription of its target genes.[15] This p53 target site was identified in regions upstream to the RNA start site of genes and was identified and confirmed functionally in introns.

Recent developments in human genome sequencing efforts made it possible to identify such p53 binding sites on a genomic scale, leading to the global identification of p53 target genes and cellular functional networks controlled by p53. An algorithm that enables the identification of such DNA sequences in the genome database was used to scan 2583 genes for p53 target sites.[17] Using this algorithm, about 300 genes (out of 2583) received a score greater than the cut off value (a score of 93) and qualified as potential p53 targets. The directory of p53 sites (http://linkage.rockefeller.edu/p53/) is a useful tool for the identification of new targets. Among the 2583 genes analyzed, 226 (8.7%) were found to have a perfect match with the consensus p53 binding sequence and 304 had a similarity score above 93. The variability of the consensus sequence of the DNA target, based on a collection of 37 motifs, was qualified by assigning a proportional weight for each position in the target DNA sequence. The highest weight was given to the fourth C and the seventh G within the internal tetramer (CA/TA/TG) of the palindrome, which are conserved in most target sites, whereas the rest of the sequence shows more variability.[17] A different attempt to scan all p53 target sites on chromosome 21 and 22 was performed recently in silico,[18] and 48 high-confidence sites were found.

Extrapolation of this analysis to the entire genome predicts 1600 p53 target sites; previous estimates placed the number at 200-400 sites.[15] Surprisingly a significant number of these sites (36%) are located at the 3'end of the genes, a larger number were located at the 5' of genes, and several genes showed p53 targets at both ends.[18] This raises the possibility of simultaneous paired-transcription of coding and non-coding strands of the same gene that could contribute to a new level of regulation of gene expression, specifically in the case of p53 or more generally by other transcription factors.

The direct identification of gene targets of p53 was initially done using subtractive hybridization, and had to rely on the identification of differentially induced genes by p53. Some of the early candidates detected by this method were not very meaningful considering the then known activity of p53. The first identified target of p53 that was in line with the growth suppression effect of p53 was *p21*^{waf}. A cell line that expresses the aforementioned human p53 cDNA[8] under a steroid inducible promoter[19] was used to clone the cDNA of the differentially enriched mRNA from the induced p53-containing cells; a major new cDNA clone was identified as *p21*^{waf}.[20] The differentially expressed cDNA clones showed overwhelming excess of *p21* clones and the promoter of the cloned *p21* gene contained two p53 target sites as far as 2.2 kb upstream from the RNA start. At the same time, using a different approach, *p21* was isolated as the gene responsible for senescence[21] and was also identified as a universal inhibitor of CDK which blocked the cell cycle at the G1/S transition and inhibited cell growth.[22-23] Interestingly, even before the discovery of *p21* as a mediator of growth arrest, the apoptotic function of p53 was demonstrated,[24] but there was no clue as to whether or how *p21* can be involved in apoptosis. The first identified mediators of p53 induced apoptosis were Bax[25] and Fas/APO1.[26] Hence, the transcription activation by p53 was split into two functional avenues with separate sets of genes, one that involves cell cycle regulation leading to growth arrest and can be explained by transactivation of *p21*, and another one that must lead to apoptosis and involves genes that should also be turned on by p53. The choice between transcription activation of either of these pathways continues to be the main dilemma of p53 function in tumor suppression. This choice is tightly linked to the question of how the cell makes the decision whether to enter growth arrest or to undergo cell death upon p53 induction. It is also relevant to the mechanism of tumor suppression by p53 since the present dogma prefers apoptosis as the dominant way by which p53 protects against cancer. In this chapter we will discuss the systems that were selected for analysis of p53 targets that use inducible or infection vehicles, the methods used for analysis with special emphasis on microarrays, and the main results that expand our knowledge on the regulation of cell response to stress.

The p53 Network

Systems of Analysis

The analysis of p53 downstream targets needed appropriate systems that, upon induced expression of p53, lead either to cell cycle arrest or to the induction of apoptosis. Two different switches were employed to turn p53 on or off: a **conformational switch** based on the temperature-sensitive (ts) p53 that is independent of transcription, and a **transcriptional switch** based on induced p53 transcription by a variety of inducible promoters (Table 1). A number of studies were carried out where overexpression of wild-type p53 resulted in cell cycle arrest, thereby providing an attractive strategy to pan out genes that regulate cell cycle in a p53 dependent manner.[27-35] To an equal extent, extensive analyses have been done on the p53-mediated apoptotic pathway as well.[29,36-46]

A broad range of cell lines representing all major tumor types were used in an effort to maximize the identification of p53 regulated genes in all the tumor types (Table 1). Kannan et al[28] have used a temperature sensitive p53 (ts p53Val135) expressed in the human lung cancer cell line H1299 to analyze the p53 mediated transcriptional profile. This ts-p53 protein changes from a mutant to wild-type conformation by a temperature shift from 37°C

Table 1. Systems of p53 activation and cell lines utilized to determine p53 modulated gene expression profile

Mode of Activation	Cell Line	Functional Endpoint	Refs.
Transcriptional Switch			
Retroviral rADp53	*U373MG* Glioblastoma cell line	Cell Cycle Arrest	33-34
Retroviral rADp53	*ECV, ECV304 or DECV* Bladder Carcinoma Cell line	Apoptosis	38-39
Retroviral rADp53	*2774* or *2774QW1* Ovarian Cancer	Apoptosis	41, 46
Retroviral rADp53	*M3 and vm10 p53 ts murine cell lines HCT116 or PA1*	Apoptosis	42
Ecdysone	*H1299 Lung Cancer cells*	Cell Cycle Arrest	27-29
tet-Off	*DLD-1 Colon Cancer cell line*	Apoptosis	32
Conformational Switch			
ts-p53	*H1299 Lung Cancer cells*	Cell Cycle Arrest	27-29
ts-p53	*M1* Myeloid Leukemic Cells	Apoptosis	29, 37
ts-p53	*MCF* Breast Cancer cell line	Cell Cycle Arrest	31
ts-p53	*M3 and vm10 p53 ts murine cell lines HCT116 or PA1*	Apoptosis	42
Chemical Switch			
Adriamycin and Etoposide	*M3 and vm10 p53 ts murine cell lines HCT116 or PA1*	Apoptosis	42
Doxorubicin	*TK6 and WTK1 EBV immortalized Lymphoblastoid*	Apoptosis	44
5-FU	*HCT116* Colorectal cell line	Apoptosis	56
Etoposide	*U2OS* Osteosarcoma cell line	Apoptosis	30, 45
Camptothecin	*Primary Neuronal Cells*	Apoptosis	43
5-FU and Adriamycin	*DLD-1 Colon Cancer cell line*	Apoptosis	32
Zinc Chloride	*EB1* Colon Carcinoma Cell line	Apoptosis	36
Physical Switch			
UV or γ irradiation	*MCF7* Breast Cancer cell line *Tera-1* Teratocarcinoma cell line *IMR32* Neuroblastoma *HCT116* Colon Carcinoma	Apoptosis	36
UV irradiation	*M3 and vm10 p53 ts murine cell lines HCT116 or PA1*	Apoptosis	42
IR irradiation	*Primary Human B-CLL*	Apoptosis	40
γ irradiation	*VhD* Murine Fibroblast Cell line and *MEFs*	Cell Cycle Arrest	35
γ irradiation	*TK6 and WTK1 EBV immortalized Lymphoblastoid*	Apoptosis	44

(non-permissive) to 32°C (permissive temperature) and induces either cell cycle arrest[27] or apoptosis,[29,37] depending on the cell line that is being used. As this conformational switch does not require protein synthesis, it allowed us to distinguish the primary targets that are directly induced by p53 from the secondary, indirectly activated genes, by using cycloheximide to inhibit protein synthesis where the levels of p53 remained unchanged.[28] In later studies, the opposite approach was taken by screening genes during inhibition of the p53 induced apoptosis by cytokines and other factors.[37] A similar conformational switch was used in other cell types as well.[31,42]

Different from the aforementioned conformational switch, overexpression of wild-type p53 was achieved by using transcriptional switches, employing recombinant viral methodologies, chemotherapeutic drugs or irradiation (Table 1). Cells which lack wild-type p53 were infected with adenoviruses encoding wild-type p53 under a CMV promoter (Ad5CMV-p53) and RNA was harvested at different time points.[33-34,38-39,41,46] On the other hand, studies that were aimed at identifying the genes involved in p53 mediated apoptosis used mainly chemotherapeutic agents such as Etoposide, a p53 activating topoisomerase II inhibitor,[30,45] Doxorubicin,[44] and Camptothecin, a DNA damaging agent that functions through p53-dependent mechanism.[43] Zhao et al,[36] have used both UV and Gamma irradiation to induce p53 expression, and were able to detect distinct sets and subsets of gene expression patterns specific to each mode of activation. Kannan et al[27] have employed a muristerone inducible transcriptional switch to over-express p53 and p21. Similarly, a tet-off system was used to study the identification and classification of p53 regulated genes[32] and Zn induced metallothionein promoter was used by Zhao et al.[36]

Heterogeneity of the Microarray Based p53 Expression Profile Data

Microarray based transcription profiling data indicated that genes induced by gamma radiation, UV radiation and Zinc induced p53 resulted in the identification of distinct sets of genes. Also, not surprisingly, the gene expression pattern thus observed was cell- type or cell-line specific[36] and showed different expression kinetics as determined by clustering and expression profiles (Fig. 1). For example, cluster 2 contained the early genes, expressed as early as 2 hours after induction and include *p21* and *gadd45*, and cluster 4 contained late genes

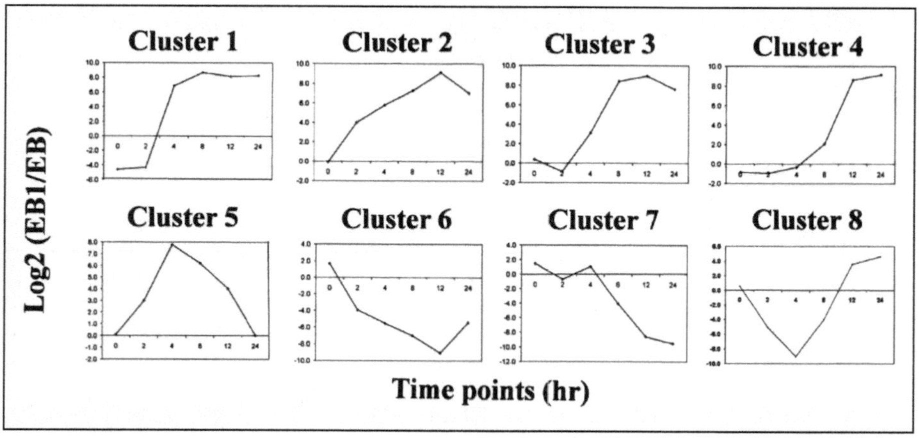

Figure 1. Gene expression patterns identified from a custom p53 chip (with selected known p53 target genes) after induction of p53. A total of 69 genes were selected for this analysis and were clustered using hierarchical clustering analysis. Average expression profiles for eight clusters with different induction kinetics were plotted showing the complexity of gene expression regulation mediated by p53.[36] EB1, cells with induced p53. EB, parental cells. (Taken from ref. 36).

that were expressed only after 8 hours. Similar clusters were also observed for the down regulated genes (Fig.1). This and other results led to the conclusion that the p53 dependent gene expression pattern depends on the levels of p53 protein in the cell, the type of induction of p53, and more importantly, the cell type examined.[36] Similar genomic approach by SAGE analysis revealed that every gene induced by p53 through retroviral infection of DLD-1 cells was also induced by another transcriptional switch used (e.g., the rTA system), indicating that the cell type rather than mode of p53 induction determines the downstream p53 target genes. Interestingly, they were able to identify only six genes upregulated by Adriamycin and 5-FU, though identical levels of p53 were found after the treatment, alluding to the fact that in addition to cell type, the nature of the induction signal is also important.[32] This observation on the cell-type specificity is correlated with the study of p53-induced genes in vivo in response to irradiation. By comparing expressed genes in $p53^{+/+}$ or $p53^{-/-}$ mice, striking tissue specificity with distinct regulation of p53-induced genes in different tissue compartments was observed.[47] For example in the liver the major up regulated gene was p21. In the spleen, PUMA was induced in the white pulp, but Noxa and Bid were upregulated in the red pulp; all three were induced in the intestine. This selectivity in transactivation by p53 following DNA damage in vivo correlates with either growth arrest, or a variable pro-apoptotic response to γ-irradiation, and it is extremely important for cancer therapy. This analysis indicates that to a certain extent p53 behaves like a tissue specific gene when activated by various stress conditions.

Primary and Secondary Targets

The tumor suppressor function of p53 mainly stems from its ability to act as a transcription factor that could sense and integrate growth, oncogenic and stress signals in a concerted manner to maintain cell growth. It is very well documented that activation of p53 leads to modulation in the expression of downstream p53 target genes that in turn controls either cell cycle arrest or apoptosis. Since p53 is a transcription factor, a first hand knowledge about its primary targets is an important step in understanding the p53 network. Use of cycloheximide (CHX, a protein synthesis inhibitor) and ts-p53, along with microarray technology on a cell culture system helped us to study this problem, since the p53 was stable during the time course of the experiment and because RNA synthesis was not affected by the inhibitor. We were able to identify 38 and 24 primary target genes that were up and down regulated, respectively, by p53 directly (Fig. 2). The results revealed altered expression of many novel genes that controlled other aspects of cell function, including adhesion, DNA repair and cell signaling,[28] in addition to well known p53 targets such as p21, *Fos*, *Bax* and *Bak* (Table 2).

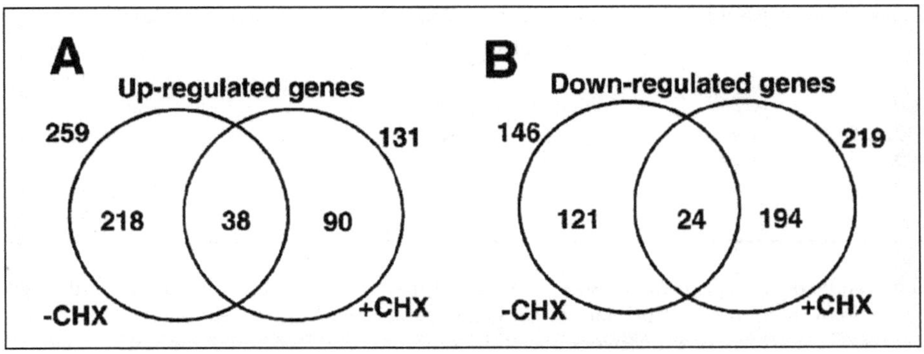

Figure 2. Venn diagram of the number of p53 regulated genes in the presence or absence of CHX. Only 38 of the up-regulated and 24 of the down-regulated genes were modulated irrespective of CHX addition indicating that those were the primary direct p53 target genes.[28] (Taken from ref. 28).

Table 2. Partial list of genes identified as trans-activated by p53 through microarray analysis

Accession Number	Gene Symbol	Description	Function
U03106	p21	Cyclin Dependent Kinase Inhibitor 1A	Cell Cycle
U82987	PUMA	Bcl2 binding component 3	Apoptosis
L19871	ATF3	Activating Transcription Factor 3	Transcription
NM_006763	BTG2	BTG family member 2	DNA Repair
M60974	GADD45	Growth Arrest and DNA Damage Inducible, Alpha	Cell Cycle
AB000584	MIC-1	TGF-B superfamily protein	Growth Factor Inhibitor
M15796	PCNA	Proliferating Cell Nuclear Antigen	DNA Repair
M31159	IGFBP3	Insulin Growth Factor Binding Protein 3	Growth Factor Inhibitor
NM_004419	DUSP5	Dual Specificity Phosphatase 5	Signal Transduction
X57348	SFN	14-3-3	Cell Cycle
AV109962	TRAF4	TNFR Associated Factor	Apoptosis
X63717	Fas/Apo1	TNFR6	Apoptosis
AF064071	APAF-1	Apoptotic Protease Activating Factor 1	Apoptosis
AF010309	PIG3	Quinone Oxidoreductase homolog	Oxidative Stress
A1853375	MDM2	Transformed 3T3 double minute 2	Cell Cycle
AF016266	Killer/DR5	TRAIL Receptor 2	Apoptosis
U23765	BAK	Bcl2 Antagonist/Killer 1	Apoptosis
X70340	TGF-alpha	Transforming Growth Factor Alpha	Growth Factor Inhibitor
U18300	DDB2	Damage specific DNA binding Protein 2	DNA Repair
L22474	Bax	Bcl2-associated X Protein	Apoptosis
M36067	LIG1	DNA Ligase I	DNA Repair
L20046	ERCC5	DNA excision repair-related gene	DNA Repair
M63488	RPA1	Replication Factor A protein 1	DNA Repair
U16811	Bcl-6	Bcl-6	Apoptosis
X72012	Eng	Endoglin	ECM/Receptors
U24389	LOXL1	Lysyl oxidase-like protein	Metabolism
X62535	DGKA	Diacylglycerol Kinase (Alpha)	Signal Transduction
L19871	ATF3	Activating Transcription Factor 3	Transcription
AF012923	Wig1	Wild type p53 induced gene 1	Apoptosis
M83649	TNFRSF6	TNF Receptor Super Family member 6	Apoptosis
M65199	EDN2	Endothelin 2	Cell-Cell signaling

The results of this experiment also indicated that upon p53 activation at least 10 fold more genes than the number of primary target genes are activated in the cell, placing p53 at the center of a gene expression network of cellular response (Fig. 3). Similar approaches towards finding the primary targets in the in vivo context were carried out by treating mice (p53 null or wild-type p53) with cycloheximide; irradiation resulted in the identification of marked

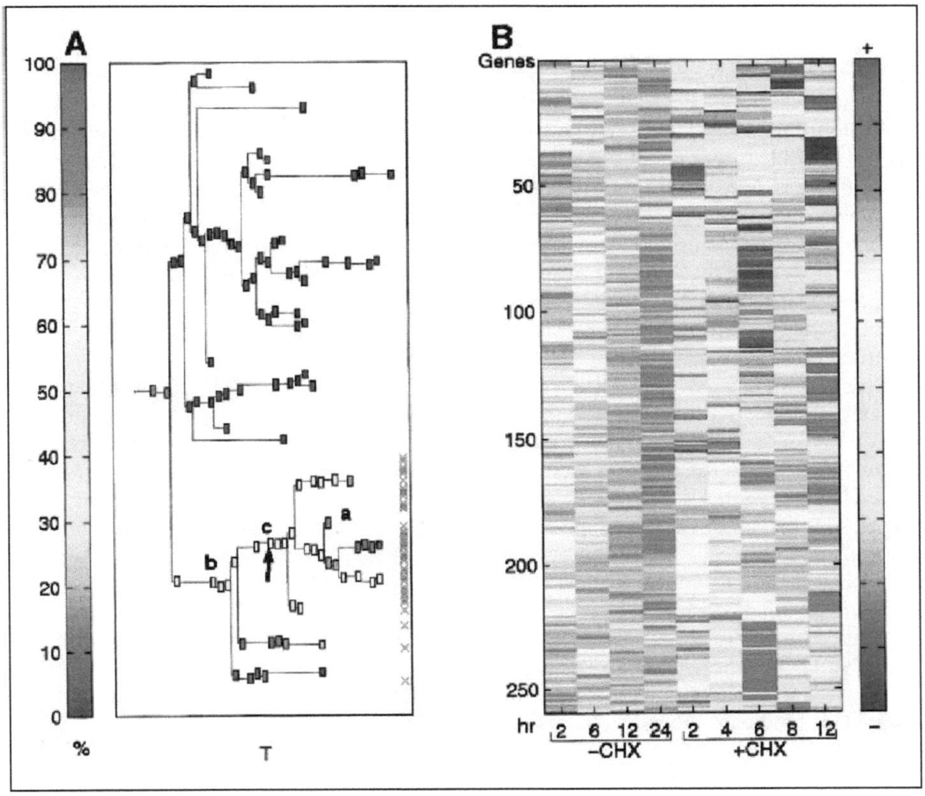

Figure 3. SPC (Super Paramagnetic Clustering) of 259 genes up-regulated by p53. A) Dendrogram of the p53 regulated genes, arrow showing the cluster containing 87% of the 38 primary target genes shown in Figure 2 (the boxes are colored according to the percentage of primary target genes). The distribution of the 38 primary target genes is marked by red crosses on the right. B) The normalized log ratio of the experiments with and without CHX is plotted. The color represents induction (red) or repression (blue). The genes are ordered according to the dendrogram on the left.[28] (Taken from ref. 28).

differences between the gene expression profile in intestine, spleen and thymus (Margalit O and Rechavi G, unpublished).

Major p53 Targets and Alternate Pathways Identified by Microarrays: Who Makes the Decisions?

Upon activation, the ability of the ts-p53 (val135) to induce cell cycle arrest but not apoptosis in the lung cancer H1299 cells, and vice versa in the myeloid LTR6 cells, provided us with an opportunity to dissect the distinct pathways that are key to the tumor suppressor function of p53. This further supports the notion that the p53 transcription program is dependent on the cell type more so than the mode of activation. Comparison of the p53 regulated gene expression profile in LTR6 cells to that of H1299 cells revealed that only 15% of the genes are common to both systems, indicating the presence of two distinct transcriptional programs in response to p53 signaling: one leading to growth arrest and the other to apoptosis, depending on the cellular context. The proapoptotic genes that were induced only in LTR6 cells, such as *Apaf-1*, *Sumo1* and *gelsolin*, may suggest a possible explanation for apoptosis in LTR6 cells.[29]

Genetic and cellular data indicated previously that *Apaf-1* deficiency (e.g., by promoter methylation) abrogates the apoptotic effect of p53 and substitutes for p53 loss in promoting tumor formation, particularly in melanoma.[48] Apaf-1 associates with Cytochrome C and caspase 9 to form the ***apoptosome***, a crucial intermediate in the mitochondrial pathway of apoptosis and activation of caspase 3. The identification of a p53 target site in the promoter sequence of *Apaf-1*[29,49] added to our understanding of the role of p53 in this pathway. In addition, we found that p53 trans-activation of *Apaf-1* was enhanced by Zac-1, a transcription factor that has previously been shown to inhibit cell proliferation. It appears that Zac-1 functions as a co-activator of p53. Furthermore, it was demonstrated that *Zac-1* itself is a possible direct target of p53, since the sequence upstream to the first coding exon of *Zac-1* contains a p53 recognition site and that the luciferase construct containing this region was activated by p53. These results suggest the existence of a tightly controlled self-amplifying mechanism of transcriptional activation leading to apoptosis by p53.[50]

This is a part of a more general emerging picture that suggests that p53 is not working alone in transactivation, but that it is colaborating with a wide group of co-activators or co-repressors in its effect on gene transcription. Important examples are a group of acetyl transferase enzymes like p300 and CBP that were shown to associate with p53 as co-activators. The deacetylating enzyme HDAC may be a co-repressor for p53. A very recent study demonstrated that p300 is the dominant factor in determining p53 dependent apoptosis.[51] Absence of p300 in colorectal cancer cells increases apoptosis in response to DNA damage, (e.g., UV, etoposide, doxorubicin) and this is accompanied by the increase in PUMA and decrease in p21^{waf} expression. On the other hand, the presence of p300 reverses the situation, increases p21 expression, and drives the cells to growth arrest. The expression ratio of PUMA over p21^{waf} is determined by p300. These results place a question mark on what decides between growth arrest and apoptosis: Is it p53 itself or its associate activators? Obviously this will open new directions for research on candidate chemotherapeutic agents against cancer.

Induction of neuronal cell death by camptothecin, a DNA damaging agent that functions through a p53 dependent mechanism, resulted in increased Apaf-1 expression. Isogenic cell lines that undergo p53-mediated apoptosis after treatment with ionizing radiation or Doxorubicin were also found to have *Apaf-1* up regulation through microarray analysis.[44] Results from these studies indicate that Apaf-1 can be a key mediator of p53-induced apoptosis. Some of the previously known and unknown key p53 target genes that were identified in microarrays are listed in Table 2. Another candidate gene that was up-regulated in these studies was the Bcl2-binding component 3 (bbc3),[29] that later on was cloned, renamed as PUMA and found to be a major factor in p53 induced apoptosis[52-53] (Fig. 4). The global analysis of gene expression by the SAGE (serial analysis of gene expression) screening method identified a group of genes, denoted PIGs (p53 Induced Genes) predicted to encode proteins that could generate or respond to oxidative stress. Perhaps the formation of reactive oxygen species and the subsequent oxidative degradation of mitochondrial components culminating in cell death are related to the p53 pathway of apoptosis.[54] The microarray experiments always detected some, if not all of the PIG genes, such as *PIG3*[27] and *PIG8*.[29]

Inhibition of p53 Induced Apoptosis without Affecting p53 Transactivation

The study of p53-regulated apoptosis is important for cancer research as it may lead to design strategies for cancer therapy. The genes or factors involved in the inhibition of p53-induced apoptosis could be informative for the treatment of cancer; in ~50% of cancer, the p53 is in the wild-type form despite the fact that p53 activated apoptosis is inhibited. It was shown previously[55] that apoptosis induction by p53 in the myeloid leukemia cells M1 could be inhibited by various compounds including cytokines such as interleukin-6 (IL-6) and calcium-mobilizing compounds such as the Ca^{2+}ATPase inhibitor thapsigargin (TG). This inhibition can be due to either inhibition of p53 transcription function, or due to p53 degradation. Other means are also possible, and a microarray analysis was carried out to study this effect[37] (Fig. 5). Clustering

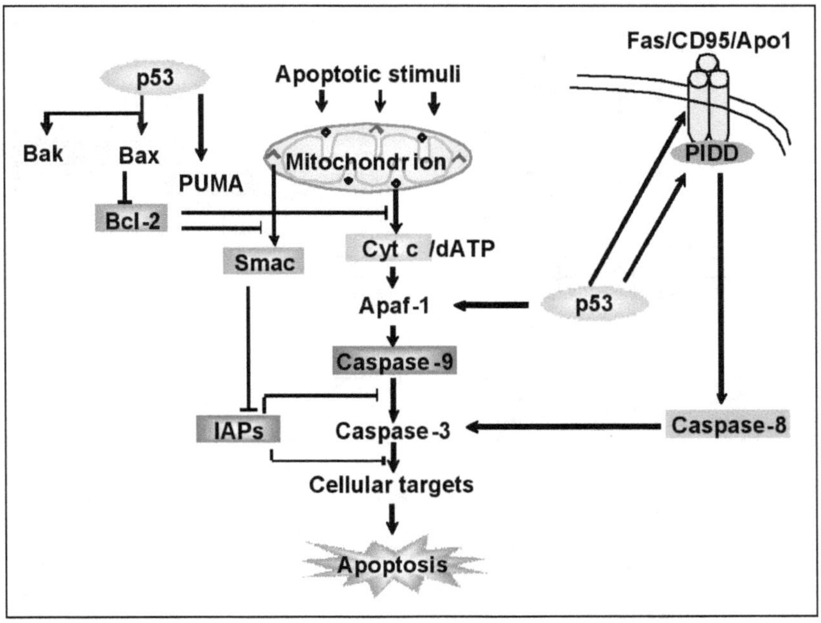

Figure 4. A scheme that integrates the genes in the mitochondrial and cell surface receptor apoptotic pathways that are the targets of activation by p53, as depicted by microarrays.

analysis of 1,786 genes, the expression level of which changed after activation of wild-type p53 in either the absence or presence of IL-6 or TG, showed that neither IL-6, nor TG caused a general inhibition of the ability of p53 to up- or down-regulate gene expression. IL-6 or TG also did not affect expression of various p53 targets implicated as mediators of p53-induced apoptosis; nevertheless apoptosis was inhibited effectively. These compounds thus can bypass the effect of wild-type p53 on gene expression and inhibit apoptosis. Hence IL-6 and TG activated different p53-independent pathways of gene expression that include up-regulation of antiapoptotic genes, thereby facilitating tumor development as well as tumor resistance to radiation and chemotherapy in cells that retain wild-type p53.[37] This is important for designing therapy that can maximize the anti-tumor effect of p53-induced apoptosis by inhibiting the effect of such cytokines. Cytokines decrease apoptosis in normal cells, and therefore inhibition of cytokine activity may improve cancer therapy by enhancing apoptosis in cancer cells. Anti-cytokine therapy against cytokines that inhibit apoptosis can increase the susceptibility of cancer cells to cytotoxic agents such as radiation and chemotherapy. Extending the same line of thinking, there could also be inhibitors that can improve cancer therapy by inhibiting the antiapoptotic function of other compounds.[37]

New p53 Regulated Pathways Identified by Microarrays

Gene Suppression

Equally important to the identification of up-regulated genes by p53 is the discovery of pattern of downregulated genes upon p53 induction (Table 3). Kannan et al[28] reported 24 genes that were downregulated by p53, including *BRCA1* and *cyclin E.* There were several genes reported by various groups to be downregulated by p53 that control cell cycle, DNA replication and signal transduction.[27-28,36] Recently, a genome-wide study by Kho et al[56] has identified 189 genes repressed compared to 41 genes activated upon p53 induction during

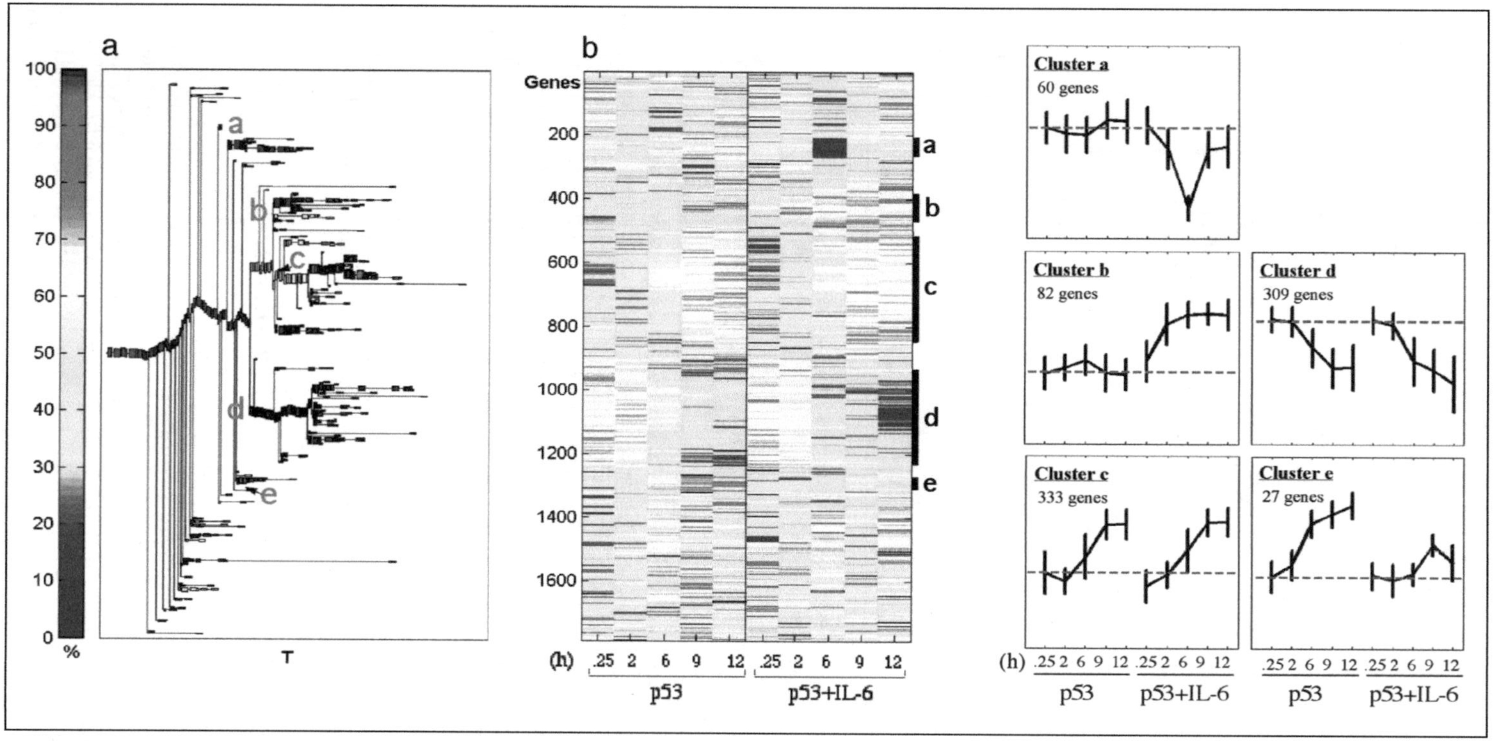

Figure 5. Inhibition of p53-driven apoptosis by IL-6. Patterns of gene expression upon p53 activation alone (leading to apoptosis) and in combination with IL-6 (leading to inhibition of cell death). A) Dendrogram of the regulated genes showing the clusters a-e. (B-left) The matrix of the up- and down-regulated genes. (B-right) The profiles of the main clusters a-e. Clusters 'c' and 'd' show similar pattern with or without IL-6 and contain most of the p53 induced genes demonstrating that p53 transactivation is not impaired, whereas the small cluster 'e' with only 27 genes shows IL-6 inhibition of p53 up regulated genes. The antiapoptotic cytokine IL-6 did not affect the expression pattern of 93% of the p53 up-regulated genes indicating that IL-6 can inhibit p53 induced apoptosis by regulating a p53 independent gene expression pathway.[37] (Adapted from ref. 37).

Table 3. Partial list of genes identified as trans-repressed by p53 through microarray analysis

Accession Number	Gene Symbol	Description	Function
M25753	Cyclin B1	Cyclin	Cell Cycle
X59798	Cyclin D1	Cyclin	Cell Cycle
M74093	Cyclin E	Cyclin	Cell Cycle
A1932735	Cyclin B2	Cyclin	Cell Cycle
NM_001759	Cyclin D2	Cyclin	Cell Cycle
X51688	Cyclin A	Cyclin	Cell Cycle
U79269	CDK4	Cyclin Dependent Kinase 4	Cell Cycle
M68520	CDK2	Cyclin Dependent Kinase 2	Cell Cycle
X05360	CDK1	Cyclin Dependent Kinase 1	Cell Cycle
J04088	TOPO-IIAlpha	Topoisomerase II Alpha	DNA Replication
U15642	E2F-TF5	E2F Transcription Factor 5	Cell Cycle
L11370	Protocadherin-1	Cell to Cell Adhesion and Communication	Extracellular Matrix
M34065	CDC25C Phosphatase	Tyrosine Protein Phosphatase	Signal Transduction
U01038	PLK	Polo/cdc5 like Kinase	Cell Cycle
U70310	XRCC9	DNA post-replication repair and cell cycle checkpoint control	DNA Repair/ Replication
L33264	PISSLRE	Cdc2 related protein kinase	Cell Cycle
NM007610	Caspase 2	Apoptosis related Cysteine Protease	Apoptosis
NM009810	Caspase 3	Apoptosis related Cysteine Protease	Apoptosis
AB019600	Caspase 9	Apoptosis related Cysteine Protease	Apoptosis
M87339	RFC4	Replication Factor 4	DNA Repair/ Replication
U45878	MIHB	Inhibitor of Apoptosis Protein 1	Apoptosis
U37546	MIHC	Inhibitor of Apoptosis Protein 2	Apoptosis
S78085	PDCD2	Programmed Cell Death 2 Gene	Apoptosis
L78833	BRCA-1	Breast Cancer 1	Cell Cycle
U77949	Cdc6	Cdc6-related protein	Cell Cycle
M72885	GOS2	Putative G0/G1 switch gene	Cell Cycle
U80017	BTF2	Basic Transcription Factor 2	Transcription

5-FU induced apoptosis. They have reported, similar to upregulated primary target genes, that there were primary repressed genes such as *PTTG1* and *CHEK1* that contained the p53 target site, and secondary repressed genes such as *PLK* that may be repressed by indirect mechanisms. These results also mirror the repression pattern observed by us.[28] Table 3 summarizes p53 trans-repressed genes that are involved in cell cycle arrest and apoptosis mediated by p53.

Taken together, all these results suggest that p53 regulates concerted opposing signals (e.g., activation of pro-apoptotic, and suppression of anti-apoptotic genes) and exerts its effect through a network of transcriptional changes, which collectively alter the cell phenotype in response to stress. The power of microarray analysis has yielded an opportunity to look at alternate pathways that were previously not known to involve p53. A cursory look at the p53 mediated transcription profile resulted in the identification of alteration in genes involved in the cell cycle control circuitry, along with other functions such as cell adhesion, signaling, transcription, neuronal growth, DNA repair, oxidative stress, ECM, cell migration and angiogenesis[27-29,36] (Tables 2 and 3). The new p53 regulated pathways identified by microarrays include metabolic enzymes, DNA repair, signal transduction and secreted factors. We will describe several examples to illustrate the relevance of these newly uncovered pathways for p53 function.

Secreted Factors

Two different studies, one aimed at identifying p53 mediated growth arrest responsive genes,[27,29] and another aimed at identifying p53 mediated apoptosis responsive genes upon Etoposide treatment,[45] have identified MIC1 gene, a member of the TGF-β family, as a novel p53 target gene. The promoter region of this gene was later found to contain the putative p53 target site.[30] The *MIC1* product was found to be secreted; the conditioned medium from cells expressing this protein can suppress the growth of certain TGF-β receptor and smad-4 expressing tumor cells. This implies a connection between p53 and the growth-inhibitory TGF-β family of genes, and underscores a potentially important novel paracrine mechanism of growth suppression by p53. An extended study of such p53 targets using irradiation and cDNA arrays revealed that in addition to its activity on targets of intracellular growth arrest mediators, p53 induces several secreted growth inhibitory factors with anti-cancer effects on the near-by cells that can also potentiate the cytotoxic effect of chemotherapeutic drugs on tumor cells.[57] Another type of secreted factors, such as thrombospondin 1, inhibit vascularization and angiogenesis and therefore cut off blood supply to the tumor cells thus preventing tumor growth and metastasis.[58]

Metabolic Pathways

Microarray analysis of differential gene expression in p53 mediated apoptosis resistant and apoptosis sensitive tumor cell line identified several metabolic enzymes involved in p53 mediated growth arrest and apoptosis; one example is proline oxidase, a transcription target for p53.[39] A proline oxidase antisense vector repressed the p53-induced up-regulation of proline oxidase, release of cytochrome *c* from mitochondria, and apoptosis in renal carcinoma cells. The localization of proline oxidase to mitochondria suggests that it can directly influence the apoptotic pathway and thus tumorigenesis.[59] In another study, an orphan adaptor protein TRAF4, that mediates cellular signaling by binding to members of TNFR superfamily and IL-1/Toll like receptor superfamily, was found to be overexpressed and involved in p53 mediated proapoptotic signaling.[42] Other metabolic enzymes and processes such as ribosomal protein synthesis and histone modifications are also found as targets by the analysis of p53 induced genes.

Signal Transduction

Transcription profiling analysis of a glioblastoma cell line showed up-regulation by p53 induction of the expression of hCDC4b, one of the four subunits of SCF (ubiquitin Ligase) complex responsible for degradation of cyclin E.[33] This indicates that a previously unrecognized transcriptional target of p53 was being employed to negatively regulate cyclin E, which might stop the cell-cycle progression at G1-S. This would represent a novel p53 dependent mechanism that controls cell division in addition to the well-known p21 pathway. Using the same cell line, another study has identified upregulation of the phosphatase DUSP5 gene by p53 induction that resulted in dephosphorylation of Erk1/2 followed by growth suppression of human cancer cells. This represents a novel mechanism by which p53 might negatively regulate cell cycle progression by down regulating mitogen or stress activated protein kinases.[34] In another study, expression array analysis identified heparin-binding epidermal growth factor-like growth factor (HB-EGF) as being markedly up-regulated by p53. In response to DNA damage, HB-EGF was induced in wild-type, but not in mutant p53-containing cells. This in turn activated the MAPK cascade through pathways involving Ras and Raf. This expression of HB-EGF protects cells from H_2O_2 induced apoptosis through MAPK activation. Additionally, the PI3K/Akt pathway was activated in response to HB-EGF induction, whereas inhibition of MAPK and Akt activation after DNA damage decreased cell survival in wild-type p53-containing cells. These findings point to a novel aspect of p53 function: p53-induced growth factors such as HB-EGF, which activate MAPK and Akt signaling, may be involved in a compensatory mechanism to alleviate adverse effects of cellular stresses. Hence, p53 induction counteracts p53 growth suppression through activation of MAPK and PI3K/Akt signaling cascades.[60] Perhaps the outcome of p53 induction is the balance between effects of pro and anti apoptotic genes and that this balance can be modulated by small changes due to factors like cytokines.

Tumor Suppression Function of p53 by Growth Arrest: Reassessment

Although in vitro studies and knock-out mice (mainly *p21* deficient mice) point to the pro-apoptotic function of p53 as the dominant factor in tumor suppression, it is still possible that the growth arrest function is important in vivo. The major theme of this chapter was that the dilemma of p53 transactivation in mediating tumor suppression is biased towards preference for apoptosis as the mechanism of cancer prevention, and therefore effort should be made to discover the p53- induced apoptotic genes. This is partly due to the fact that very few systems can separate the growth arrest function from the apoptotic one. However, as was shown recently, the p53 mutant R172P (R175P in human) probably could. This mutant lost the capacity to activate apoptosis, but retained the growth arrest function, and is able to up-regulate p21 much like wild-type p53.[61] Recent experiments renew interest in the growth inhibitory function of p53 in vivo as an important factor in suppressing cancer. Lozano and colleagues[62] prepared a knocked-in mouse containing this mutant (R172P) and compared it to the $p53^{-/-}$ mouse for the spontaneous appearance of tumors. Remarkably at 7 month of age, 85% of the p53R172P mice remained alive and tumor free, whereas 90% of the p53 deficient mice died, mainly from lymphoma.[62] Comparison of the MEFs from the two mice strains showed that while both were resistant to apoptosis, only those derived from the R172P mutant mice showed elevated levels of p21. These results suggest that the growth arrest function of p53 is an important part of its tumor suppression, and the currently held dogma emphasizing the dominance of apoptosis in tumor suppression should be reconsidered.[63] This raises the question as to the mechanism for this anti tumor effect of this mutant p53. One possibility is that it results from the generation of replicative senescence as was shown, for example, by p53's effect in H1299[64-65] and other cells,[66] and perhaps may be important in cancer prevention in various tissues, in particular in the intestine.[67-68] An additional unexpected observation was that the tumors derived from the $p53^{-/-}$ mice showed aneuploid karyotypes where the few tumors that developed late in the R172P mutant showed a diploid or sometimes tetraploid but not aneuploid karyotypes. This was due to extra number of centrosomes in the tumors from the $p53^{-/-}$ mice that cause unequal segregation of chromosomes and genome instability.[62] It is suggested that $p21^{waf}$ regulates the involvement of cyclin E in centrosome duplication and prevents the generation of more than two centrosomes per cell. Since most cancer cells show chromosomal aberration and aneuploidy, it is possible that p53 gained its title as "guardian of the genome" because of its role in preventing aneuploidy, perhaps through the activity of p21. In view of the marked heterogeneity in gene expression profile in various organs as a result of p53 activation, more information using tissue specific microarray analysis after in vivo p53 induction is needed.

Conclusion and Perspectives

The analysis of p53-induced genes had begun with methods of subtractive hybridization and differential display and isolation of individual relevant genes in various systems.[20] Many important genes were identified in this way, and their relevance was verified by Northern blots and by functional analysis using transfections. The main functions of such genes were cell cycle inhibition, apoptosis and genome stability. The introduction of microarray analysis to the p53 field in 1999 allowed the screening of the equivalent of thousands of Northern blots at once, and obtaining a gene expression profile that is meaningful in the context of the p53 system. Obviously the major genes that were previously identified were picked up by the microarray methodology as well, and new key genes in the pathway of p53 activation were also discovered (Tables 2 and 3). In addition, the microarray method extended the pattern of p53 gene activation to many other cell functions previously unknown to be related to p53. These include: gene suppression, upregulation of secreted factors, inhibitors of angiogenesis, genes for DNA repair enzymes and nucleotide metabolism, genes for ribosomal and protein synthesis activity, genes related to cytoskeletal functions and cell division, and the determination of the presence of a second envelope of co-activators or co-repressors that assist in the transactivation by p53. Hence p53 is one of the most highly connected nodes in the cellular network, and as such, an attack on p53 (by mutation) will disrupt basic cellular functions, particularly the responses to DNA damage and tumor-predisposing stresses.[69] On the other hand, this also indicates the possibility of mechanisms for correcting defects in p53 functions by employing by-passes at various points of the network in order to overcome tumor growth.

It should, however, be noted that the microarray gene expression profile is like a snapshot of global gene expression that lacks the functional dimension. It will be desirable to have an analytical system that enables the global detection of gene expression combined with functional selection of the relevant genes. A recent study points to such a complementary approach by using libraries of siRNA to select for genes that are functionally involved in the p53 effect on cells. Berns et al[70] used RNA interference (RNAi) to perform loss-of-function genetic screens of cellular signaling pathways in mammalian cells. They developed expression vectors to direct the synthesis of short hairpin RNAs (shRNAs) that act as short interfering RNA (siRNA)-like molecules to stably suppress gene expression. In their recent report, they constructed a set of retroviral vectors encoding 23,742 distinct shRNAs, which target 7,914 different human genes for suppression, and used this RNAi library in human cells to identify new modulators of p53-dependent proliferation arrest. The cells express a temperature sensitive SV40 T antigen that will lose its inhibition of p53 at 39°C. The ts T antigen binds to p53 at 32°C and therefore cell proliferation is not inhibited by p53. However, at 39°C (where T antigen is inactive) proliferation is inhibited by p53 and the cells develop colonies only if p53 growth inhibition is by- passed or blocked by the gene knockdown due to the siRNA. This knockdown confers resistance to p53- dependent growth arrest and allows for the isolation of the genes involved from the selected colonies. In their experiment they identified five new targets that confer resistance to p53 dependent growth arrest. These are *RPS6KA6* (ribosomal S6 kinase 4, RSK4), *HTATIP* (histone acetyl transferase TIP60), *HDAC4* (histone deacetylase 4), KIAA0828 (a putative *S*-adenosyl-L-homocysteine hydrolase, SAH3) and *CCT2* (T-complex protein 1, β-subunit). These new tools are similar to the global gene expression profile detected by microarrays combined with a built-in functional output, and therefore will greatly facilitate a large-scale loss-of-function genetic screen in mammalian cells. In view of the many decisions facing p53 function, as well as the variability and cell specificity pointed out in this chapter, the new gene expression knock-down/cell selection methodology is a promising addition to the field and may facilitate the discovery of new p53 targets with functional implications.

In addition, methods that combine microarray results with promoter identification of co-regulated genes and proteomics promise to decipher cellular networks and to shed more light on the p53 network.

Acknowledgment

We thank Prof. Eytan Domany for helpful discussion.

References

1. Zakut-Houri R, Oren M, Bienz B SL et al. A single gene and a pseudogene for the cellular tumor antigen p53. Nature 1983; 306:594-597.
2. Bienz B, Zakut-Houri R, Givol D SL et al. Analysis of the gene coding for the murine cellular tumor antigen p53. EMBO J 3:2179-2183.
3. Eliyahu D, Raz A, Gruss P SL et al. Participation of p53 cellular tumor antigen in transformation of normal embryonic cells. Nature 1984; 312:646-649.
4. Eliyahu D, Michalovitz D, Eliyahu S SL et al. Wild-type p53 can inhibit oncogene-mediated focus formation. Proc Natl Acad Sci 1989; 86:8763-8767.
5. Michalovitz D, Halevy O, Oren M. Conditional inhibition of transformation and of cell proliferation by a temperature-sensitive mutant of p53. Cell 1990; 62:671-680.

6. McBride OW, Merry DE, Oren M et al. Human p53 cellular antigen is on chromosome 17p13. Cytogenet. and Cell Genet 1985; 40:694.
7. Zakut-Houri R, Bienz-Tadmor B SL et al. Human p53 cellular tumor antigen : cDNA sequence and expression in COS cells. EMBO J 1985; 4:1251-1255.
8. McBride OW, Merry D, Givol D. The gene for human p53 cellular tumor antigen is located on chromosome 17 short arm (17p13). Proc Natl Acad Sci 1986; 83:130-134.
9. Vogelstein B, Fearon ER, Hamilton SR SL et al. Genetic alterations during colorectal-tumor development. New Engl J Med 1988; 319:525-532.
10. Baker SJ, Fearon ER, Nigro JM et al., Chromosome 17 deletions and p53 gene mutations in colorectal carcinomas. Science, 1989; 244:217-221.
11. Knudson AG Jr. Mutation and cancer: statistical study of retinoblastoma. Proc Natl Acad Sci 1971; 68:820-823.
12. Knudson AG. Two genetic hits (more or less) to cancer. Nat Rev Cancer 2001; 1:157-162.
13. Takahashi T, Nau MM, Chiba I SL et al. p53: a frequent target for genetic abnormalities in lung cancer. Science 1989; 246:491-494.
14. Finlay CA, Hinds PW, Levine AJ. The p53 proto-oncogene can act as a suppressor of transformation. Cell 1989; 57:1083-1093.
15. El-Deiry WS, Kern SE, Pietenpol JA SL et al. Definition of a consensus binding site for p53. Nat Genet 1992; 1:45-49.
16. Cho Y, Gorina S, Jeffrey SL et al. Crystal structure of a p53 tumor suppressor-DNA complex : understanding tumorigenic mutations. Science 1994; 265:346-355.
17. Hoh J, Jin S, Parrado T SL et al. the p53MH algorithm and its application in detecting p53-responsive genes. Proc Natl Acad Sci 2002; 99:8467-8472.
18. Cawley S, Bekiranov S, Ng HH SL et al. Unbiased mapping of transcription factor binding sites along human chromosomes 21 and 22 points to widespread regulation of noncoding RNAs. Cell 2004; 116:499-509.
19. Mercer WE, Shields MT, Amin M SL et al. Negative growth regulation in a glioblastoma tumor cell line that conditionally expresses human wild-type p53. Proc Natl Acad Sci 1990; 87:6166-6170.
20. El-Deiry WS, Tokino T, Velculescu VE, Levy DB, Parsons R, Trent JM, Lin D, Mercer WE, Kinzler KW and Vogelstein B. WAF1, a potential mediator of p53 tumor suppression. Cell, 1993; 75:817-825.
21. Noda A, Ning Y, Venable SF SL et al. Cloning of senescent cell derived inhibitors of DNA synthesis using an expression screen. Exp Cell Res 1994; 211:90-98.
22. Harper JW, Adami GR, Wei N SL et al. The p21 cdk-interacting protein Cip1 is a potent inhibitor of G1 cyclin dependent kinases. Cell 1993; 75:805-816.
23. Xiong Y, Hannon GJ, Zhang H SL et al. p21 is a universal inhibitor of cyclin kinases. Nature 1983; 366:634-635.
24. Yonish-Rouach E, Resnitzky D, Lotem J SL et al. Wild-type p53 induces apoptosis of myeloid leukaemic cells that is inhibited by interleukin-6. Nature 1991; 352:345-347.
25. Miyashita T, Reed JC. Tumor suppressor p53 is a direct transcriptional activator of the human bax gene. Cell 1995; 80:293-299.
26. Owen-Schaub LB, Zhang W, Cusack JC et al., Wild-type human p53 and a temperature-sensitive mutant induce Fas/APO-1 expression. Mol Cell Biol 1995; 15:3032-3040.
27. Kannan K, Amariglio N, Rechavi G SL et al. Profile of gene expression regulated by induced p53: connection to the TGF-B family. FEBS Lett., 2000; 470:77-82.
28. Kannan K, Amariglio N, Rechavi G SL et al. DNA microarrays identification of primary and secondary target genes regulated by p53. Oncogene 2001; 20:2225-2234.
29. Kannan K, Kaminski N, Rechavi G SL et al. DNA microarray analysis of genes involved in p53 mediated apoptosis activation of apaf-1. Oncogene 2001; 20:3449-3455.
30. Tan M, Wang Y, Guan K SL et al. PTGF-B, a type B transforming growth factor (TGF-B) superfamily member, is a p53 target gene that inhibits tumor cell growth via TGF-B signaling pathway. Proc Natl Acad Sci 2000; 97:109-114.
31. Li C, Shridhar K, Liu J. Molecular characterization of oncostatin M-induced growth arrest of MCF-7 cells expressing a temperature sensitive mutant of p53. Breast Caner Res And Treatment 2003; 80:23-37.
32. Yu J, Zhang L, Hwang PM, Rago C, Kinzler KW and Vogelstein B. Identification and classification of p53-regulated genes. Proc Natl Acad Sci 1999; 96:14517-14522.
33. Kimura T, Gotoh M, Nakamura Y SL et al. hCDC4b, a regulator of cyclin E, as a direct transcriptional target of p53. Cancer Sci 2003; 94:431-436.
34. Ueda K, Arakawa H, Nakamura Y. Dual-specificity phosphatase 5 (DUSP5) as a direct transcriptional target of tumor suppressor p53. Oncogene 2003; 22:5586-5591.

35. Ongusaha PP, Ouchi T, Kim K SL et al. BRCA1 shifts p53-mediated cellular outcomes towards irreversible growth arrest. Oncogene 20032; 22:3749-3758.
36. Zhao R, Gish K, Murphy M SL et al. Analysis of p53-regulated gene expression patterns using oligonucleotide arrays. Genes and Dev 2000; 14:981-993.
37. Lotem J, Gal H, Kama R SL et al. Inhibition of p53-induced apoptosis without affecting expression of p53 regulated genes. Proc Natl Acad Sci 2003; 100:6718-6723.
38. Maxwell SA, Davis GE. Differential gene expression in p53-mediated apoptosis-resistant vs. apoptosis-sensitive tumor cell lines. Proc Natl Acad Sci 2000; 97:13009-13014.
39. Maxwell SA, Davis GE. Biological and molecular characterization of an ECV-304-derived cell line resistant to p53 mediated apoptosis. Apoptosis 2000; 5:277-290.
40. Stankovic T, Hubank M, Cronin D SL et al. Microarray analysis reveals that TP53- and ATM-mutant B-CLLs share a defect in activating proapoptotic responses after DNA damage but are distinguished by major differences in activating prosurvival responses. Blood 2004; 103:291-300.
41. Mirza A, Wu Q, Wang L SL et al. Global transcriptional program of p53 target genes during the process of apoptosis and cell cycle progression. Oncogene 2003; 22:3645-3654.
42. Sax JK, El-Deiry WS. Identification and characterization of the cytoplasmic protein TRAF4 as a p53-regulated proapoptotic gene. J Biol Chem 2003; 278:36435-36444.
43. Fortin A, Cregan SP. MacLaurin JG SL et al. APAF1 is a key transcriptional target for p53 in the regulation of neuronal cell death. J Cell Biol 2001; 155:207-216.
44. Robles AI, Bemmels NA. Foraker AB SL et al. APAF-1 is a transcriptional target of p53 in DNA damage-induced apoptosis. Cancer Res., 2001; 61:6660-6664.
45. Wang Y, Rea T, Bian J SL et al. Identification of the genes responsive to etoposide-induced apoptosis:application of DNA chip technology. FEBS Lett 1999; 445:269-273.
46. Wu Q, Kirschmeier P, Hockenberry T SL et al. Transcriptional regulation during p21waf1/cip1 induced apoptosis in human ovarian cancer cells. J Biol Chem 2002; 277:36329-36337.
47. Fei P, Bernhard EJ, El-Deiry WS. Tissue-specific induction of p53 targets in vivo. Cancer Res 2002;62:7316-7327.
48. Soengas MS, Alarcon RM, Yoshida H SL et al. Apaf-1 and caspase-9 in p53-dependent apoptosis and tumor inhibition. Science, 1999; 284:156-159.
49. Moroni MC, Hickman ES, Denchi EL SL et al. Apaf-1 is a transcriptional target for E2f and p53. Nat Cell Biol 2001; 3:552-558.
50. Rozenfeld-Granot G, Krishnamurthy J SL et al. A positive feedback mechanism in the transcriptional activation of Apaf-1 by p53 and the coactivator Zac-1. Oncogene 2002; 21:1469-1476.
51. Iyer NG, Chin S, Ozdag H, Daigo Y, Hu D, Cariati M, Brindle K, Aparicio S and Caldas C. p300 regulates p53 dependent apoptosis after DNA damage in colorectal cancer cells by modulation of PUMA/p21 levels. Proc Natl Acad Sci 2004; 101:7386-7391.
52. Yu J, Wang Z, Kinzler KW SL et al. PUMA mediates the apoptotic response to p53 in colorectal cancer cells. Proc Natl Acad Sci 2003; 100:1931-1936.
53. Nakano K, Vousden KH. PUMA, a novel proapoptotic gene, is induced by p53. Mol Cell 2001; 7:683-694.
54. Polyak K, Xia Y, Zweier JL SL et al. A model for p53 induced apoptosis. Nature 1997; 389:300-305.
55. Lotem J, Sachs L. Different mechanisms for suppression of apoptosis by cytokines and calcium mobilizing compounds. Proc Natl Acad Sci 1998; 95:4601-4606.
56. Kho PS, Wang Z, Zhuang L SL et al. p53-regulated transcriptional program associated with genotoxic stress-induced apoptosis. J Biol Chem 279:21183-21192.
57. Komarova EA, Diatchenko L, Rokhlin OW SL et al. Stress-induced secretion of growth inhibitors: a novel tumor suppressor function of p53. Oncogene 1998; 17:1089-1096.
58. Dameron KM, Volpert OV, Tainsky MA SL et al. Control of angiogenesis in fibroblasts by p53 regulation of thrombospondin-1. Science 1994; 265:1582-1584.
59. Maxwell SA, Rivera A. Proline oxidase induces apoptosis in tumor cells, and its expression is frequently absent or reduced in renal carcinomas. J Biol Chem 2003; 278:9784-9789.
60. Fang L, Li G, Liu G SL et al. P53 induction of heparin-binding EGF-like growth factor counteracts p53 growth suppression through activation of MAPK and PI3K/Akt signaling cascades. EMBO J 2001; 20:1931-1939.
61. Rowan S, Ludwig RL, Haupt Y SL et al. Specific loss of apoptotic but not cell cycle arrest function in a human tumor derived p53 mutant. EMBO J 1996; 15:827-838.
62. Liu G, Parant JM, Lang G SL et al. Chromosome stability, in the absence of apoptosis, is critical for suppression of tumorigenesis in Trp53 mutant mice. Nat Genet 2003; 36:63-68.
63. Attardi LD, de Vries A, Jacks T. Activation of the p53 dependent G1 checkpoint response in mouse embryo fibroblasts depends on the specific DNA damage inducer. Oncogene 2004; 23:973-980.

64. Wang Y, Blandino G, Oren M SL et al. Induced p53 expression in lung cancer cell line promotes cell senescence and differentially modifies the cytotoxicity of anti-cancer drugs. Oncogene 1998; 17:1923-1930.
65. Wang Y, Blandino G, Givol D. Induced p21^{waf} expression in H1299 cell line promotes cell senescence and protects against cytotoxic effect of radiation and doxorubicin. Oncogene 1999; 18:2643-2649.
66. Sugrue MM, Shin DY, Lee SW SL et al. Wild-type p53 triggers a rapid senescence program in human tumor cells lacking functional p53. Proc Natl Acad Sci 1997; 94:9648-9653.
67. Fazeli A, Steen RG, Dickinson SL et al. Effects of p53 mutations on apoptosis in mouse intestinal and human colonic adenomas. Proc Natl Acad Sci 1997; 94:10199-10204.
68. Campisi J. Cellular senescence as a tumor suppressor mechanism. Trends in Cell Biol 2001; 11:S27-S31.
69. Vogelstein B, Lane D, Levine AJ. Surfing the p53 network. Nature 2000; 408:307-10.
70. Berns K, Hijmans EM, Mullenders J et al. A large scale RNAi screen in human cells identifies new components of the p53 pathway. Nature 2004; 428:431-7.

CHAPTER 11

Tumor Viruses and p53

Nobuo Horikoshi*

Abstract

Viruses are potent oncogenic agents and are considered to link to approximately one-fifth of all human malignancies. The study of tumor viruses has provided fundamental bases of how normal cells become cancer cells. A number of virus-encoded oncogenes have been identified and their functions were characterized. One of the striking functional similarities of such virus oncogenes, especially the ones encoded by DNA tumor viruses, is that they bind to and inactivate tumor suppressor genes, such as *p53* and *pRb*. In this chapter, I am going to summarize the function of virus oncogenes focusing on their interactions with p53.

Introduction

Historically, small DNA viruses have been ideal tools to study mechanisms of "life" including cellular transformation and tumorigenesis (for virus classification: http://www.nlv.ch/Virologytutorials/Classification.htm, http://www.ncbi.nlm.nih.gov/ICTVdb). There are several advantages to using such small DNA viruses as model systems; they have relatively small genome sizes and are easy to genetically manipulate, the viruses are easy to grow in a laboratory setting and to induce cellular transformation in a tissue culture system, and transformed cells can form tumors in laboratory animals. In fact, p53 was first discovered in 1979 as a cellular protein tightly bound to the large T antigen of the simian virus 40 (SV40) DNA tumor virus.

Once viruses infect cells, they use all necessary cellular components to replicate themselves. Immediate early (IE) gene products will be expressed using cellular transcription and translational machineries. These products play essential roles for the expression of other virus early gene products, and the modification of cellular gene transcriptions, the translocation of mRNAs, and translations. In many cases, these IE products also stimulate cellular DNA synthesis suitable for the replication of virus genomes. The process of modification of the host cell metabolisms includes the activation of cellular proto-oncogenes and the inactivation of tumor suppressor genes.[1] For example, adenovirus, human papillomaviruses (HPVs) and SV40 belong to different virus families, all three have adopted very similar strategies to deregulate cell growth. Each virus encodes oncoproteins which interact with the same cellular targets; tumor suppressor proteins pRb and p53 (Table 1). Since somatic mutations result in the loss of pRb and p53 functions in many cancers, the contribution of these viruses to tumor development appear to reflect their ability to inactivate these tumor suppressors. From this sense, the discovery of the physical and functional interaction between adenovirus E1A and tumor suppressor pRb was a milestone for the understanding of the relationship of DNA tumor virus and tumor suppressors.[2] The p53 protein negatively regulates the cell cycle and induces apoptosis in cells. Therefore, the action of the viral oncoproteins that inactivate p53 has the central importance in virus reproduction and tumorigenesis of host cells.

*Nobuo Horikoshi—Division of Molecular Radiation Biology, Department of Radiation Oncology, University of Texas Southwestern Medical School, Dallas, TX 75390, USA. Email: nobuo.horikoshi@utsouthwestern.edu

p53, edited by Ayeda Ayed and Theodore Hupp. ©2010 Landes Bioscience and Springer Science+Business Media.

Table 1. Tumor virus oncoproteins and their interacting tumor suppressors

Tumor Viruses	Viral Oncoproteins	Tumor Suppressors
Adenovirus type 2, 5	E1A E1B55K E4orf6	pRb, p107, p130 p53 p53
SV40 (simian virus 40)	Large T antigen	p53, pRb, p107, p130
HPV-16, 18 (human papillomavirus 16, 18)	E6 E7	p53 pRb
HBV (hepatitis B virus)	HBx	p53
EBV (Epstein-Barr virus)	EBNA-5 BXLF1	p53 p53
HCMV (human cytomegalovirus)	IE72 IE86	p107 p53, pRb

Tumor Viruses

Adenovirus

There are many types of adenoviruses for different hosts, including human, monkey, bovine, horse, mouse, chicken, and many of them have transformation activity. There are 51 different serotypes of human adenovirus so far identified. Human adenoviruses are known as flu viruses and do not relate to human cancers so far investigated, however, they have a potential to transform cultured cells and some of them are able to form tumors in the rodent. Adenovirus encodes three oncogenes within its early region genes, E1A, E1B, E4, which are expressed immediately after infection (Fig. 1). E1A is an essential transcription factor for the efficient transcription from other virus gene promoters as well as is able to immortalize and/or transform primary cultured cells. In some genotypes of adenoviruses, E1A-transformed cells grow in nude mice. Both E1B and E4 do not have immortalization/transformation

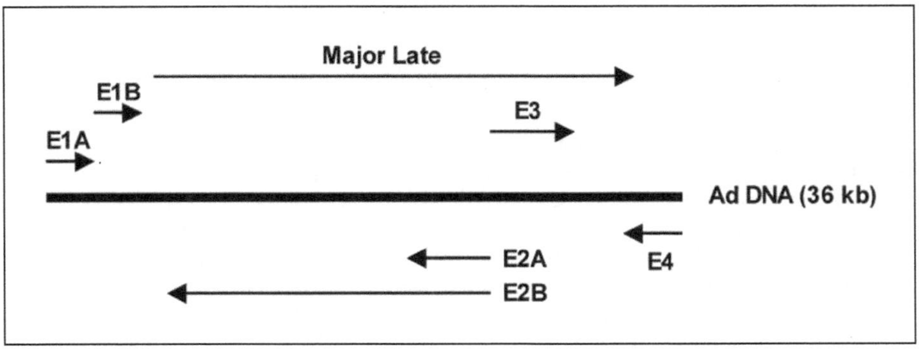

Figure 1. Genomic structure of human adenovirus. Immediate after infection, transcriptions from E1 and E4 promoters are initiated. Arrows indicate RNA transcripts from each virus gene.

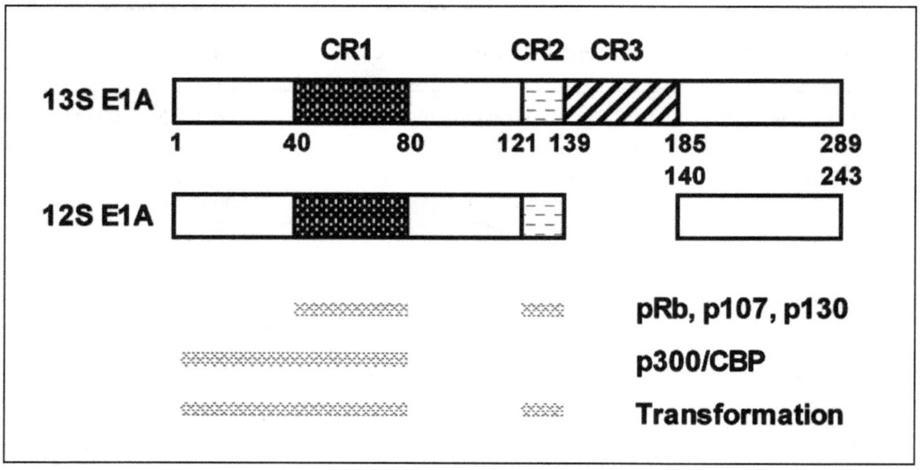

Figure 2. The structure of E1A proteins. E1A protein contains three conserved regions (CR1-3). The interaction domains for Rb family proteins and p300/CBP are shown.

activity alone, instead they can enhance the transformation efficiency by E1A and increase tumor malignancy.

E1A region produces several different sizes of proteins. The major products are 289 amino acids (289R, or 13S E1A) and 243 amino acids (243R or 12S E1A) in the Type 5 adenovirus generated by alternative splicing to give the so-called 13S and 12S mRNAs. 13S E1A protein contains three conserved regions (CR1, CR2, and CR3) throughout different genotypes (Fig. 2). The difference between 13S E1A and 12S E1A is the CR3 region which is only present in13S E1A. The essential domains for transformation have been mapped in the N-terminal nonconserved region, CR1, and CR2. The N-terminal region and CR1 are responsible for the interaction with p300/CBP binding protein, and the CR1 and CR2 are essential for the binding to Rb family proteins. The Rb-binding domain in CR2 contains a LXCXE motif which is found in other cellular and virus Rb-interacting proteins (Fig. 3). The stimulation of DNA synthesis by virus oncogenes induces cell cycle arrest or apoptosis in host cells. In adenovirus-encoded oncogenes, E1A protein stabilizes the p53 protein and induces apoptosis.[3,4] Both 13S and 12S E1A proteins have this effect and CR1 is required. Since CR1 is involved in the interaction with Rb family and p300/CBP proteins, E1A-induced p53-dependent apoptosis might involve these proteins. 13S E1A has also been shown to induce a p53-independent apoptosis.[5,6]

E1B region produces two major products, 19 kDa (19K) and 55 kDa (55K) proteins. E1B 19K protein shares homology to the cellular anti-apoptotic protein Bcl-2 (BH domain). Similar to the Bcl-2 protein, E1B 19K protein localizes to the mitochondrial inner membrane.[7] Therefore, E1B 19K enhances E1A-dependent transformation by preventing apoptosis.[8] E1B 55K directly binds to the transcription activation domain of p53 (amino acid 1-123) and inhibits its transactivation function. Since most of p53-dependent apoptosis requires the transactivation function of p53, it is assumed that E1B 55K inhibits p53-dependent apoptosis by inhibiting the activation of the downstream target genes of p53.[6] Posttranslational modifications (for example phosphorylations and acetylations) play a critical role in the activation of transactivation function of p53. Adenovirus has been shown to alter p53 acetylation at two different levels: E1B 55K interacts with p53 and inhibits its acetylation induced by PCAF (p300/CBP-associated factor),[9] whereas E1A interacts with p300/CBP and PCAF and represses their acetylation activity for p53.[10,11]

E4 gene encodes 7 open reading frames (ORFs). E4orf6 (E4 open reading frame 6) has been shown to bind the oligomerization domain of p53.[12] E4orf6 inhibits the interaction of

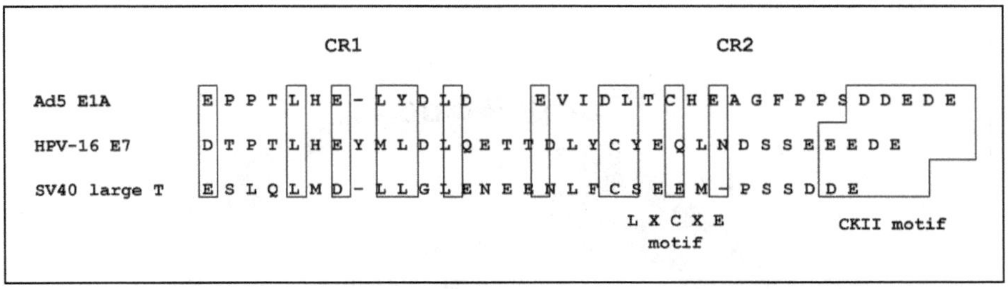

Figure 3. Sequence similarity of pRb binding region within viral oncoproteins. The LXCXE motif is critical for the interaction with pRb. Casein kinase II target sequence is also found at the C-terminus of CR2.

the transcription activation domain of p53 and $TAF_{II}31$, and inhibits its transactivation function.[12] E4orf6 also reduces cellular concentration of p53 by reducing stability of p53.[13,14] The inhibition of E1A-induced p53-independent apoptosis by E4orf6 has also been reported.[15] By these mechanisms, E4orf6 inhibits p53-dependent apoptosis and therefore enhances E1A-dependent transformation and tumorigenesis. Adenovirus Type 9 (Ad9) elicits exclusively estrogen-dependent mammary tumors in female rats,[16] and the transforming genes encoded in the viral E1 region are not required for tumorigenicity.[17] Three different histological types of mammary tumors (benign fibroadenomas, phyllodes tumors, and malignant solid sarcomas) have been described in Ad9-infected animals, and benign fibroadenomas are seen most frequently. Using hybrid viruses generated between Ad9 and the closely related nontumorigenic virus Ad26, E4orf1 and E4orf2 have been identified as essential oncogenes for the induction of mammary tumors.[18] Tumorigenic Ad9 E4orf1, but not nontumorigenic E4orf1, proteins encoded by adenovirus Types 5 and 12 binds to cellular PDZ domain (http://pawsonlab.mshri.on.ca)-containing protein ZO-2, a candidate tumor suppressor protein.[19] Ad9-induced rat mammary tumors have been studied as a unique model system for understanding the molecular mechanism of developing fibroadenoma, a common human breast tumor.

SV40 and Polyomavirus

SV40, a polyomavirus of rhesus macaque origin, was discovered in 1960 as a contaminant of polio vaccines that were distributed to millions of people from 1955 through early 1963.[20-23] SV40 is a potent DNA tumor virus that induces tumors in rodents and transforms many types of cells in culture, including those of human origin. The virus could be responsible for human mesotheliomas in the United States, at least in part.[24,25] Mesotheliomas are malignant tumors usually associated with occupational asbestors exposure.

The early region of SV40 produces 2 types of mRNAs for large T antigen (97 kDa) and small t antigen (17 kDa) by alternative splicing (Fig. 4). The large T antigen has a transformation activity in cultured cells. It is a multifunctional protein which holds all necessary functions for the replication of virus DNA, including binding activities to DNA polymerase α and SV40's replication origin (*ori*) DNA, ATPase activity, and helicase activity. However, most of these functions are not necessary for the transformation activity. The essential region for the transformation of rodent cells are located within the first 147 amino acids of the large T antigen[26] (Fig. 5). This region contains a domain for binding with the Rb family proteins[27] (pRb, p107, p130) including a LXCXE motif (aa 102-115), and a domain essential for the virus DNA replication and transformation (aa 1-82). The large T antigen binds the hypophosphorylated form of pRb[28] and thus disrupts the role of Rb in coordinating cell cycle progression. Rb protein normally binds transcription factor E2F in early G1 phase of the cell cycle. E2F is a family of transcription factors containing E2F1-5 and pRb preferentially interacts and inactivates E2F1-3, whereas p130 has a

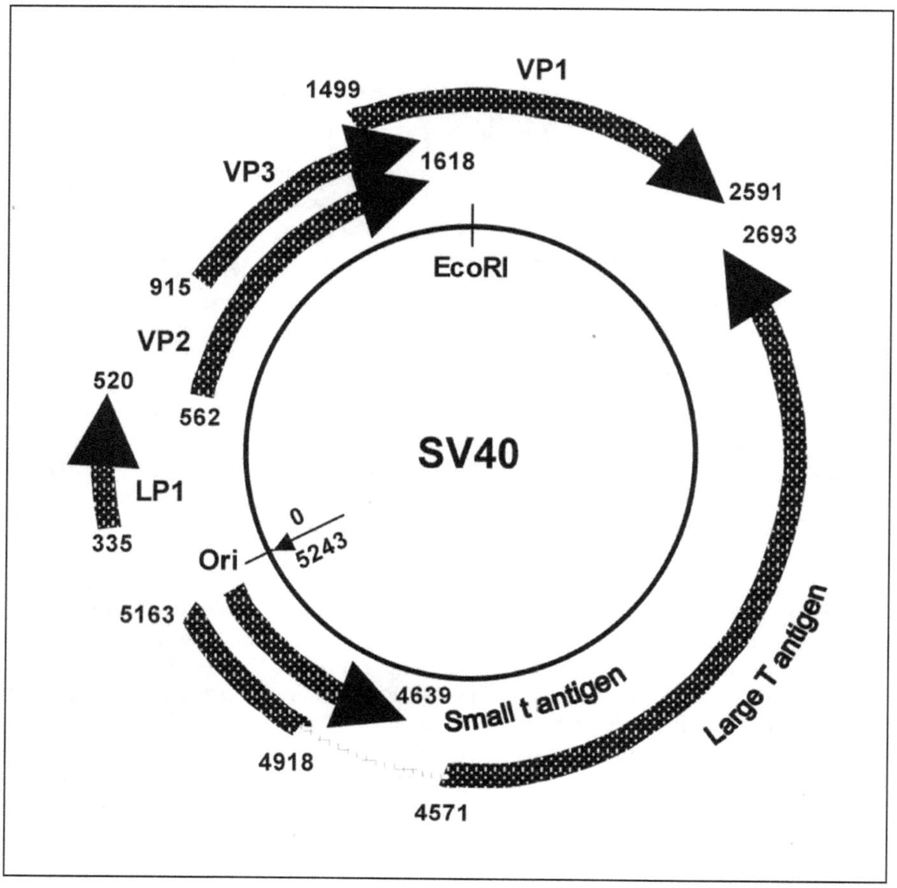

Figure 4. Genetic map of simian virus 40 (SV40). The circular SV40 DNA genome is represented with the unique *Eco*RI site. Nucleotide numbers based on reference strain SV40-776 begin and end at the origin (Ori) of viral DNA replication (0/5243). The open reading frames that encode viral proteins are indicated. Arrowheads point in the direction of transcription; the beginning and end of each open reading frame are indicated by nucleotide numbers.

preference to bind E2F4 and 5.[29,30] When pRb is phosphorylated by cyclin-dependent kinases, E2F is released and functions to activate expression of growth-stimulatory genes. The large T antigen causes dissociation of Rb-E2F complexes and releases active E2F, and as a result, induces unregulated growth of host cells. For the transformation of primary culture cells, the N-terminal 147 amino acids region and additional C-terminal region which binds to p53 are required. The regions responsible for p53-binding are mapped in amino acids 350-450 and 533-626 of the large T antigen. The large T antigen binds amino acids 94-293 of p53, the central DNA binding domain of p53 and inhibits the function of p53 to induce apoptosis in cells.[31]

The closest relatives are two polioviruses recovered from humans, JC virus (JCV) and BK virus (BKV). They share about 69% genomic similarity at the DNA sequence level. JCV possesses an oncogenic potential and induces development of various neuroectodermal origin tumors including medulloblastomas and glioblastomas in experimental animals. The virus early gene product, large T antigen, has been shown to bind to p53 and Rb and inhibits their functions. JCV has been also linked to human brain tumors.[32]

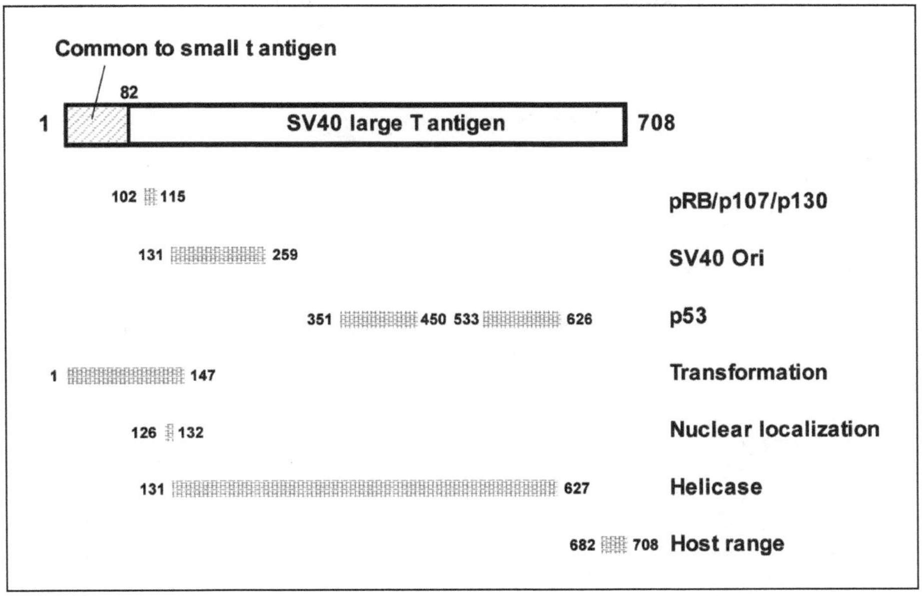

Figure 5. Functional domains of simian virus 40 (SV40) large T antigen. The numbers given are the amino acid residues using the numbering system for SV40-776.

Papillomavirus

More than 100 subtypes have been identified in human papillomaviruses (HPVs).[33] The viruses are small double-stranded DNA viruses with a genome of approximately 8000 bp (http://cinvestav.mx/genetica/MyFiles/Papillomavirus/PAPgeno.htm). HPVs infect either cutaneous or mucosal epithelia where they can cause warts or epithelial tumors.[34,35] Approximately 30 HPV types preferentially infect anogenital tract mucosa. These HPVs are also detected in the oral mucosa, and further classified into high-risk (HPV-16, -18) and low-risk (HPV-6, -11) types based on the clinical prognosis of the lesions that they cause.[36] Low-risk HPVs induce benign genital warts and very rarely associated with malignancy, whereas high-risk HPVs cause lesions that more frequently progress to cervical carcinoma. Even though the rate of carcinogenic progression is relatively low, infections with high-risk HPVs account for more than 90% of human cervical cancers,[37] a leading cause of cancer death in younger women worldwide.[38,39] Moreover, approximately 20% of oral cancers, particularly oropharyngeal carcinoma in patients that often lack risk factors such as alcohol and tobacco abuse, are also associated with high-risk HPVs.[40,41]

Less is known about the cutaneous HPV types. However, a subset of cutaneous HPVs, including HPV-5 and HPV-8, is linked with the development of skin cancers, particularly squamous cell carcinoma at sun exposed sites in immunocompromised individuals.[35,42]

Two viral oncogenes have been identified from high-risk HPVs, *E6* and *E7*. The products of those genes are essential in the process of HPV-induced cellular immortalization and transformation. Indeed, in cervical tumors and derived cell lines, HPV DNA is invariably retained and the viral early proteins E6 and E7 are continually expressed and their expressions are required for maintenance of the transformed phenotype.[43-45]

HPV E7 proteins are small, rather acidic polypeptides composed of approximately 100 amino acid residues (Fig. 6). HPV E7 proteins are phosphoproteins and are predominantly localized in the nucleus.[46,47] The high-risk HPV E7 protein contains an LXCXE motif, similar to adenovirus E1A and SV40 large T antigen, and binds to Rb family proteins such as pRb,

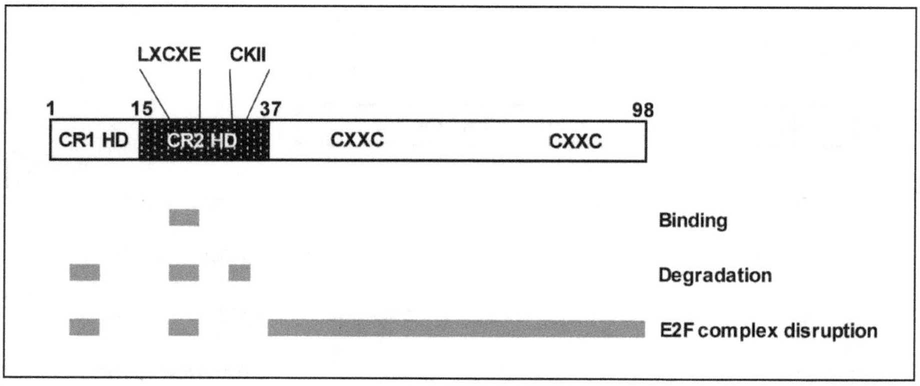

Figure 6. Interaction of Rb family proteins (pRb, p107, p130) with HPV E7. A schematic representation of the HPV-16 E7 domain structure is shown. CR1 HD and CR2 HD indicate the homology domains to conserved region 1 and 2 of the adenovirus E1A protein, respectively. CR2 HD contains the pRb binding motif (LXCXE) and a consensus sequence for casein kinase II phosphorylation (CKII). The metal binding domain consisting of two Cys-X-X-Cys (CXXC) sequences spaced 29 amino acid residues in the carboxyl terminus is also shown. Gray bars indicate domains required for binding or degradation of Rb family proteins and for the dissociation of pRb-E2F complexes.

p107, and p130[48] (Table 1). This interaction increases the active form of E2F and thus upregulating gene expressions required for G1/S transition and DNA synthesis. The inactivation of pRb seems to be essential for the E7 protein's for its transforming function since mutations in high-risk HPV E7 that interfere with pRb binding failed to transform cells,[49] and mutations enhancing the pRb binding efficiency of low-risk HPV-6 E7 could increase its transforming activity as well.[50,51] However, the fact that additional sequences in the carboxyl terminal dimerization domain of E7 are also required for disruption of Rb/E2F complexes[52,53] but not for transformation implies that the abilities of E7 to disrupt Rb/E2F complexes and cellular transformation are distinguishable.[54,55] It was also demonstrated that the binding of high-risk HPV E7 protein destabilizes Rb family proteins[56-59] possibly through the proteosome-mediated protein degradation.[57,60]

HPV E6 proteins are approximately 160 amino acid polypeptides with the molecular weight of approximately 18 kD (Fig. 7). High-risk HPV E6 proteins bind and functionally inactivate the p53 protein. A major mechanism by which these E6 proteins inhibit p53 function is that E6 protein stimulates the degradation of p53 through the E6-associated protein (E6AP, molecular weight is about 100 kD) mediated ubiquitin-dependent pathway.[61] The ubiquitin-dependent protein degradation pathway plays a major role in selective protein degradation in cells, and requires the sequential processes of the ubiquitin activating enzyme E1, ubiquitin conjugating enzymes E2, and ubiquitin protein ligases E3. Ubiquitin ligases E3 are the key components to show substrate specificities and therefore contain a large number of family enzymes. E6AP belongs to the HECT-domain family of ubiquitin ligases, whose large and divergent N-terminal domains mediate substrate recognition, while ubiquitination of bound substrate is catalyzed by the conserved C-terminal HECT domain.[62] High-risk HPV E6 binds E6AP within its N-terminal substrate recognition domain.[63] While E6AP alone is unable to bind p53, formation of a stable E6-E6AP complex precedes association with p53, thereby redirecting the substrate specificity of E6AP toward p53.[64] E3 ubiquitin ligase activity of E6AP is independent of E6, and it does not appear to be normally involved in the degradation of p53. Indeed, the blockage of E6AP activity, either by the use of antisense oligonucleotides[65] or dominant negative mutants,[66] increased the levels of p53 in HPV-positive, but not in HPV-negative cells, confirming that E6AP plays an essential role in E6-directed degradation of

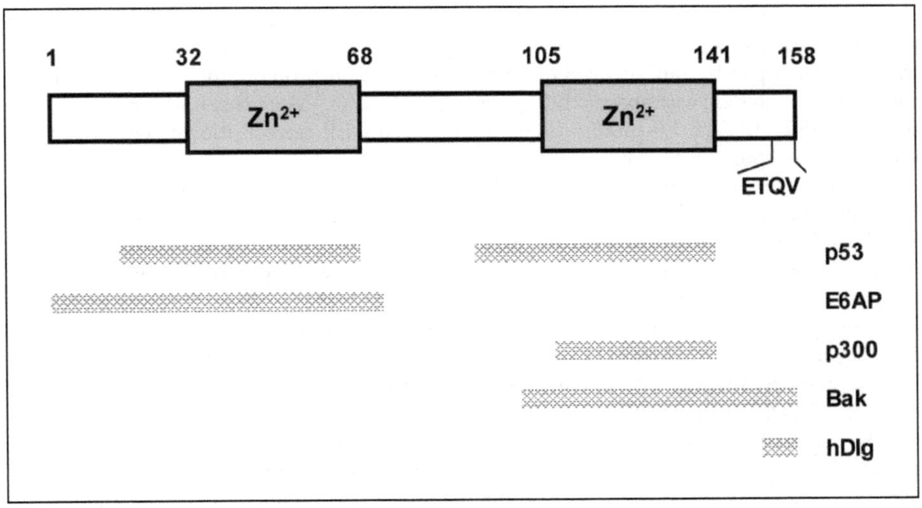

Figure 7. The high-risk HPV E6 protein. Schematic diagram of E6 protein including two zinc fingers (Zn^{2+}) and the PDZ-binding motif (ETQV) at the C-terminus is shown. The regions involving in the binding with cellular proteins are shown as gray boxes.

p53 in vivo, but has no effect on p53 levels in cells lacking E6. The degradation of p53 in normal, HPV-uninfected cells is pursued by MDM2 cellular oncoprotein, which also belongs to an E3 ubiquitin ligase family.[67,68]

The efficiency of E6 proteins to stimulate degradation of p53 correlates to their binding affinity to the central DNA biding region of p53; the high-risk HPV E6 proteins bind to this part of p53 more tightly than low-risk HPV E6 proteins, and this interaction is enhanced by the presence of E6AP and thus induce efficient degradation of p53.[69] Both high- and low-risk HPV E6 proteins are able to bind the C-terminal region of p53. However, such interactions do not induce the degradation of p53. In addition, high-risk HPV-16 E6 binds E6AP more strongly, and concomitantly degrades p53 more effectively, than another high-risk HPV-18 E6. Low-risk HPV-11 E6 has minimal level of binding to E6AP,[64] and degrades p53 only weakly in vivo.[70] HPV E6 also stimulates the E6AP-mediated degradation of c-Myc[71] and Bak[72] but, in contrast to p53, these proteins are also normal targets of E6AP, thus E6 only enhances the degradation processes of such cellular proteins.

Although p53 protein is extensively modified at different sites by phophorylation and acetylation, there is no report whether certain modified forms of p53 are better or weaker targets for E6-induced E6AP-mediated degradation at present. On the other hand, it has been reported that a polymorphic change in p53 protein alters susceptibility to HPV E6-dependent degradation. A common p53 polymorphism at proline 72 (72P) substituted to arginine (72R) is found in approximately 30% of humans.[73] The arginine variant disrupts one of the five PXXP SH3 motifs in the proline-rich region of p53.[74] Interestingly, 72R variant is significantly more susceptible than 72P wild-type to HPV-18 E6-dependent degradation,[70] and it can be also degraded by the low-risk HPV-11 E6-mediated proteolysis. Indeed, it was shown that individuals homozygous for 72R had a 7 times higher chance to bear an HPV-associated cervical carcinoma. However, similar analyses performed in other populations did not confirm the association between such polymorphism of the p53 gene and the risk to develop HPV-associated lesions,[75,76] it still remains unclear whether such association exists. Although p53 72P is a stronger in transcription activation function, the 72R variant is somehow twice as effective in preventing immortalization of primary rodent cells.

Since some E6 mutants defective for the degradation of p53 still retain the ability to inhibit transcriptional activation by p53, it has been suggested that high-risk HPV E6 may also inhibit the function of p53 by a mechanism not involving the stimulation of protein degradation. E6 was shown to interfere with specific DNA binding by p53.[77,78] Moreover, it has been suggested that high-risk HPV-16 E6 protein binds to the p300/CBP transcriptional coactivator and inhibits p53-responsive promoter activities.[79,80]

It is now clear that, similar to many other viral oncoproteins, E6 is multifunctional. E6 has been shown to inhibit NF-κB-dependent transcription[81] and to activate the promoter for the telomerase catalytic subunit, hTERT.[82,83] Furthermore, genetic analysis using transgenic mice expressing E6 has revealed that E6-induced hyperproliferation is seen in both $p53^{+/+}$ and $p53^{-/-}$ mice at similar levels, indicating that such hyperproliferation effects are not p53-dependent.[84] Studies regarding such p53-independent functions of HPV E6 are focusing on its highly conserved C-terminal domain (Fig. 7), which is not involved in p53 binding and degradation[85,86] but contributes to transformation activity of E6.[87] This region contains a PDZ-binding motif (XT/SXV), a short stretch of amino acids which mediates the specific interaction with proteins containing PDZ domains,[88,89] and several candidate proteins which bind to the C-terminal domain of E6 have been identified including hDlg (human disk large).[87,90] This is the human homologue of the developmentally essential Drosophila tumor suppressor Dlg, and the hDlg protein localizes at cell junctions functioning in the control of cell-cell contact and cell polarity.[91,92] High-risk HPV E6 proteins bind hDlg and stimulate degradation by ubiquitin-dependent manner.[93] HPV-18 E6 binds and thus degrades hDlg more efficiently than HPV-16 E6,[93,94] which correlates well to the clinical outcomes: cervical tumors associated with HPV-18 are more invasive and recurrent than those caused by HPV-16.[95-97] The search for the PDZ domain-containing proteins which are involved in E6-mediated transformation is still on going.

Others

Hepatitis B Virus

Hepatocellular carcinoma (HCC) is the most common malignant tumor of males with an incidence of one million new cases each year. Chronic infection with hepatitis B virus (HBV) is one of the major causes of HCC[98] (see more hepatitis virus: http://mmbr.asm.org/cgi/content/full/64/1/51). HBV has an approximately 3.2 kb double stranded DNA as a genome. HBV encodes the X protein (HBx) which is 154 amino acids in length, and has been suspected to be oncogenic,[99,100] although the precise function(s) remain to be elucidated.

HBx has been reported to interact with a wide variety of both cytoplasmic, nuclear proteins[101]. Nuclear target proteins of HBx include transcriptional factors, basal transcription factors, positive and negative transcriptional cofactors, RNA polymearse II, and DNA repair proteins.[100] Direct interaction between p53 and HBx has been demonstrated.[102,103] HBx binds the C-terminal region of p53 and inhibits its specific DNA binding, transcriptional activation, and apoptosis.[103-106] It has been shown that HBx inhibits the function of p53 by squelching p53 from the nucleus since HBx localizes mainly in the cytoplasm.[102,107,108] In addition, it has been reported that the promoter activity of p53 is downregulated by HBx possibly through the E-box in the promoter,[109] which is the binding site of the transcription factors that belong to the bHLH family (helix-loop-helix), including Myc, Mad, Max.

Epstein-Barr Virus

Epstein-Barr virus (EBV) is a member of the herpesvirus family and one of the most common human viruses. In the United States, approximately 95% of adults between 35-40 years of age have been infected. Transmission of EBV through the air or blood does not normally occur but usually requires intimate contact with the saliva of the infected individual. EBV is known to be closely associated with several human cancers, including Burkitt's lymphoma, nasopharyngeal carcinoma, Hodgkin's disease, and lymphoproliferative disorders.[110,111]

The EBV genome in the virus particle is a linear double-stranded 172kb DNA encoding more than 80 open reading frames (http://www.nature.com/onc/journal/v22/n33/full/1206556a.html). EBV predominantly infects resting B lymphocytes and induces the continuous proliferation of the B cells.[110-112] In the resulting lymphoblastoid cells, where EBV exists in a latent state, EBV expresses only 10 of the viral genes. Those include the transformation-associated proteins, such as EBV nuclear antigens (EBNA1-6) and three latent membrane proteins (LMP-1, -2A and-2B). LMP-1 acts as an oncogene and plays essential role in growth and tranformation of B cells. LMP-1 binds and activates cytoplasmic TNF receptor-associated factors (TRAFs), leading to activation of the transcription factor NF-κB, culminating in cellular proliferation.[113] EBNA-2 is an essential transcriptional activator for the expression of viral genes, including LMP-1, and some other host genes.[114-116] EBNA-1 binds the *oriP*, a cis-acting element required for EBNA1-dependent episome maintenance, in the viral genome.[117,118] EBNA-1 contains Gly-Ala repeat motif thus avoids proteosomal degradation for the antigen presentation by MHC class I, and therefore EBNA-1 expressing cells escape from host immune system.[119]

Two EBV-encoded viral proteins have been shown to interact with p53: EBNA-5[120] and BZLF1[121] (also designated ZEBRA, Zta, EB1). EBNA-5 and EBNA-2 are the first viral genes expressed after B cell infection. Therefore, it is assumed that EBNA-5 may play a role at the initial stage of B cell transformation. Indeed, EBNA-5 can cooperate with EBNA-2 in the activation of viral promoters.[122] EBNA-5 also binds to pRb, and p53 competes for EBNA-5 binding with pRb.[120] The physiological significance of these interactions is not clear at present. EBNA-5 has been shown to localize in PODs (promyelocytic leukemia protein (PML) oncogenic domains, or ND10) in EBV infected B cells.[123,124]

The EBV lytic program is initiated by the expression of BZLF1.[125-128] The BZLF1 protein is a transcription factor which binds to AP-1 like sequences present in the promoters of early lytic genes, and eventually leads to virus reproduction.[129-134] Binding between BZLF1 and p53 occurs in part through sequences in the C-terminus dimerization domain of the BZLF1 protein and the C-terminus of p53.[121]

Expression of the BZLF1 protein in an EBV-negative epithelial cell system induced G1/S cell cycle arrest with increased expression of p53 and cyclin-dependent kinase inhibitors, such as p21/Waf-1 and p27/Kip-1.[135,136] Although p53 activates p21 expression, inhibition of cell cycle progression by BZLF1 has been shown to be independent of p53.[136] Instead, it was reported that BZLF1 inhibits p53 transcriptional function in reporter gene assays.[121] Infection of cells with the BZLF1-expressing recombinant adenviruses increased the level of cellular p53 but prevented the induction of p53-dependent cellular target genes, such as p21 and MDM2.[137] Inhibition of p53 function appears to be important for efficient lytic herpesvirus infection since a number of herpesviruses encode lytic proteins that inhibit p53 function. For example, the HCMV IE86 protein,[138-141] Kaposi's sarcoma-associated herpesvirus (KSHV or HHV-8) ORF K8 protein,[142] and human herpesvirus 6 (HHV-6) ORF 1 protein[143] all stabilize p53, thus increasing overall level of p53 protein while inhibiting its transactivation function.

However, the effect of BZLF1 on p53 function remains somewhat controversial. While other groups similarly reported that BZLF1 increases the level of cellular p53, they found that this effect was accompanied by increased p53 transactivation function.[135,144,145] Moreover, in epithelial cell lines, the coexpression of p53 and BZLF1 leads to an increase in the transcription of target genes by p53.[135] It is possible that BZLF1 has both potentially activating and inhibitory effects on p53. The reason for these different results is not clear but could reflect cell type-dependent effects of BZLF1 on p53 function or the effect of BZLF1 on p53 could be modified by the presence of other viral proteins.

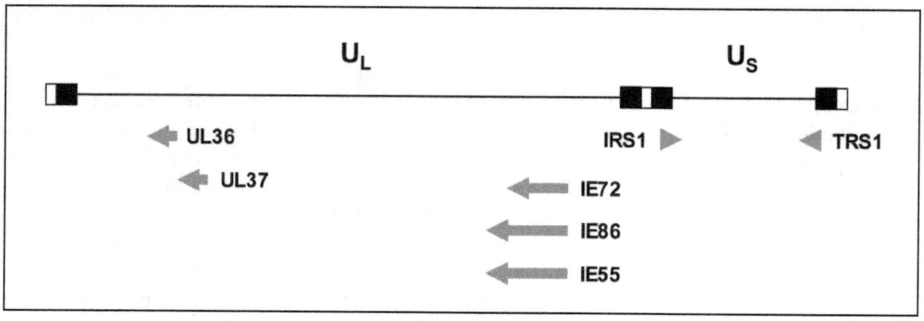

Figure 8. HCMV IE proteins and their location within the genome. The HCMV genome is divided into two segments, designated U_L (unique long) and U_S (unique short), bounded by inverted repeats. Arrows represent transcripts from IE genes.

Cytomegalovirus

Human cytomegalovirus (HCMV) belongs to the Herpesviridae family and has an approximately 230 kb double-stranded DNA genome containing approximately 200 open reading frames (Fig. 8). HCMV infection rarely causes symptomatic disease in healthy, immunocompetent individuals. However, it becomes hazardous in immunocompromised and immunosuppressed individuals such as organ transplant recipients and HIV-positive individuals.[146,147] Likewise, HCMV infection can be a potential threat for pregnant women and infants.

Several reports have shown an association of HCMV infection with the overproliferation and migration of arterial smooth muscle cells along the vessel wall intima, both of which are characteristic of restenosis following balloon angioplasty.[139,148] Although HCMV does not appear to be oncogenic, many studies have demonstrated that HCMV can transform both human[149] and rodent fibroblasts.[150,151] Several regions of HCMV genome can transform cells, including some of the IE genes. Although both IE86 and IE72 cooperate with adenovirus E1A to transform primary rodent cells, the transformation requires E1A protein. Therefore, IE86 and IE72 are considered as potentially mutagenic proteins, and these proteins may contribute to cellular transformation only under the appropriate conditions.[138]

It has been shown that IE86 binds p53 through its N-terminal region (aa 1-135) and inhibits its transcriptional activation function of p53[139,141,152] (Fig. 9). Expression of IE86 has been

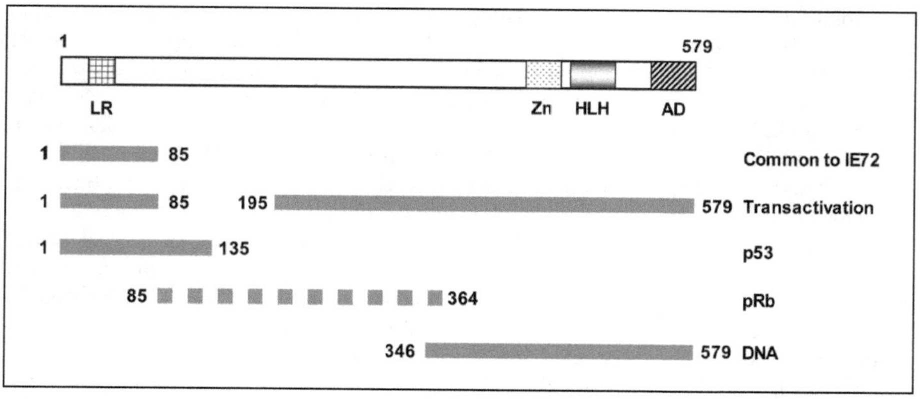

Figure 9. Schematic diagram of the HCMV IE86 protein. Gray boxes indicate the functional and protein binding domain. The pRb-binding domain is still unclear. LR; leucine-rich domain, Zn; zinc finger domain, HLH; helix-loop-helix domain, AD; acidic domain.

shown to inhibit the activation of p53-responsive promoter and its endogenous target gene in cultured smooth muscle cells.[139,140] In addition, IE86 also binds to pRb,[153-156] whereas another IE protein IE72 binds p107 but not pRb.[157,158] The inactivation of these tumor suppressor genes is considered to be responsible for the overproliferation of arterial smooth muscle cells in vivo. These studies have revealed that IE proteins encoded by HCMV have similar functions as IE equivalents encoded by small DNA tumor viruses, such as adenovirus, in terms of inactivation function of Rb and p53 families.

Conclusion and Perspectives

Infection of tumor viruses plays critical roles in developing certain human cancers, and the inactivation of tumor suppressor proteins such as p53 and pRB by virus-encoded oncogenes is one of the major mechanisms to induce unregulated growth of host cells. Therefore, reactivation of tumor suppressors in cancer cells would be one strategy for treatment. For example, if a small molecular weight compound which binds to virus oncogenes and prevents the interaction with p53 or pRB, it can be considered as a potential anticancer drug for the virus-induced cancers. Note that virus oncogenes that bind to pRB share a common "pocket" structure (Fig. 3), however, no common sequence homology has been identified for the interaction with p53. This implies that a compound to reactivate p53 should be custom-made for each virus oncogene product. Since most of the mutations in the p53 gene result in the production of point-mutant p53s which generally have lost transactivation function of p53, the recovery of such activity to mutant p53s by compounds would be another approach.[159,160] Alternatively, gene therapies to express ectopic p53 or pRB would also be effective if ectopically expressed large amounts of tumor suppressor proteins overcome virus oncogenes. Another type of gene therapy would be to target a virus oncogene itself by antisense oligonucleotide or by siRNA against virus oncogenes.

More fundamentally, virus infections would be inhibited by vaccinations. A recent clinical study using HPV-16 vaccine involving 2392 young women showed a complete protection of HPV-16 (100% efficacy) after 17.4 months follow-up period.[161] Moreover, all nine cases of HPV-16-related cervical intraepithelial neoplasia occurred among the placebo recipients. Therefore, immunization of HPV16-negative women is likely to reduce their risk of cervical cancer. Effective immunization with HPV-16, -18 or HBV, hepatitis C virus (HCV) may drastically reduce virus-related cervical and hepatitis cancer incident.

Acknowledgement

I would like to thank Ms Regan Nanz for helpful comments to the manuscript.

References

1. Efferth T, Grassmann R. Impact of viral oncogenesis on responses to anti-cancer drugs and irradiation. Crit Rev Oncog 2000; 11(2):165-187.
2. Whyte P, Buchkovich KJ, Horowitz JM et al. Association between an oncogene and an anti-oncogene: The adenovirus E1A proteins bind to the retinoblastoma gene product. Nature 1988; 334(6178):124-129.
3. Lowe SW, Ruley HE. Stabilization of the p53 tumor suppressor is induced by adenovirus 5 E1A and accompanies apoptosis. Genes Dev 1993; 7(4):535-545.
4. Debbas M, White E. Wild-type p53 mediates apoptosis by E1A, which is inhibited by E1B. Genes Dev 1993; 7(4):546-554.
5. Subramanian T, Tarodi B, Chinnadurai G. p53-independent apoptotic and necrotic cell deaths induced by adenovirus infection: Suppression by E1B 19K and Bcl-2 proteins. Cell Growth Differ 1995; 6(2):131-137.
6. Teodoro JG, Shore GC, Branton PE. Adenovirus E1A proteins induce apoptosis by both p53-dependent and p53-independent mechanisms. Oncogene 1995; 11(3):467-474.
7. Boyd JM, Malstrom S, Subramanian T et al. Adenovirus E1B 19 kDa and Bcl-2 proteins interact with a common set of cellular proteins. Cell 1994; 79(2):341-351.
8. White E. Life, death, and the pursuit of apoptosis. Genes Dev 1996; 10(1):1-15.
9. Liu Y, Colosimo AL, Yang XJ et al. Adenovirus E1B 55-kilodalton oncoprotein inhibits p53 acetylation by PCAF. Mol Cell Biol 2000; 20(15):5540-5553.

10. Lundblad JR, Kwok RP, Laurance ME et al. Adenoviral E1A-associated protein p300 as a functional homologue of the transcriptional coactivator CBP. Nature 1995; 374(6517):85-88.
11. Sang N, Avantaggiati ML, Giordano A. Roles of p300, pocket proteins, and hTBP in E1A-mediated transcriptional regulation and inhibition of p53 transactivation activity. J Cell Biochem 1997; 66(3):277-285.
12. Dobner T, Horikoshi N, Rubenwolf S et al. Blockage by adenovirus E4orf6 of transcriptional activation by the p53 tumor suppressor. Science 1996; 272(5267):1470-1473.
13. Moore M, Horikoshi N, Shenk T. Oncogenic potential of the adenovirus E4orf6 protein. Proc Natl Acad Sci USA 1996; 93(21):11295-11301.
14. Steegenga WT, Riteco N, Jochemsen AG et al. The large E1B protein together with the E4orf6 protein target p53 for active degradation in adenovirus infected cells. Oncogene 1998; 16(3):349-357.
15. Marcellus RC, Teodoro JG, Wu T et al. Adenovirus type 5 early region 4 is responsible for E1A-induced p53-independent apoptosis. J Virol 1996; 70(9):6207-6215.
16. Javier R, Raska Jr K, Macdonald GJ et al. Human adenovirus type 9-induced rat mammary tumors. J Virol 1991; 65(6):3192-3202.
17. Thomas DL, Shin S, Jiang BH et al. Early region 1 transforming functions are dispensable for mammary tumorigenesis by human adenovirus type 9. J Virol 1999; 73(4):3071-3079.
18. Thomas DL, Schaack J, Vogel H et al. Several E4 region functions influence mammary tumorigenesis by human adenovirus type 9. J Virol 2001; 75(2):557-568.
19. Glaunsinger BA, Weiss RS, Lee SS et al. Link of the unique oncogenic properties of adenovirus type 9 E4-ORF1 to a select interaction with the candidate tumor suppressor protein ZO-2. EMBO J 2001; 20(20):5578-5586.
20. Lewis AJ. Experience with SV40 and adenovirus-SV40 hybrids. Biohazards in biological research. New York: Cold Spring Harbor Laboratory Press, 1973.
21. Fraumeni JJ, Ederer F, Miller R. An evaluation of the carcinogenicity of simian virus 40 in man. JAMA 1963; 185:713-718.
22. Geissler E. SV40 and human brain tumors. Prog Med Virol 1990; 37:211-222.
23. Shah K, Nathanson N. Human exposure to SV40: Review and comment. Am J Epidemiol 1976; 103(1):1-12.
24. Carbone M, Rizzo P, Grimley PM et al. Simian virus-40 large-T antigen binds p53 in human mesotheliomas. Nat Med 1997; 3(8):908-912.
25. Cristaudo A, Vivaldi A, Sensales G et al. Molecular biology studies on mesothelioma tumor samples: Preliminary data on H-ras, p21, and SV40. J Environ Pathol Toxicol Oncol 1995; 14(1):29-34.
26. Sompayrac L, Danna KJ. The amino-terminal 147 amino acids of SV40 large T antigen transform secondary rat embryo fibroblasts. Virology 1991; 181(1):412-415.
27. Stubdal H, Zalvide J, DeCaprio JA. Simian virus 40 large T antigen alters the phosphorylation state of the RB-related proteins p130 and p107. J Virol 1996; 70(5):2781-2788.
28. DeCaprio JA, Ludlow JW, Figge J et al. SV40 large tumor antigen forms a specific complex with the product of the retinoblastoma susceptibility gene. Cell 1988; 54(2):275-283.
29. Vairo G, Livingston DM, Ginsberg D. Functional interaction between E2F-4 and p130: Evidence for distinct mechanisms underlying growth suppression by different retinoblastoma protein family members. Genes Dev 1995; 9(7):869-881.
30. Hijmans EM, Voorhoeve PM, Beijersbergen RL et al. E2F-5, a new E2F family member that interacts with p130 in vivo. Mol Cell Biol 1995; 15(6):3082-3089.
31. Bargonetti J, Reynisdottir I, Friedman PN et al. Site-specific binding of wild-type p53 to cellular DNA is inhibited by SV40 T antigen and mutant p53. Genes Dev 1992; 6(10):1886-1898.
32. Del Valle L, Gordon J, Assimakopoulou M et al. Detection of JC virus DNA sequences and expression of the viral regulatory protein T-antigen in tumors of the central nervous system. Cancer Res 2001; 61(10):4287-4293.
33. de Villiers EM, Lavergne D, McLaren K et al. Prevailing papillomavirus types in nonmelanoma carcinomas of the skin in renal allograft recipients. Int J Cancer 1997; 73(3):356-361.
34. zur Hausen H. Viruses in human cancers. Science 1991; 254(5035):1167-1173.
35. Shah K, Howley PM. Papillomaviruses. In: Fields B, Knipe D, Howley P, eds. Fields Virology. Philadelphia: Lippincott-Raven Publishers, 1996:2077-2109.
36. Lowy DR, Kirnbauer R, Schiller JT. Genital human papillomavirus infection. Proc Natl Acad Sci USA 1994; 91(7):2436-2440.
37. Munoz N, Bosch FX. The causal link between HPV and cervical cancer and its implications for prevention of cervical cancer. Bull Pan Am Health Organ 1996; 30(4):362-377.
38. zur Hausen H. Papillomavirus infections—a major cause of human cancers. Biochim Biophys Acta 1996; 1288(2):F55-78.

39. Pisani P, Parkin DM, Bray F et al. Estimates of the worldwide mortality from 25 cancers in 1990. Int J Cancer 1999; 83(1):18-29.
40. Sugerman PB, Shillitoe EJ. The high risk human papillomaviruses and oral cancer: Evidence for and against a causal relationship. Oral Dis 1997; 3(3):130-147.
41. Gillison ML, Koch WM, Capone RB et al. Evidence for a causal association between human papillomavirus and a subset of head and neck cancers. J Natl Cancer Inst 2000; 92(9):709-720.
42. de Villiers EM, Ruhland A, Sekaric P. Human papillomaviruses in nonmelanoma skin cancer. Semin Cancer Biol 1999; 9(6):413-422.
43. Schwarz E, Freese UK, Gissmann L et al. Structure and transcription of human papillomavirus sequences in cervical carcinoma cells. Nature 1985; 314(6006):111-114.
44. Smotkin D, Wettstein FO. Transcription of human papillomavirus type 16 early genes in a cervical cancer and a cancer-derived cell line and identification of the E7 protein. Proc Natl Acad Sci USA 1986; 83(13):4680-4684.
45. Munger K, Phelps WC, Bubb V et al. The E6 and E7 genes of the human papillomavirus type 16 together are necessary and sufficient for transformation of primary human keratinocytes. J Virol 1989; 63(10):4417-4421.
46. Sato H, Watanabe S, Furuno A et al. Human papillomavirus type 16 E7 protein expressed in Escherichia coli and monkey COS-1 cells: Immunofluorescence detection of the nuclear E7 protein. Virology 1989; 170(1):311-315.
47. Greenfield I, Nickerson J, Penman S et al. Human papillomavirus 16 E7 protein is associated with the nuclear matrix. Proc Natl Acad Sci USA 1991; 88(24):11217-11221.
48. Lee JO, Russo AA, Pavletich NP. Structure of the retinoblastoma tumor-suppressor pocket domain bound to a peptide from HPV E7. Nature 1998; 391(6670):859-865.
49. Phelps WC, Munger K, Yee CL et al. Structurefunction analysis of the human papillomavirus type 16 E7 oncoprotein. J Virol 1992; 66(4):2418-2427.
50. Heck DV, Yee CL, Howley PM et al. Efficiency of binding the retinoblastoma protein correlates with the transforming capacity of the E7 oncoproteins of the human papillomaviruses. Proc Natl Acad Sci USA 1992; 89(10):4442-4446.
51. Sang BC, Barbosa MS. Single amino acid substitutions in "low-risk" human papillomavirus (HPV) type 6 E7 protein enhance features characteristic of the "high-risk" HPV E7 oncoproteins. Proc Natl Acad Sci USA 1992; 89(17):8063-8067.
52. Huang PS, Patrick DR, Edwards G et al. Protein domains governing interactions between E2F, the retinoblastoma gene product, and human papillomavirus type 16 E7 protein. Mol Cell Biol 1993; 13(2):953-960.
53. Wu EW, Clemens KE, Heck DV et al. The human papillomavirus E7 oncoprotein and the cellular transcription factor E2F bind to separate sites on the retinoblastoma tumor suppressor protein. J Virol 1993; 67(4):2402-2407.
54. Braspenning J, Marchini A, Albarani V et al. The CXXC Zn binding motifs of the human papillomavirus type 16 E7 oncoprotein are not required for its in vitro transforming activity in rodent cells. Oncogene 1998; 16(8):1085-1089.
55. Mavromatis KO, Jones DL, Mukherjee R et al. The carboxyl-terminal zinc-binding domain of the human papillomavirus E7 protein can be functionally replaced by the homologous sequences of the E6 protein. Virus Res 1997; 52(1):109-118.
56. Berezutskaya E, Yu B, Morozov A et al. Differential regulation of the pocket domains of the retinoblastoma family proteins by the HPV16 E7 oncoprotein. Cell Growth Differ 1997; 8(12):1277-1286.
57. Boyer SN, Wazer DE, Band V. E7 protein of human papilloma virus-16 induces degradation of retinoblastoma protein through the ubiquitin-proteasome pathway. Cancer Res 1996; 56(20):4620-4624.
58. Jones DL, Munger K. Analysis of the p53-mediated G1 growth arrest pathway in cells expressing the human papillomavirus type 16 E7 oncoprotein. J Virol 1997; 71(4):2905-2912.
59. Smith-McCune K, Kalman D, Robbins C et al. Intranuclear localization of human papillomavirus 16 E7 during transformation and preferential binding of E7 to the Rb family member p130. Proc Natl Acad Sci USA 1999; 96(12):6999-7004.
60. Gonzalez SL, Stremlau M, He X et al. Degradation of the retinoblastoma tumor suppressor by the human papillomavirus type 16 E7 oncoprotein is important for functional inactivation and is separable from proteasomal degradation of E7. J Virol 2001; 75(16):7583-7591.
61. Scheffner M, Werness BA, Huibregtse JM et al. The E6 oncoprotein encoded by human papillomavirus types 16 and 18 promotes the degradation of p53. Cell 1990; 63(6):1129-1136.
62. Schwarz SE, Rosa JL, Scheffner M. Characterization of human hect domain family members and their interaction with UbcH5 and UbcH7. J Biol Chem 1998; 273(20):12148-12154.

63. Huibregtse JM, Scheffner M, Howley PM. Cloning and expression of the cDNA for E6-AP, a protein that mediates the interaction of the human papillomavirus E6 oncoprotein with p53. Mol Cell Biol 1993; 13(2):775-784.
64. Huibregtse JM, Scheffner M, Howley PM. Localization of the E6-AP regions that direct human papillomavirus E6 binding, association with p53, and ubiquitination of associated proteins. Mol Cell Biol 1993; 13(8):4918-4927.
65. Beer-Romero P, Glass S, Rolfe M. Antisense targeting of E6AP elevates p53 in HPV-infected cells but not in normal cells. Oncogene 1997; 14(5):595-602.
66. Talis AL, Huibregtse JM, Howley PM. The role of E6AP in the regulation of p53 protein levels in human papillomavirus (HPV)-positive and HPV-negative cells. J Biol Chem 1998; 273(11):6439-6445.
67. Honda R, Tanaka H, Yasuda H. Oncoprotein MDM2 is a ubiquitin ligase E3 for tumor suppressor p53. FEBS Lett 1997; 420(1):25-27.
68. Shimizu H, Burch LR, Smith AJ et al. The conformationally flexible S9-S10 linker region in the core domain of p53 contains a novel MDM2 binding site whose mutation increases ubiquitination of p53 in vivo. J Biol Chem 2002; 277(32):28446-28458.
69. Li X, Coffino P. High-risk human papillomavirus E6 protein has two distinct binding sites within p53, of which only one determines degradation. J Virol 1996; 70(7):4509-4516.
70. Storey A, Thomas M, Kalita A et al. Role of a p53 polymorphism in the development of human papillomavirus-associated cancer. Nature 1998; 393(6682):229-234.
71. Gross-Mesilaty S, Reinstein E, Bercovich B et al. Basal and human papillomavirus E6 oncoprotein-induced degradation of Myc proteins by the ubiquitin pathway. Proc Natl Acad Sci USA 1998; 95(14):8058-8063.
72. Thomas M, Banks L. Inhibition of Bak-induced apoptosis by HPV-18 E6. Oncogene 1998; 17(23):2943-2954.
73. Matlashewski GJ, Tuck S, Pim D et al. Primary structure polymorphism at amino acid residue 72 of human p53. Mol Cell Biol 1987; 7(2):961-963.
74. Walker KK, Levine AJ. Identification of a novel p53 functional domain that is necessary for efficient growth suppression. Proc Natl Acad Sci USA 1996; 93(26):15335-15340.
75. Minaguchi T, Kanamori Y, Matsushima M- et al. No evidence of correlation between polymorphism at codon 72 of p53 and risk of cervical cancer in Japanese patients with human papillomavirus 16/18 infection. Cancer Res 1998; 58(20):4585-4586.
76. Rosenthal AN, Ryan A, Al-Jehani RM et al. p53 codon 72 polymorphism and risk of cervical cancer in UK. Lancet 1998; 352(9131):871-872.
77. Thanos D, Maniatis T. Virus induction of human IFN beta gene expression requires the assembly of an enhanceosome. Cell 1995; 83(7):1091-1100.
78. Lechner MS, Laimins LA. Inhibition of p53 DNA binding by human papillomavirus E6 proteins. J Virol 1994; 68(7):4262-4273.
79. Patel D, Huang SM, Baglia LA et al. The E6 protein of human papillomavirus type 16 binds to and inhibits coactivation by CBP and p300. EMBO J 1999; 18(18):5061-5072.
80. Zimmermann H, Degenkolbe R, Bernard HU et al. The human papillomavirus type 16 E6 oncoprotein can down-regulate p53 activity by targeting the transcriptional coactivator CBP/p300. J Virol 1999; 73(8):6209-6219.
81. Hu Y, Baud V, Delhase M et al. Abnormal morphogenesis but intact IKK activation in mice lacking the IKKalpha subunit of IkappaB kinase. Science 1999; 284(5412):316-320.
82. Gewin L, Galloway DA. E box-dependent activation of telomerase by human papillomavirus type 16 E6 does not require induction of c-myc. J Virol 2001; 75(15):7198-7201.
83. Veldman T, Horikawa I, Barrett JC et al. Transcriptional activation of the telomerase hTERT gene by human papillomavirus type 16 E6 oncoprotein. J Virol 2001; 75(9):4467-4472.
84. Song S, Pitot HC, Lambert PF. The human papillomavirus type 16 E6 gene alone is sufficient to induce carcinomas in transgenic animals. J Virol 1999; 73(7):5887-5893.
85. Pim D, Storey A, Thomas M et al. Mutational analysis of HPV-18 E6 identifies domains required for p53 degradation in vitro, abolition of p53 transactivation in vivo and immortalisation of primary BMK cells. Oncogene 1994; 9(7):1869-1876.
86. Crook T, Tidy JA, Vousden KH. Degradation of p53 can be targeted by HPV E6 sequences distinct from those required for p53 binding and trans-activation. Cell 1991; 67(3):547-556.
87. Kiyono T, Hiraiwa A, Fujita M et al. Binding of high-risk human papillomavirus E6 oncoproteins to the human homologue of the Drosophila discs large tumor suppressor protein. Proc Natl Acad Sci USA 1997; 94(21):11612-11616.
88. Doyle DA, Lee A, Lewis J et al. Crystal structures of a complexed and peptide-free membrane protein-binding domain: Molecular basis of peptide recognition by PDZ. Cell 1996; 85(7):1067-1076.

89. Songyang Z, Fanning AS, Fu C et al. Recognition of unique carboxyl-terminal motifs by distinct PDZ domains. Science 1997; 275(5296):73-77.
90. Lee SS, Weiss RS, Javier RT. Binding of human virus oncoproteins to hDlg/SAP97, a mammalian homolog of the Drosophila discs large tumor suppressor protein. Proc Natl Acad Sci USA 1997; 94(13):6670-6675.
91. Woods DF, Hough C, Peel D et al. Dlg protein is required for junction structure, cell polarity, and proliferation control in Drosophila epithelia. J Cell Biol 1996; 134(6):1469-1482.
92. Bilder D, Li M, Perrimon N. Cooperative regulation of cell polarity and growth by Drosophila tumor suppressors. Science 2000; 289(5476):113-116.
93. Gardiol D, Kuhne C, Glaunsinger B et al. Oncogenic human papillomavirus E6 proteins target the discs large tumor suppressor for proteasome-mediated degradation. Oncogene 1999; 18(40):5487-5496.
94. Pim D, Thomas M, Javier R et al. HPV E6 targeted degradation of the discs large protein: Evidence for the involvement of a novel ubiquitin ligase. Oncogene 2000; 19(6):719-725.
95. Burnett AF, Barnes WA, Johnson JC et al. Prognostic significance of polymerase chain reaction detected human papillomavirus of tumors and lymph nodes in surgically treated stage IB cervical cancer. Gynecol Oncol 1992; 47(3):343-347.
96. Kurman RJ, Schiffman MH, Lancaster WD et al. Analysis of individual human papillomavirus types in cervical neoplasia: A possible role for type 18 in rapid progression. Am J Obstet Gynecol 1988; 159(2):293-296.
97. Zhang J, Rose BR, Thompson CH et al. Associations between oncogenic human papillomaviruses and local invasive patterns in cervical cancer. Gynecol Oncol 1995; 57(2):170-177.
98. Robinson WS. Molecular events in the pathogenesis of hepadnavirus-associated hepatocellular carcinoma. Annu Rev Med 1994; 45:297-323.
99. Shirakata Y, Kawada M, Fujiki Y et al. The X gene of hepatitis B virus induced growth stimulation and tumorigenic transformation of mouse NIH3T3 cells. Jpn J Cancer Res 1989; 80(7):617-621.
100. Murakami S. Hepatitis B virus X protein: Structure, function and biology. Intervirology 1999; 42(2-3):81-99.
101. Murakami S. Hepatitis B virus X protein: A multifunctional viral regulator. J Gastroenterol 2001; 36(10):651-660.
102. Elmore LW, Hancock AR, Chang SF et al. Hepatitis B virus X protein and p53 tumor suppressor interactions in the modulation of apoptosis. Proc Natl Acad Sci USA 1997; 94(26):14707-14712.
103. Lin Y, Nomura T, Yamashita T et al. The transactivation and p53-interacting functions of hepatitis B virus X protein are mutually interfering but distinct. Cancer Res 1997; 57(22):5137-5142.
104. Feitelson MA, Zhu M, Duan LX et al. Hepatitis B x antigen and p53 are associated in vitro and in liver tissues from patients with primary hepatocellular carcinoma. Oncogene 1993; 8(5):1109-1117.
105. Wang XW, Forrester K, Yeh H et al. Hepatitis B virus X protein inhibits p53 sequence-specific DNA binding, transcriptional activity, and association with transcription factor ERCC3. Proc Natl Acad Sci USA 1994; 91(6):2230-2234.
106. Ueda H, Ullrich SJ, Gangemi JD et al. Functional inactivation but not structural mutation of p53 causes liver cancer. Nat Genet 1995; 9(1):41-47.
107. Dandri M, Petersen J, Stockert RJ et al. Metabolic labeling of woodchuck hepatitis B virus X protein in naturally infected hepatocytes reveals a bimodal half-life and association with the nuclear framework. J Virol 1998; 72(11):9359-9364.
108. Zhang Z, Torii N, Furusaka A et al. Structural and functional characterization of interaction between hepatitis B virus X protein and the proteasome complex. J Biol Chem 2000; 275(20):15157-15165.
109. Lee SG, Rho HM. Transcriptional repression of the human p53 gene by hepatitis B viral X protein. Oncogene 2000; 19(3):468-471.
110. Rickinson A, Kieff E. Epstein-barr virus. In: Fields B, Knipe D, Howley P, eds. Fields Virology. Philadelphia: Lippincott-Raven Publishers, 1996:2397-2446.
111. Kieff E. Epstein-barr virus and its replication. In: Fields B, Knipe D, Howley P et al., eds. Fields Virology. Philadelphia: Lippincott-Raven Publishers; 1996:2343-2395.
112. Li Q, Spriggs MK, Kovats S et al. Epstein-barr virus uses HLA class II as a cofactor for infection of B lymphocytes. J Virol 1997; 71(6):4657-4662.
113. Liebowitz D. Epstein-barr virus and a cellular signaling pathway in lymphomas from immunosuppressed patients. N Engl J Med 1998; 338(20):1413-1421.
114. Chen W, Cooper NR. Epstein-barr virus nuclear antigen 2 and latent membrane protein independently transactivate p53 through induction of NF-kappaB activity. J Virol 1996; 70(7):4849-4853.
115. Woisetschlaeger M, Jin XW, Yandava CN et al. Role for the Epstein-barr virus nuclear antigen 2 in viral promoter switching during initial stages of infection. Proc Natl Acad Sci USA 1991; 88(9):3942-3946.

116. Lengyel P. Tumor-suppressor genes: News about the interferon connection. Proc Natl Acad Sci USA 1993; 90(13):5893-5895.
117. Yates JL, Warren N, Sugden B. Stable replication of plasmids derived from Epstein-barr virus in various mammalian cells. Nature 1985; 313(6005):812-815.
118. Rawlins DR, Milman G, Hayward SD et al. Sequence-specific DNA binding of the Epstein-barr virus nuclear antigen (EBNA-1) to clustered sites in the plasmid maintenance region. Cell 1985; 42(3):859-868.
119. Levitskaya J, Sharipo A, Leonchiks A et al. Inhibition of ubiquitin/proteasome-dependent protein degradation by the Gly-Ala repeat domain of the Epstein-barr virus nuclear antigen 1. Proc Natl Acad Sci USA 1997; 94(23):12616-12621.
120. Szekely L, Selivanova G, Magnusson KP et al. EBNA-5, an Epstein-barr virus-encoded nuclear antigen, binds to the retinoblastoma and p53 proteins. Proc Natl Acad Sci USA 1993; 90(12):5455-5459.
121. Zhang Q, Gutsch D, Kenney S. Functional and physical interaction between p53 and BZLF1: Implications for Epstein-Barr virus latency. Mol Cell Biol 1994; 14(3):1929-1938.
122. Harada S, Kieff E. Epstein-barr virus nuclear protein LP stimulates EBNA-2 acidic domain-mediated transcriptional activation. J Virol 1997; 71(9):6611-6618.
123. Szekely L, Pokrovskaja K, Jiang WQ et al. Resting B-cells, EBV-infected B-blasts and established lymphoblastoid cell lines differ in their Rb, p53 and EBNA-5 expression patterns. Oncogene 1995; 10(9):1869-1874.
124. Szekely L, Pokrovskaja K, Jiang WQ et al. The Epstein-Barr virus-encoded nuclear antigen EBNA-5 accumulates in PML-containing bodies. J Virol 1996; 70(4):2562-2568.
125. Chevallier-Greco A, Manet E, Chavrier P et al. Both Epstein-Barr virus (EBV)-encoded trans-acting factors, EB1 and EB2, are required to activate transcription from an EBV early promoter. EMBO J 1986; 5(12):3243-3249.
126. Countryman J, Miller G. Activation of expression of latent Epstein-Barr herpesvirus after gene transfer with a small cloned subfragment of heterogeneous viral DNA. Proc Natl Acad Sci USA 1985; 82(12):4085-4089.
127. Rooney CM, Rowe DT, Ragot T et al. The spliced BZLF1 gene of Epstein-Barr virus (EBV) transactivates an early EBV promoter and induces the virus productive cycle. J Virol 1989; 63(7):3109-3116.
128. Takada K, Shimizu N, Sakuma S et al. Trans activation of the latent Epstein-Barr virus (EBV) genome after transfection of the EBV DNA fragment. J Virol 1986; 57(3):1016-1022.
129. Chang YN, Dong DL, Hayward GS et al. The Epstein-Barr virus Zta transactivator: A member of the bZIP family with unique DNA-binding specificity and a dimerization domain that lacks the characteristic heptad leucine zipper motif. J Virol 1990; 64(7):3358-3369.
130. Farello GA, Cerofolini A, Zardini C et al. [Mechanical duodenopancreatectomy]. G Chir 1990; 11(3):127-128.
131. Flemington E, Speck SH. Autoregulation of Epstein-Barr virus putative lytic switch gene BZLF1. J Virol 1990; 64(3):1227-1232.
132. Kenney S, Holley-Guthrie E, Mar EC et al. The Epstein-Barr virus BMLF1 promoter contains an enhancer element that is responsive to the BZLF1 and BRLF1 transactivators. J Virol 1989; 63(9):3878-3883.
133. Lieberman PM, Hardwick JM, Hayward SD. Responsiveness of the Epstein-Barr virus NotI repeat promoter to the Z transactivator is mediated in a cell-type-specific manner by two independent signal regions. J Virol 1989; 63(7):3040-3050.
134. Urier G, Buisson M, Chambard P et al. The Epstein-Barr virus early protein EB1 activates transcription from different responsive elements including AP-1 binding sites. EMBO J 1989; 8(5):1447-1453.
135. Cayrol C, Flemington EK. The Epstein-Barr virus bZIP transcription factor Zta causes G0/G1 cell cycle arrest through induction of cyclin-dependent kinase inhibitors. EMBO J 1996; 15(11):2748-2759.
136. Rodriguez A, Armstrong M, Dwyer D et al. Genetic dissection of cell growth arrest functions mediated by the Epstein-Barr virus lytic gene product, Zta. J Virol 1999; 73(11):9029-9038.
137. Mauser A, Saito S, Appella E et al. The Epstein-Barr virus immediate-early protein BZLF1 regulates p53 function through multiple mechanisms. J Virol 2002; 76(24):12503-12512.
138. Castillo JP, Kowalik TF. Human cytomegalovirus immediate early proteins and cell growth control. Gene 2002; 290(1-2):19-34.
139. Speir E, Modali R, Huang ES et al. Potential role of human cytomegalovirus and p53 interaction in coronary restenosis. Science 1994; 265(5170):391-394.

140. Tanaka K, Zou JP, Takeda K et al. Effects of human cytomegalovirus immediate-early proteins on p53-mediated apoptosis in coronary artery smooth muscle cells. Circulation 1999; 99(13):1656-1659.
141. Tsai HL, Kou GH, Chen SC et al. Human cytomegalovirus immediate-early protein IE2 tethers a transcriptional repression domain to p53. J Biol Chem 1996; 271(7):3534-3540.
142. Park J, Seo T, Hwang S et al. The K-bZIP protein from Kaposi's sarcoma-associated herpesvirus interacts with p53 and represses its transcriptional activity. J Virol 2000; 74(24):11977-11982.
143. Doniger J, Muralidhar S, Rosenthal LJ. Human cytomegalovirus and human herpesvirus 6 genes that transform and transactivate. Clin Microbiol Rev 1999; 12(3):367-382.
144. Chen H, Lee JM, Wang Y et al. The Epstein-Barr virus latency BamHI-Q promoter is positively regulated by STATs and Zta interference with JAK/STAT activation leads to loss of BamHI-Q promoter activity. Proc Natl Acad Sci USA 1999; 96(16):9339-9344.
145. Dreyfus DH, Nagasawa M, Kelleher CA et al. Stable expression of Epstein-Barr virus BZLF-1-encoded ZEBRA protein activates p53-dependent transcription in human Jurkat T-lymphoblastoid cells. Blood 2000; 96(2):625-634.
146. Mocarski E, Courcelle C. Cytomegaloviruses and their replication. In: Knipe D, Howley P, Griffin D, Lamb R, Martin M, Straus S, eds. Fields Virology. Philadelphia: Lippincott Williams & Wilkins, 2001:2629-2673.
147. Pass R. Cytomegalovirus. In: Knipe D, Howley P, Griffin D, Lamb R, Martin M, Straus S, eds. Fields Virology. Philadelphia: Lippincott Williams & Wilkins, 2001:2675-2705.
148. Zhou YF, Leon MB, Waclawiw MA et al. Association between prior cytomegalovirus infection and the risk of restenosis after coronary atherectomy. N Engl J Med 1996; 335(9):624-630.
149. Geder KM, Lausch R, O'Neill F et al. Oncogenic transformation of human embryo lung cells by human cytomegalovirus. Science 1976; 192(4244):1134-1137.
150. Albrecht T, Rapp F. Malignant transformation of hamster embryo fibroblasts following exposure to ultraviolet-irradiated human cytomegalovirus. Virology 1973; 55(1):53-61.
151. Boldogh I, Gonczol E, Vaczi L. Transformation of hamster embryonic fibroblast cells by UV-irradiated human cytomegalovirus. Acta Microbiol Acad Sci Hung 1978; 25(4):269-275.
152. Bonin LR, McDougall JK. Human cytomegalovirus IE2 86-kilodalton protein binds p53 but does not abrogate G1 checkpoint function. J Virol 1997; 71(8):5861-5870.
153. Hagemeier C, Caswell R, Hayhurst G et al. Functional interaction between the HCMV IE2 transactivator and the retinoblastoma protein. EMBO J 1994; 13(12):2897-2903.
154. Sommer MH, Scully AL, Spector DH. Transactivation by the human cytomegalovirus IE2 86-kilodalton protein requires a domain that binds to both the TATA box-binding protein and the retinoblastoma protein. J Virol 1994; 68(10):6223-6231.
155. Choi KS, Kim SJ, Kim S. The retinoblastoma gene product negatively regulates transcriptional activation mediated by the human cytomegalovirus IE2 protein. Virology 1995; 208(2):450-456.
156. Fortunato EA, Sommer MH, Yoder K et al. Identification of domains within the human cytomegalovirus major immediate-early 86-kilodalton protein and the retinoblastoma protein required for physical and functional interaction with each other. J Virol 1997; 71(11):8176-8185.
157. Poma EE, Kowalik TF, Zhu L et al. The human cytomegalovirus IE1-72 protein interacts with the cellular p107 protein and relieves p107-mediated transcriptional repression of an E2F-responsive promoter. J Virol 1996; 70(11):7867-7877.
158. Johnson RA, Yurochko AD, Poma EE et al. Domain mapping of the human cytomegalovirus IE1-72 and cellular p107 protein-protein interaction and the possible functional consequences. J Gen Virol 1999; 80(Pt 5):1293-1303.
159. Hupp TR, Sparks A, Lane DP. Small peptides activate the latent sequence-specific DNA binding function of p53. Cell 1995; 83(2):237-245.
160. Nikolova PV, Henckel J, Lane DP et al. Semirational design of active tumor suppressor p53 DNA binding domain with enhanced stability. Proc Natl Acad Sci USA 1998; 95(25):14675-14680.
161. Koutsky LA, Ault KA, Wheeler CM et al. A controlled trial of a human papillomavirus type 16 vaccine. N Engl J Med 2002; 347(21):1645-1651.

Chapter 12

p53 and Immunity

Vikram Narayan, Sarah E. Meek and Kathryn L. Ball*

Introduction

Since its discovery in 1979, many different roles for the tumor suppressor protein p53 in tumorigenesis have been described. Correct p53 function is required for proper regulation of cell division, apoptosis, senescence, and the responses to cellular stresses such as DNA damage and hypoxia. Indeed, mutations in p53 are observed in as many as 50% of human cancers.[1] However, recent reports have highlighted an emerging role for p53 in anti-viral immunity. This chapter reviews the available literature on p53 and the body's immune response, and how p53 may link immunity and cancer.

Overview of the Immune System

The immune system comprises an innate and an adaptive component.[2] Upon exposure to pathogens or foreign molecules (antigens), the innate response is immediately triggered. If the innate component is unable to clear the pathogen and infection persists, the adaptive response is activated.

The innate immune system comprises structural barriers against infection, such as the skin, and phagocytic cells such as macrophages and neutrophils. The innate response has been therefore believed to be largely nonspecific. However, specialized cell surface receptors, toll-like receptors (TLRs), respond to particular pathogen-derived molecules, such as lipopolysaccharides, CpG deoxynucleotides and double-stranded (ds) RNA (reviewed in ref. 3). Viral dsRNA is also recognized by cytoplasmic receptors such as RIG-I. Overall, particular pathogen-associated molecular patterns (PAMPs) will generate particular patterns of signaling pathway activation. This leads to the production of Type I interferons and pro-inflammatory cytokines. Interestingly, p53 appears to be involved in the TLR response, as described later in this chapter.

The adaptive component is the "specific" response of the immune system. It comprises T and B lymphocytes, and antigen presenting cells (APCs). During development these lymphocytes undergo a series of gene rearrangements and selections, culminating in the expression of a unique cell surface receptor (T cell receptor or B cell receptor ('antibody')). APCs internalize antigenic molecules, process them and express antigens on their surface with molecules known as MHCs (major histocompatibility complex). T cells, through their specific T cell receptors, recognize these MHC-antigen complexes, and initiate the cell-mediated response. Activated T cells give rise to various T cell sub-groups having distinct functions, including cytotoxic T_C cells, which secrete toxic molecules to kill pathogens, and helper T_H cells, which secrete cytokines including IL-4 (T_H2 cells) and Type II interferons (IFNγ, T_H1 cells). B cells, through their B cell receptor recognize and bind to free antigens that are not usually presented by MHCs. The binding of a B cell to an antigen causes it to divide rapidly, generating clones secreting the same

*Corresponding Author: Kathryn L. Ball—Cell Signalling Unit, Division of Cancer Biology, IGMM, University of Edinburgh, Crewe Rd South, Edinburgh, EH4 2XR, UK.
Email: kathryn.ball@ed.ac.uk

p53, edited by Ayeda Ayed and Theodore Hupp. ©2010 Landes Bioscience and Springer Science+Business Media.

specific antibody. T_H cells are required for this process. Adaptive immune cells which remain in the bloodstream give the immune system a component of 'memory'.

Recently, it has become clear that the innate and adaptive immune systems are not entirely distinct. For instance, TLRs link the two systems: TLRs are required both for activation of T_H cells,[4] and for B-cell activation.[5]

p53 in the Antiviral Response

In addition to its role in ensuring genome stability as a key activator of the cells DNA damage responses, the tumor suppressor protein p53 appears to play a role in the response to viral infections. It is well-known that p53 is targeted by many oncogenic viruses. Examples of this are human Papillomavirus protein E6 binding to and inactivating p53 and SV40 cellular transformation where p53 is inactivated by large T-antigen,[6,7] both mechanisms allow host cell survival and proliferation, which correlates with tumorigenesis. However, interestingly, the anti-viral role of p53 is not limited to tumor viruses and this is highlighted by the fact that many viruses have evolved proteins that bind to and inactivate or degrade p53, highlighting the role of p53 in general defense, in addition to protection against cancer.

p53 Cooperation with IFNα/β

The first link between p53 and the general antiviral response was demonstrated by Takaoka et al.[8] The authors showed that treating cells with Interferon (IFN) α and β caused an increase in both p53 mRNA and protein levels. This induction was mediated by Interferon Stimulated Gene Factor 3 (ISGF3), a heterotrimer comprising Stat1 (signal transducer and activator of transcription factor), Stat2 and IRF-9 (interferon regulatory factor 9), binding to interferon-stimulated response element (ISRE) consensus sequences in the p53 promoter.

Interestingly, this induction did not correlate with activating phosphorylations or production of p53 target genes. Therefore, IFN induction of p53 may sensitize cells to subsequent stresses, but may not by itself be sufficient to fully activate the p53 pathway. In addition to direct activation by the Type 1 interferons p53 can also be indirectly activated by IFIXα-1 an IFN-inducible protein which inhibits MDM2-dependent degradation of p53.[9] Further, p53 has been shown to cooperate with a number of IFN-inducible proteins to regulate gene expression including STAT1, IRF-1 and PML.[10-12]

p53 Protects against Virus Infection

In addition to a role in the response to DNA tumor viruses such as SV40, human papillomavirus, Kaposi's sarcoma herpesvirus and adenivirus,[7] p53 function is also activated by nontransforming virus such as the DNA virus HHV-6B and RNA viruses such as influenza virus, poliovirus and vesicular somatitis virus (VSV). One of the ways in which cells respond to viral infection is by activating signaling pathways that initiate a protective programme leading to cell death, thus limiting the spread of the viral infection.[13] The role of p53 in viral infection has been linked to this cell death arm of the anti-viral response. For example, using MEFs (mouse embryonic fibroblasts) from 'super' p53 mice which exhibit enhanced cancer resistance,[14] a 10-fold decrease in virus yield is seen in VSV infected cells when compared to MEFs from wild-type animal.[9] This corresponds with enhanced apoptosis of infected super p53 MEFs, which is likely to be p53-mediated: the proapoptotic p53 target gene *Puma* was upregulated in these cells, while *p21*, a p53 cell cycle arrest target, was not. Studies on influenza virus also supported the link between p53 status and virus induced cell death as dominant negative p53 inhibited cell death in influenza infected lung cell lines.[15] More recently, it has been suggested that p53 is activated at a much earlier stage in the infection process suggesting that p53-mediated gene expression plays a role in the early as well as the later stages of viral infection. Thus, influenza infection produces a bi-phasic p53 response with an early peak in protein levels 6-8 hours post infection that is not associated with viral replication and which also occurs in the presence of UV-inactivated viral particles.[16] At later time points a second p53 peak is observed which has been shown to require live replication competent virus.[15]

Mechanisms of p53 Induction in Response to Viral Infection

As discussed above, the presence of p53 in cells appears to give them an advantage in fighting viral infection. This section attempts to summarize some of the responses that activate p53 upon viral infection.

The E2F family of transcription factors function primarily in regulating cell cycle progression, particularly through the G1/S phase. Of the eight members, E2F1 is unique in that in addition to promoting cell cycle progression, it is also involved in inducing apoptosis. E2F1-mediated apoptosis may be p53-dependent or p53-independent.[17] The tumor suppressor protein retinoblastoma (RB) regulates G1/S progression by binding to E2Fs and inhibiting their function. Loss of the RB protein thus leads to uncontrolled cell division, giving rise to malignancies (reviewed in ref. 18).

Upon infection by certain types of virus, E2Fs are released from an inactive complex with retinoblastoma, thereby inducing cell division.[19] However, this deregulated growth results in the E2F induction of the alternative reading frame protein (ARF). The tumor suppressor ARF has been shown to have antiviral properties independent of p53.[20] Like p53, ARF is often mutated in human cancers.[21] Interestingly, while p53 responds to viral infection by inducing apoptosis, ARF appears to respond by increasing cell viability through reduced production of viral progeny.[20] ARF may also respond to viruses independently of E2Fs: ARF is induced by both virus infection and IFNβ, and its promoter contains an interferon-responsive ISRE element.[20]

ARF induction in response to virus infection induces p53.[20] This may occur in two ways: on the one hand, ARF binds to nucleophosmin (NPM), a protein frequently over-expressed in cancers, thereby relieving the NPM-mediated repression of PKR (protein kinase, dsRNA dependent).[22] PKR in turn is involved in shutting down protein synthesis in the cell, and, incidentally, plays a role in the activation of p53. The role of PKR in the dsRNA antiviral response is described later in this chapter.

Secondly, the oncoprotein MDM2 is an E3 ligase belonging to the RING family, that ubiquitinates p53 and targets it for degradation. ARF induction inhibits MDM2-mediated degradation of p53.[23] Thus, by inhibiting the degradation of p53, ARF expression results in increased levels of the p53 protein, which can then enhance the antiviral response, for example by inducing apoptosis.

Interestingly, E2F1 itself binds p53 and has been shown to enhance p53 mediated apoptosis, albeit in response to DNA damage.[24] In addition, E2F1 can induce p53 even in the absence of ARF, by activating the ATM (ataxia telangiectasia mutated) signaling pathway.[25,26] In response to viral infection, however, E2F1-mediated p53-induced apoptosis has only been demonstrated in the presence of ARF. It is interesting to speculate whether viral DNA/RNAs might induce the cellular DNA damage response.

The polypeptide cytokine TNFα (tumor necrosis factor alpha) is induced by a variety of molecules including other cytokines, bacteria, and viruses.[27] In addition to its role in necrosis, TNFα has been shown to have a variety of other functions—it promotes cell survival through NF-κB (nuclear factor for κ chain of B cells), and causes cell death through apoptosis and necrosis.[28-30] Interestingly, TNFα has been shown to induce p53, albeit transiently.[31] TNFα also induces PKR, an antiviral kinase (see below). The induction of p53 by TNFα lagged behind the induction of PKR, suggesting that p53 activation might be downstream of PKR activation in response to TNFα induction. Consistent with this model, PKR has been shown to phosphorylate IκB (inhibitor of NF-κB), though recent evidence seems to suggest that this phosphorylation is indirect.[32,33] IκB binds to NF-κB and retains it in the cytoplasm, preventing it from functioning as a transcription factor. Phosphorylation of IκB signals for its ubiquitination and subsequent degradation, blocking its inhibition of NF-κB, and thus allowing NF-κB to move into the nucleus. NF-κB has been shown to directly induce p53 transcription, and is also required for the induction of IFNβ.[34,35]

As mentioned previously p53 gene expression can also be directly activated by Type I IFN through an IRSE in the p53 promoter or through the inhibition of MDM2 mediated degradation.[8,9] However, evidence exists to suggest that post-translational modification is also an important step on the pathway to activation of p53 in response to viral infection. For example, in cells infected with polio virus, activation of p53-dependent gene expression is dependent on phosphorylation at the Ser[15] site. Thus, although total p53 protein levels decrease shortly after infection, the protein is rapidly phosphrylated at Ser[15] and phosphorylaiton is accompanied by the PML-dependent recruitment of p53 to PML nuclear bodies.[11] Takaoka et al suggest that activation of p53 in viral infected cells may be dependent on the activity of the ATM kinase.[8] Better characterized is the post-translational activation of p53 in response to HHV-6B, a DNA virus that activates a growth arrest rather than an apoptotic response.[36] HHV-6B infection leads to the phosphorylaiton of p53 at multiple sites including the well characterized activating sites of Ser[15], Ser[20] and Ser[392].[37] Furthermore the CK1α kinase has recently been shown to phosphorylate the Ser[20] site in response to HHV-6B infection.[38]

Once activated, p53 has been shown to induce apoptosis in two ways: through the release of cytochrome c from the mitochondria, resulting in the formation of the apoptosome (intrinsic pathway), and through the activation of death receptors (extrinsic pathway). Both pathways ultimately lead to the activation of caspases, which mediate apoptosis.[39] Thus, it is likely that, upon viral infection, TNFα is induced, which in turn induces PKR. This PKR phosphorylates IκB, causing its degradation. This frees NF-κB to move into the nucleus and activate transcription of *p53* and other NF-κB responsive genes. Induced p53 becomes modified post-translationally and ultimately causes caspase-dependent apoptosis.

p53 and the dsRNA Response

dsRNA (double stranded RNA) is a by-product of transcription in most viruses. In the case of RNA viruses, during RNA replication for viral proliferation, the formation of these dsRNA molecules is essential. However, some DNA viruses also synthesize dsRNA—since ORFs (open reading frame) in some viruses are present in opposite orientations, overlapping mRNA transcripts are formed, which have the potential to then fold into dsRNA molecules.[22] Treatment of cells with synthetic dsRNAs such as polyinosine-polycytidylic acid (pIpC) is commonly used to mimic the dsRNA synthesized during viral replication. Cells posess several different dsRNA sensors, including the cytoplasmic kinase PKR (protein kinase, dsRNA dependent), the cytosolic adaptor proteins Rig-I and mda5, and toll-like receptors (TLRs) at the cell surface and endosomal membrane.

PKR is activated by direct binding to dsRNA, which relieves its autoinhibition.[40] PKR is also induced by Interferons (ref.41 and references therein). Notable among its targets is the translation initiation factor eIF2α (eukaryotic initiation factor alpha). As its name suggests, eIF2α is essential for the initiation of translation, and when phosphorylated, is unable to function.[42] As a result, protein synthesis is arrested.

In addition to its role in arresting protein synthesis, PKR has also been implicated in apoptosis (reviewed in ref. 22). As discussed above, through molecules like TRAF, PKR activates the IKKα/IKKβ complex, which in turn phosphorylates IκB. Phosphorylated IκB is unable to bind NF-κB, which can then move into the nucleus and activate its target genes, including genes involved in apoptosis. PKR is also involved in the release of cytochrome c from the mitochondria and the subsequent formation of the apoptosome, through the activation of FADD, which triggers the activation of caspase-8. PKR$^{-/-}$ mice show impaired antiviral responses and impaired stress-induced apoptosis.[43,44]

PKR has been shown to physically interact with p53 both in vitro and in coimmunoprecipitation experiments. Endogenous p53 was coimmunoprecipitated with endogenous PKR in HeLa S3 cells, and this coimmunoprecipitation was enhanced by IFNβ.[45] The authors also provide evidence for better interaction between the two proteins (the N terminus of PKR with the C terminus of p53) when p53 is in its wild-type conformation. Furthermore, PKR was able to phosphorylate p53 in vitro at serine 392. The consequences of p53 phosphorylation at S^{392} are as yet unknown. Given as phosphorylation of p53 at its N-terminus has been implicated in its activity as a transcription factor, the effect of PKR on S^{15} p53 phosphorylation was studied.[45] While PKR was unable to phosphorylate p53 at S^{15} either in vivo or in vitro , it was observed that basal S^{15} p53 levels were consistently higher in $PKR^{+/+}$ cells than in $PKR^{-/-}$ cells. DNA damage studies revealed that S^{15} p53 phosphorylation upon DNA damage was higher in $PKR^{+/+}$ cells than in $PKR^{-/-}$ cells. Moreover, the specific PI-3 kinase inhibitor LY294002 inhibited DNA-damage-induced S^{15} p53 phosphorylation to a higher extent in $PKR^{+/+}$ cells than in $PKR^{-/-}$ cells. Collectively, these results led the authors to believe that PKR is involved in p53 phosphorylation at S^{15}, albeit not directly, but instead possibly through a member of the PI-3 kinase family. Known PI-3 kinase family members that phosphorylate p53 include ATM (at S^{15}; refs. 46, 47), DNA-PK (at S^{15}; refs. 48, 49) and casein kinase I (CKI; multiple sites in the N terminus of p53; ref. 50). It is therefore possible that any of these kinases is involved in the PKR-induced phosphorylation of p53 at S^{15}.

As described earlier in this chapter, p53 has been shown to be involved in the antiviral response against the nononcogenic virus VSV. Upon infection with VSV, proapoptotic p53 target genes were shown to be activated, suggesting that p53 functions in the antiviral response by inducing apoptosis. However, recently, it was demonstrated that in the absence of mechanisms that can activate p53, some nononcogenic viruses such as encephalomyocarditis virus (EMCV) and human parainfluenza virus Type 3 (HPIV3) induce a down regulation of p53.[51] In the same paper, dsRNA was shown to induce a similar reduction of p53 protein levels. Surprisingly, in this study, the down regulation of p53 marked the onset of dsRNA induced apoptosis, and p53 knockout cells were shown to have an increased sensitivity to dsRNA induced apoptosis, implying that p53 inhibits antiviral responses. The half life of p53, and its binding to the ubiquitin E3 ligase MDM2 (which leads to proteasome-mediated degradation), remained unchanged upon dsRNA treatment. Because dsRNA is known to activate PKR, the authors speculate that the down-regulation of p53 is due to the inhibition of protein synthesis by PKR mediated phosphorylation of eIF2α, and the degradation of the already synthesized protein by the ubiquitin proteasome machinery, which is unaffected (MDM2 binding to p53 was unaffected by dsRNA). How specific this down-regulation is for p53 remains to be established. Nonetheless, it is worth noting that the authors demonstrate how p53 responds differently to different viruses—in its absence, VSV replication was enhanced, while EMCV and HPIV3 replication was hindered.

dsRNA also induces the expression of several interferon regulatory factors (IRFs), via toll-like receptor 3 (see below) and Rig-I/Mda5 mediated pathways.[52] The IRFs then activate various target genes including proapoptotic genes and IFNs, and play a critical role in the antiviral response.[53] In one of many examples of feedback loops in anti-viral signaling, IRF-1 is key in PKR induction by IFNβ.[54]

The toll-like receptor (TLR) proteins are a family of 11 transmembrane signaling proteins that respond to a variety of biomolecules present in invading pathogens.[3] The 11 family members respond to different ligands, giving the innate immune response an element of specificity thus far attributed only to the adaptive response. Of the TLRs, TLR3 recognizes dsRNA (reviewed in ref. 55). dsRNA binds to the N terminus of TLR3 in the endosomal lumen (and on the cell surface in some cell types ref. 56), and activates its C-terminal signaling domain residing in the cytosol. The TLR3 signaling cascade is mediated by the adaptor protein TRIF, and activates downstream targets including IRF-1 and NF-κB, which respond by the activation of genes involved in apoptosis and defense (reviewed in ref. 52). The role of TLR3 in the antiviral response is controversial—while studies have shown that TLR3 is induced upon infection by some viruses and is required for the prevention of systemic infection,[57] other studies have demonstrated that TLR3 is not required for the response to infection by several viruses,[58] possibly due to the presence of multiple redundant antiviral signaling pathways. In vitro transcribed mRNA and endogenous mRNA from necrotic cells have been shown to activate TLR3 extracellularly,[59] suggesting that TLR3 may have physiological roles in addition to a role in the response to viral infection, such as detection of self dsRNA released from dying cells within a tumor.

There is in fact some evidence to suggest that p53 itself is important for normal TLR responses. Data from the Ball group has identified a link between p53 and IRF-1 in response to dsRNA (Eckert and Ball, unpublished observation) and this is supported by a study showing that TLR3 has a p53[60] responsive element in its promoter. This data suggests that p53 associates directly with the TLR3 promoter, and is responsible for regulating basal levels of TLR3. In the absence of p53, the TLR3 mediated dsRNA response was not activated.

p53 and Interferon Signaling

IFN and dsRNA signaling pathways converge in the regulation of ISG15 (IFN stimulated gene with 15kDa product). ISG15 is an ubiquitin-like protein that is induced by Type I IFNs.[61] p53 has been shown to induce ISG15 in TS (temperature sensitive) HeLa cells.[62] Interestingly, while dsRNA induced gene transcription was shown to be dependent on p53 (p53 null cells showed a weakened response), the induction of ISG15 by the Newcastle disease virus (NDV) and EMCV was shown to be independent of p53 status. Thus, in addition to a p53 dependent dsRNA mediated signal, a p53-independent viral signal is also generated in virus-infected cells. A study on influenza virus lends further support for a p53 role in the regulation of interferon responsive genes.[15] This study shows that an ISRE reporter was induced 14-fold following infection of A549 cells, whereas no activation was seen in the presence of a p53 dominant negative construct. Thus, in A549 cells viral activation of genes through the ISRE is dependent on p53.[15]

Additional support comes for studies on the HPV E6 protein. E6 in complex with E6AP acts as an E3-ligase that targets p53 for degradation in infected cells. Microarray analysis of E6 expressing cells showed that a cluster of interferon responsive genes were down regulated by E6 and a high proportion of these were also down regulated by dominant negative p53.[63]

An additional potential mechanism of p53 involvement in the antiviral response is through interaction with the classical STAT (signal transducer and activator of transcription) mediated interferon signaling pathways.[64] Firstly, IFN-stimulated gene production was attenuated in cells with a mutant compared to wild-type p53.[65] In these cells, p53 and IRF9, a component of the ISGF3 complex, physically interact. Thus, both dsRNA and IFN response pathways may account for the inhibition of Hepatitis C virus replication seen in the p53-deficient cells. More recently, it has been demonstrated that IRF-9 is under the transcriptional control of p53 in cells infected with virus or treated with IFN/dsRNA leading to an overall enhancement of the IFN response.[66]

Secondly, another potential point of interaction between antiviral signaling pathways and p53 is through STATs themselves. STATs are crucial mediators of IFN signaling, and STAT1 and p53 physically interact. STAT1 activation by genotoxic agents is p53-dependent; conversely, STAT1 stabilizes p53.[67]

Thus, p53 appears to be intricately linked to the dsRNA and viral response—not only does it regulate the expression of sensors of viral infection (TLR3), but in addition, it is itself activated by antiviral signaling, and plays a role in the apoptosis of infected cells. At the same time, one study demonstrated a downregulation of p53 in response to certain viruses and dsRNA, suggesting that the p53 mediated antiviral response may be activated only in response to some viruses.

Opposing Roles of the Immune System in Carcinogenesis

There is a causal link between oncogenic viruses and cancer. Here, p53's anti-viral responses will have a tumor suppressive function. Additionally, there is also a more general role of immune system activation in carcinogenesis. Inflammatory conditions predispose to multiple tumor types, and immune suppression correlates with decreased cancer risk even for tumors that are not associated with oncogenic viruses.[68] This is thought to be due to the production of proinflammatory mediators by innate immune cells, including IL6, TNFα, growth factors, and also angiogenic factors and matrix proteases, which together promote tumor initiation, promotion and progression.[69] This is perhaps best-characterized for the

NFκB signaling pathways described above. In this scenario, p53's functions in the innate immune response could play a role in tumor promotion.

In contrast, the recent demonstration of cells responsible for tumor dormancy[70] supports the long-held notion that immunosurveillance functions as an anti-cancer mechanism. In this mouse model of tumor dormancy, T-cells inhibit tumor growth; immunosuppression allowed tumors to form. Whether p53 is involved in this immunosurveillance function of the adaptive immune system is not known.

p53's Anti-Viral Response and Immunotherapy

In addition to the classical approach of attempting to restore p53 function in p53-deficient tumor cells to halt proliferation or induce apoptosis, virus responses of p53 may provide an additional route to therapy. Oncolytic viruses that replicate specifically within a tumor and kill infected and neighbouring tumor cells, such as VSV, may prove to be useful anti-cancer agents.[71] Adenoviruses lacking proteins that normally inactivate p53 have been engineered; in theory, these should replicate selectively in p53-deficient cancer cells. Indeed, one such virus is increased in p53-defective cells,[72] although other groups carrying out similar experiments have not observed selectivity for p53 defects.[71]

Conclusion—An Evolutionary Perspective

p53 is a well-known tumor suppressor protein. However, novel roles are emerging for p53 in multiple aspects of the immune response, particularly in antiviral immunity. Although the immune system does play a general role in both protection against and promotion of cancer, these novel roles of p53 are not limited to its tumor suppressor functions. Indeed, p53 protects against both oncogenic and nontumorigenic viruses. Thus, antiviral responses may be another example of p53's crucial role in coordinating cellular responses to stress. One might speculate that p53 initially evolved to evade viruses, preventing host cell proliferation and inducing host cell apoptosis, and only later acquired a tumor suppressor function, through responses to cellular DNA damage, hypoxia, and other stresses involved in tumor initiation and development.

References

1. Pfeifer GP, Holmquist GP. Mutagenesis in the P53 gene. Biochim Biophys Acta 1997; 1333(1):M1-8.
2. Kindt T, Osborne B, Goldsby R. Kuby Immunology. 6th Edition. W.H.Freeman, 2006.
3. Takeda K, Akira S. Toll-like receptors in innate immunity. Int Immunol 2005; 17(1):1-14.
4. Imanishi T, Hara H, Suzuki S et al. Cutting edge: TLR2 directly triggers Th1 effector functions. J Immunol 2007; 178(11):6715-9.
5. Pasare C, Medzhitov R. Control of B-cell responses by Toll-like receptors. Nature 2005; 438(7066):364-8.
6. Scheffner M, Werness BA, Huibregtse JM et al. The E6 oncoprotein encoded by human papillomavirus types 16 and 18 promotes the degradation of p53. Cell 1990; 63(6):1129-36.
7. Levine AJ. The common mechanisms of transformation by the small DNA tumor viruses: The inactivation of tumor suppressor gene products: p53. Virology 2009; 384(2):285-93.
8. Takaoka A, Hayakawa S, Yanai H et al. Integration of interferon-alpha/beta signalling to p53 responses in tumor suppression and antiviral defence. Nature 2003; 424(6948):516-23.
9. Ding Y, Lee JF, Lu H et al. Interferon-inducible protein IFIXalpha1 functions as a negative regulator of HDM2. Mol Cell Biol 2006; 26(5):1979-96.
10. Dornan D, Eckert M, Wallace M et al. Interferon regulatory factor 1 binding to p300 stimulates DNA-dependent acetylation of p53. Mol Cell Biol 2004; 24(22):10083-98.
11. Pampin M, Simonin Y, Blondel B et al. Cross talk between PML and p53 during poliovirus infection: implications for antiviral defense. J Virol 2006; 80(17):8582-92.
12. Townsend PA, Scarabelli TM, Davidson SM et al. STAT-1 interacts with p53 to enhance DNA damage-induced apoptosis. J Biol Chem 2004; 279(7):5811-20.
13. Barber GN. Host defense, viruses and apoptosis. Cell Death Differ 2001; 8(2):113-26.
14. Garcia-Cao I, García-Cao M, Martín-Caballero J et al. "Super p53" mice exhibit enhanced DNA damage response, are tumor resistant and age normally. EMBO J 2002; 21(22):6225-35.
15. Turpin E, Luke K, Jones J et al. Influenza virus infection increases p53 activity: role of p53 in cell death and viral replication. J Virol 2005; 79(14):8802-11.

16. Shen Y, Wang X, Guo L et al. Influenza A virus induces p53 accumulation in a biphasic pattern. Biochem Biophys Res Commun 2009; 382(2):331-5.
17. Bell LA, Ryan KM. Life and death decisions by E2F-1. Cell Death Differ 2004; 11(2):137-42.
18. Classon M, Harlow E. The retinoblastoma tumor suppressor in development and cancer. Nat Rev Cancer 2002; 2(12):910-7.
19. Swanton C, Jones N. Strategies in subversion: de-regulation of the mammalian cell cycle by viral gene products. Int J Exp Pathol 2001; 82(1):3-13.
20. Garcia MA, Collado M, Muñoz-Fontela C et al. Antiviral action of the tumor suppressor ARF. EMBO J 2006; 25(18):4284-92.
21. Sharpless NE, DePinho RA. The INK4A/ARF locus and its two gene products. Curr Opin Genet Dev 1999; 9(1):22-30.
22. Garcia MA, Gil J, Ventoso I et al. Impact of protein kinase PKR in cell biology: from antiviral to antiproliferative action. Microbiol Mol Biol Rev 2006; 70(4):1032-60.
23. Honda R, Yasuda H. Association of p19(ARF) with Mdm2 inhibits ubiquitin ligase activity of Mdm2 for tumor suppressor p53. EMBO J 1999; 18(1):22-7.
24. Hsieh JK, Yap D, O'Connor DJ et al. Novel function of the cyclin A binding site of E2F in regulating p53-induced apoptosis in response to DNA damage. Mol Cell Biol 2002; 22(1):78-93.
25. Powers JT, Hong S, Mayhew CN et al. E2F1 uses the ATM signaling pathway to induce p53 and Chk2 phosphorylation and apoptosis. Mol Cancer Res 2004; 2(4):203-14.
26. Hong S, Paulson QX, Johnson DG. E2F1 and E2F3 activate ATM through distinct mechanisms to promote E1A-induced apoptosis. Cell Cycle 2008; 7(3):391-400.
27. Tracey KJ, Cerami A. Tumor necrosis factor: a pleiotropic cytokine and therapeutic target. Annu Rev Med 1994; 45:491-503.
28. Van Antwerp DJ, Martin SJ, Kafri T et al. Suppression of TNF-alpha-induced apoptosis by NF-kappaB. Science 1996; 274(5288):787-9.
29. Wang CY, Mayo MW, Baldwin AS Jr. TNF- and cancer therapy-induced apoptosis: potentiation by inhibition of NF-kappaB. Science 1996; 274(5288):784-7.
30. Tartaglia LA, Weber RF, Figari IS et al. The two different receptors for tumor necrosis factor mediate distinct cellular responses. Proc Natl Acad Sci U S A 1991; 88(20):9292-6.
31. Yeung MC, Lau AS. Tumor suppressor p53 as a component of the tumor necrosis factor-induced, protein kinase PKR-mediated apoptotic pathway in human promonocytic U937 cells. J Biol Chem 1998; 273(39):25198-202.
32. Kumar A, Haque J, Lacoste J et al. Double-stranded RNA-dependent protein kinase activates transcription factor NF-kappa B by phosphorylating I kappa B. Proc Natl Acad Sci U S A 1994; 91(14):6288-92.
33. Leaman DW, Salvekar A, Patel R et al. A mutant cell line defective in response to double-stranded RNA and in regulating basal expression of interferon-stimulated genes. Proc Natl Acad Sci U S A 1998; 95(16):9442-7.
34. Wu H, Lozano G. NF-kappa B activation of p53. A potential mechanism for suppressing cell growth in response to stress. J Biol Chem 1994; 269(31):20067-74.
35. Cheng G, Nazar AS, Shin HS et al. IP-10 gene transcription by virus in astrocytes requires cooperation of ISRE with adjacent kappaB site but not IRF-1 or viral transcription. J Interferon Cytokine Res 1998; 18(11):987-97.
36. Oster B, Kaspersen MD, Kofod-Olsen E et al. Human herpesvirus 6B inhibits cell proliferation by a p53-independent pathway. J Clin Virol 2006; 37(Suppl 1):S63-8.
37. Oster B, Bundgaard B, Hupp TR et al. Human herpesvirus 6B induces phosphorylation of p53 in its regulatory domain by a CK2- and p38-independent pathway. J Gen Virol 2008; 89(Pt 1):87-96.
38. MacLaine NJ, Oster B, Bundgaard B et al. A central role for CK1 in catalyzing phosphorylation of the p53 transactivation domain at serine 20 after HHV-6B viral infection. J Biol Chem 2008; 283(42):28563-73.
39. Haupt S, Berger M, Goldberg Z et al. Apoptosis—the p53 network. J Cell Sci 2003; 116(Pt 20):4077-85.
40. Clemens MJ. PKR—a protein kinase regulated by double-stranded RNA. Int J Biochem Cell Biol 1997; 29(7):945-9.
41. Zhou A, Paranjape JM, Der SD et al. Interferon action in triply deficient mice reveals the existence of alternative antiviral pathways. Virology 1999; 258(2):435-40.
42. Williams BR. PKR; a sentinel kinase for cellular stress. Oncogene 1999; 18(45):6112-20.
43. Yang YL, Reis LF, Pavlovic J et al. Deficient signaling in mice devoid of double-stranded RNA-dependent protein kinase. EMBO J 1995; 14(24):6095-106.
44. Der SD, Yang YL, Weissmann C et al. A double-stranded RNA-activated protein kinase-dependent pathway mediating stress-induced apoptosis. Proc Natl Acad Sci U S A 1997; 94(7):3279-83.

45. Cuddihy AR, Wong AH, Tam NW et al. The double-stranded RNA activated protein kinase PKR physically associates with the tumor suppressor p53 protein and phosphorylates human p53 on serine 392 in vitro. Oncogene 1999; 18(17):2690-702.
46. Banin S, Moyal L, Shieh S et al. Enhanced phosphorylation of p53 by ATM in response to DNA damage. Science 1998; 281(5383):1674-7.
47. Canman CE, Lim DS, Cimprich KA et al. Activation of the ATM kinase by ionizing radiation and phosphorylation of p53. Science 1998; 281(5383):1677-9.
48. Shieh SY, Ikeda M, Taya Y et al. DNA damage-induced phosphorylation of p53 alleviates inhibition by MDM2. Cell 1997; 91(3):325-34.
49. Siliciano JD, Canman CE, Taya Y et al. DNA damage induces phosphorylation of the amino terminus of p53. Genes Dev 1997; 11(24):3471-81.
50. Steegenga WT, van der Eb AJ, Jochemsen AG. How phosphorylation regulates the activity of p53. J Mol Biol 1996; 263(2):103-13.
51. Marques JT, Rebouillat D, Ramana CV et al. Down-regulation of p53 by double-stranded RNA modulates the antiviral response. J Virol 2005; 79(17):11105-14.
52. Sen GC, Sarkar SN. Transcriptional signaling by double-stranded RNA: role of TLR3. Cytokine Growth Factor Rev 2005; 16(1):1-14.
53. Honda K, Taniguchi T. IRFs: master regulators of signalling by Toll-like receptors and cytosolic pattern-recognition receptors. Nat Rev Immunol 2006; 6(9):644-58.
54. Nguyen H, Lin R, Hiscott J. Activation of multiple growth regulatory genes following inducible expression of IRF-1 or IRF/RelA fusion proteins. Oncogene 1997; 15(12):1425-35.
55. Vercammen E, Staal J, Beyaert R. Sensing of viral infection and activation of innate immunity by toll-like receptor 3. Clin Microbiol Rev 2008; 21(1):13-25.
56. Matsumoto M, Funami K, Tanabe M et al. Subcellular localization of Toll-like receptor 3 in human dendritic cells. J Immunol 2003; 171(6):3154-62.
57. Tabeta K, Georgel P, Janssen E et al. Toll-like receptors 9 and 3 as essential components of innate immune defense against mouse cytomegalovirus infection. Proc Natl Acad Sci U S A 2004; 101(10):3516-21.
58. Edelmann KH, Richardson-Burns S, Alexopoulou L et al. Does Toll-like receptor 3 play a biological role in virus infections? Virology 2004; 322(2):231-8.
59. Kariko K, Ni H, Capodici J et al. mRNA is an endogenous ligand for Toll-like receptor 3. J Biol Chem 2004; 279(13):12542-50.
60. Taura M, Eguma A, Suico MA et al. p53 regulates Toll-like receptor 3 expression and function in human epithelial cell lines. Mol Cell Biol 2008; 28(21):6557-67.
61. Pitha-Rowe IF, Pitha PM. Viral defense, carcinogenesis and ISG15: novel roles for an old ISG. Cytokine Growth Factor Rev 2007; 18(5-6):409-17.
62. Hummer BT, Li XL, Hassel BA. Role for p53 in gene induction by double-stranded RNA. J Virol 2001; 75(16):7774-7.
63. Nees M, Geoghegan JM, Hyman T et al. Papillomavirus type 16 oncogenes downregulate expression of interferon-responsive genes and upregulate proliferation-associated and NF-kappaB-responsive genes in cervical keratinocytes. J Virol 2001; 75(9):4283-96.
64. Kisseleva T, Bhattacharya S, Braunstein J et al. Signaling through the JAK/STAT pathway, recent advances and future challenges. Gene 2002; 285(1-2):1-24.
65. Dharel N, Kato N, Muroyama R et al. Potential contribution of tumor suppressor p53 in the host defense against hepatitis C virus. Hepatology 2008; 47(4):1136-49.
66. Munoz-Fontela C, Macip S, Martínez-Sobrido L et al. Transcriptional role of p53 in interferon-mediated antiviral immunity. J Exp Med 2008; 205(8):1929-38.
67. Youlyouz-Marfak I, Gachard N, Le Clorennec C et al. Identification of a novel p53-dependent activation pathway of STAT1 by antitumor genotoxic agents. Cell Death Differ 2008; 15(2):376-85.
68. de Visser KE, Eichten A, Coussens LM. Paradoxical roles of the immune system during cancer development. Nat Rev Cancer 2006; 6(1):24-37.
69. Karin M, Greten FR. NF-kappaB: linking inflammation and immunity to cancer development and progression. Nat Rev Immunol 2005; 5(10):749-59.
70. Koebel CM, Vermi W, Swann JB et al. Adaptive immunity maintains occult cancer in an equilibrium state. Nature 2007; 450(7171):903-7.
71. Kirn D, Martuza RL, Zwiebel J. Replication-selective virotherapy for cancer: Biological principles, risk management and future directions. Nat Med 2001; 7(7):781-7.
72. Bischoff JR, Kirn DH, Williams A et al. An adenovirus mutant that replicates selectively in p53-deficient human tumor cells. Science 1996; 274(5286):373-6.

Index

I

K

L

M

N

O

P

R

S

T

U

V

FSC
www.fsc.org
MIX
Papier | Fördert gute Waldnutzung
FSC® C083411

Kolibri 360 GmbH
Ferdinand-Jühlke-Straße 7
99095 Erfurt, Deutschland
produktsicherheit@kolibri360.de